Rangeland Stewardship in Central Asia

Victor Squires
Editor

Rangeland Stewardship in Central Asia

Balancing Improved Livelihoods, Biodiversity Conservation and Land Protection

Springer

Editor
Victor Squires
College of Grassland Science
Gansu Agricultural University
Lanzhou, China

ISBN 978-94-007-5366-2 ISBN 978-94-007-5367-9 (eBook)
DOI 10.1007/978-94-007-5367-9
Springer Dordrecht Heidelberg New York London

Library of Congress Control Number: 2012953514

Printed on acid-free paper

Springer is part of Springer Science+Business Media (www.springer.com)

Foreword

The Asian Development Bank (ADB) has been an active supporter of Central Asian countries over the past decades via technical assistance and loans for rural development and other environmental initiatives. A major initiative was the Central Asian Countries Initiative for Land Management (CACILM) that was developed under the aegis of ADB in partnership with five countries in this important region.

ADB efforts have been focused on addressing the environmental challenges that many member countries in Central Asia have been facing, especially combating land degradation and desertification. Land is losing its nutrients; sediment from accelerated erosion is fouling rivers and lakes and obstructing their flows. There is widespread deforestation and the degradation of rangelands, leading to losses in plant and animal species as well as loss of globally important wild relatives of important food and forage plants. Central Asia is one of the major world centers for biodiversity, lying as it does at the boundary of major biomes and at the crossroads of major migratory routes.

Central Asia is vast, and the environmental diversity reflects the contrasts that give rise within the region of the coldest and hottest deserts, extensive steppe, grasslands and rangelands, alpine meadows, glacier fields, and mountains and plains. It is this great variation in geography, topography, and climate that provides the rich and unique diversity found in the region's ecosystems. There is great disparity too in the ethnicity of its peoples and the economic status of the various countries within the region.

The pressures on these rich natural resources (including water from the high mountains) and environmental systems generally have, however, been increasing over the past few decades. Rapid population growth, urbanization, rising economic output, and more consumptive lifestyles, coupled with increasing incidence of poverty, have all contributed to the region's struggle to adjust to the market economy and the demands of the increasingly disenfranchised citizens, especially those on or just above the poverty line.

Since 2005, under a cofinancing agreement with GEF, ADB undertook a lead role in developing and coordinating efforts to improve cooperation between five Central Asian countries and the promotion of a more integrated approach to the management of ecosystems.

Sustainable use of rangelands is considered a key component of the ADB/GEF approach. Development of a land ethic that promotes better long-term management is central to this effort. Sharing and dissemination of experiences in sustainable rangeland management gleaned over many years by local and foreign experts whose knowledge of these Central Asian countries is unsurpassed is a vital part of this effort. This book is a significant and most valuable contribution to the knowledge sharing that has been promoted and implemented through the ADB/GEF partnership. We hope that local governments, research institutes, and finally pastoral people and farmers can benefit from this wealth of knowledge and experience captured and presented here.

One of the main lessons learned is that the capacity of local decision makers and the land users themselves can be continuously improved. In this context, local decision makers need to stop making decisions based on short-term interests and, instead, look into the importance of long-term sustainability of rangelands on which a large proportion of the people in all Central Asian countries depend for their livelihoods but also for their spiritual and cultural nourishment. The livelihoods of these land users and indeed the economy of most countries in the region will depend on the rangeland resources and the goods and services they provide. Quite importantly, the future will also depend on the land users' ability to optimize income generation from their rangeland-based livestock.

Similarly, national and lower-level governments must take responsibility as policy makers and policy enforcers to provide an enabling environment for sustainably using the natural resources for the benefit of their own citizens, discouraging short-term profit seeking.

ADB would like to congratulate the editor, the authors, and all others, including the various stakeholders (mainly land users), whose input has contributed to this important publication. ADB also looks forward to further cooperation with all stakeholders to continue improving sustainable rangeland management under the various programs that are currently being executed in the Central Asian region.

Central and West Asia Department
Asian Development Bank
Manila

Randall Jones

Editor's Preface

Central Asia *sens. lat.* encompasses a vast territory extending from the Russian Federation and Mongolia in the north to the Indian subcontinent in the south and from western China (Xinjiang and Tibet) in the east to Turkmenistan in the west. This definition of Central Asia is broader than some but follows the UNESCO classification.[1] The environmental diversity is therefore vast and is a land mass of contrasts—it contains the region's coldest and hottest deserts, extensive steppes, grassland and rangelands, forested mountains, numerous lakes, and vast plains. It is this great variation in geography, topography, and climate that provides the rich and unique diversity found in the region's ecosystems. There is great disparity too in the ethnicity of its people and the economic status of its nations.

For at least the first half of the twenty-first century, most people in Central Asia will derive their living or depend on the productivity of the soil, water, and vegetation resources. Yet this region, perhaps more than all others, has real physical constraints on the productivity of the environment. These limitations have been compounded by layers of human use and mismanagement, and these pressures on these rich natural resources and environmental systems have, however, been continuously increasing over the past few decades. Rapid population growth, urbanization, rising economic output and consumptive lifestyles, coupled with an increasing incidence of poverty, have all contributed to the region's struggle to mitigate desertification and arrest and/or reverse land degradation in all of its forms.

This book reviews past and current land management practices and policies against a background of considerable upheavals wrought by changes in attitudes in recent decades in the political, social, and environmental spheres as a number of countries in Central Asia have emerged from the period of central planning, as part of the USSR, to a more market-oriented economy. However, this is not a comprehensive analysis of development problems in this vast region. Both the various contributors and myself are well aware of the complex combinations of historical, social, cultural, economic, and political factors which, interacting upon the physical

[1] Explained in more detail in Chap. 1.

environment, explain current individual country and regional situations. We have attempted to acknowledge these issues where they are especially relevant to our themes, but it is not practical to explore them at any great length in book of this size. Rather the authors of the various chapters direct themselves to the considerable task of analyzing the salient issues and through this introduction present the background for more detailed studies and analysis.

In attempting to analyze the most important features of the Central Asian situation, we have deliberately focused on those characteristics and issues that we judged to be most important in the current and future management of environmental resources. Issues arising from management of the environment are therefore woven into the text.

For hundreds of years, animal husbandry has played a crucial role in the Central Asian region (CAR—as defined in Chap. 1) economies, societies, and cultures. Pastoralism in its multiple forms and as a livelihood strategy continues to be dynamic and flexible, with changing pastoral strategies reflecting societal pressure, shifting power structures, and economic opportunities. Until the early socialist times, pastoralists exploited the spacious rangelands (too dry, too steep, and too rocky for cultivation) of the region predominantly as natural forage grounds by applying spatiotemporal mobility patterns between seasonal pastures. Seasonal and spatial fluctuations in forage quality, accessibility, and output provide strong incentives for migratory stock keeping in Central Asia. Over the past century, mobile livestock husbandry has either been suppressed or collapsed and a fragmented pattern of rangeland use has ensued. After a forced sedentarization campaign under Soviet rule, an intensified pasture utilization was introduced as part of the socialist-style, scientifically informed and large-scale agro-industrial sector, which was composed mainly of state and collective farms.

Historically nomadic pastoralism in the CAR remained relatively constant and presumably in balance with natural resources, i.e., herders and agropastoralists were moving up to hundreds of kilometers to follow the natural seasonal progression of range forage production across elevation and latitudinal gradients. These types of grazing were seasonal and plants had a chance to mature before they were grazed again. An important aspect of this type of system was that the "return time" was long enough for plants to recover their vigor, set seed, and build up root reserves. Although nomadic grazing has been seen "as primitive," making it difficult for government agencies to serve and regulate rural communities, it can actually allow greater productivity than the more "modern" grazing systems that may be rotational over a small spatial scale but are sedentary at the spatial scale of the most relevant environmental heterogeneity.

Population in five of the Central Asian countries (Kazakhstan, Kyrgyz Republic, Tajikistan, Turkmenistan, and Uzbekistan—the so-called: stans) is going to increase from 60.6 million (2008) to 79.9 million by 2050 mainly by the forecasted growth in Uzbekistan from 27.8 to 38.4 million. Increasing human population and expanding areas of cropland have resulted in heavier grazing pressures in rangelands in spite of the increasing role of crop residues and grains in livestock production.

As a result of erratic cropping in low-rainfall zones, overgrazing of the good rangelands, and cutting of shrubs by local population for firewood, the natural vegetation of these desert areas is under pressure from anthropogenic degradation factors. This leads to the eradication of useful, endemic, or rare wild animals, and desert plant species, and to the reduction of rangelands' productivity. Instead of the valuable fodder herbs, subshrubs, and bushes, plants with poor palatability for animals have appeared. The desert forests and some shrublands have completely disappeared from large areas under the pressure from grazing/browsing livestock and the collection of fuelwood.

Land degradation is one of the greatest challenges countries of the Central Asian region (CAR) face. The loss of land productivity over the past two decades due to unsustainable land-use practices not only directly affects the well-being and livelihoods of one-third of the combined population of these countries, it has also resulted in large losses of foreign exchange earnings and threatens national food security. The situation is equally serious from an environmental viewpoint. The CAR contains unique dryland, mountain, and riparian ecosystems of importance to global biodiversity. These are being degraded and lost as rural populations become more desperate to sustain their livelihoods as a result of policies favoring unsustainable agricultural practices. The resulting land degradation in the five CA countries, in particular, has led to declines in agricultural yields of 20–30% across the region since these countries achieved independence after the collapse of the Soviet Union in 1991. This has been mainly related to the significant reduction of forage supply throughout the region.

In the last few decades, there has been a sharp rise in livestock inventories. The livestock population is high in the five Central Asian "stans." Livestock [sheep (*43.923.000*), horses (1.837.000), and goats (7.463.000)] are fed in the steppes for about 6–7 months a year. This has induced the degradation of pastures as a result of overgrazing. Despite the clear dependence on rangelands in CAR, they have long been neglected by policy makers because they were misperceived as being degraded marginal areas, offering poor returns on development investment.

Livestock owners' experience and mobility spaces have undergone momentous changes over the last two generations, and today, they are faced with many critical challenges. Rangelands or permanent pastures represent 77–95 % of the agricultural area in the five Central Asian countries (Kazakhstan Kyrgyzstan, Tajikistan, Uzbekistan, and Turkmenistan) and similar proportions in NW China (Tibet and Xinjiang) and in Afghanistan. Rangelands are the main source of forage, fuelwood, wildlife, and medicinal herbs for millions of resource-poor pastoral and agropastoral communities whose livelihoods depend on rangeland-based livestock production. These sparsely populated areas play an important role in crops and livestock production and serve as home for tens of millions of families, mostly low-income farmers, shepherds, cattle herders, and agropastoralists. However, currently many rangelands are degraded, and this loss of productivity poses significant threats to human survival. Climate variability, drought, and temperature extremes will only exacerbate the already degrading rangelands. It is important to consider that rangelands have an economic importance from the macroeconomic national level down to the level of local households, as well as crucial ecological meanings.

The CAR is one of the most vulnerable to climate change with consequences for biodiversity loss and changes in carbon sequestration potential. The unique serious applicant for a role of CO_2 fixation remains the grassy and shrubby ecosystems of the moderate and arid zones of the world that are used mainly in the agricultural and pastoral use. Because they occupy upward of 2/5 of the world's drylands. Appreciable parts of these lands are located on the territory of CAR. For example, the total area of pastures of Uzbekistan, Kazakhstan, Tajikistan, Turkmenistan, and Kyrgyzstan is 262 million hectares (Mha), which represents a significant portion of the total arid and semiarid land areas of the world. It can be compared to the total area of steppes and semideserts (90–350 mm of precipitation) on whole region of western Asia and northern Africa that makes 9.4 Mha. Therefore, the processes occurring in this vast area, especially changes in carbon fluxes and reserves resulting from changes in cultivation, grazing use, and other forms of human activity (including drying up of Aral Sea and the resulting spread of desertification), have regional as well as global significance.

For these reasons, rangelands are significant for the development processes of the whole society in most Central Asian countries. In addition to their role in providing forage for the tens of millions of livestock, there are other ecosystem goods and services provided by rangelands, not the least of which is as watersheds for the rivers that provide water for irrigation and for hydropower generation. Pastures are an increasingly threatened resource in CAR. Large areas of most of these CAR countries are used for grazing or for the harvesting of bushes or hay for animal fodder or fuel, but access to certain pastures is, today, heavily contested and often the source of volatile conflict. At the same time, and adding to the problem, increasing pasture degradation facilitated by rangeland conversion for opportunistic cropping, overgrazing, and overcutting poses ecological challenges as well as contributing to the widespread poverty so endemic throughout CAR. Significant further progress in poverty reduction and natural resource conservation in the CAR will rely to a large extent on their ability to achieve growth in the agricultural sector—a major contributor to CAR economies—and, hence, on attaining sustainable land management. The change to market economies has been generally hard on the land, and this has been particularly exacerbated by uncertainty over land ownership.

Contested access to and conflicts over pastures along with ecological degradation has resulted in heightened land-tenure insecurity. Tenure insecurity refers to the degree to which land users lack confidence that neither the state nor other people will interfere with their access rights to the land for an extended period of time. Tenure insecurity has four broad dimensions: first, conflict over rights to pastures among groups of village residents and mobile groups; second, differences of opinion about the preservation of pastures between farmers with access to farmland and those without access to farmland but with a dependence on livestock; third, contradictions between governmental agencies empowered by formal law establishing state ownership of pastureland and local communities which, by custom and necessity, use the pastures; and fourth, land-grabbing by powerful elites who establish control over pastures.

The Central Asian region is huge and very diverse in agroecologies and production systems. Even though the countries are now moving into different directions of overall and policy development, the fundamental issues of sustainable agricultural development remain common for all countries of the region, thus necessitating a double-pronged approach—regional perspective for common strategic issues—and country-distinctive action plans for implementation at national levels. Better land stewardship is a foundation for this strategic approach.

The argument being pursued in this book is that formal institutions, especially top-down-initiated legal rules implemented since 1991 in the five "stans" in particular, are decisively contributing to the formation of socioecological pasture-related challenges. The causes of the difficulties faced are quite complex. Utilization practices applied by the actors can be understood as results of the interplay of economic necessities, weak legal institutions, legal uncertainty, and a related lack of reliable planning opportunities. In this way, inappropriate and unstable legal arrangements are stimulating the processes of socioeconomic stratification and disintegration of the society as well as those of rangeland degradation. Based on findings presented in this book, there should be support for the principles of an integrated sustainable development of the wider CAR society in economic, social, and ecological matters. Management responsibilities, access, and utilization rights need to be matched to the specifics of local contexts and legitimized through participatory approaches. Including the local population in the institution-building process can make a decisive contribution to the development of the Central Asian region's society by balancing different interests.

The book is written by leading authorities on the current and latent issues in the region and brings to the reader an up to date analysis and provides a window into this important region about which literature is not readily available in English.[2] It will be especially valuable to academics, scholars, to the donor community, and policy makers. The work is meant to be read by the nonprofessional in the field. It is hoped that it will serve some of the overview needs of professionals in resource management, environmental conservation, and development and *students* in these and other fields who are interested in Central Asia. The lessons learned from the transition to a more caring society, as embodied in the land-stewardship paradigm, will be of interest to others working in the developing world.

A key aspect of the work is that of achieving *balance*. Technical approaches can address improved livelihoods, biodiversity conservation, and land protection individually, but balancing these is the negotiated outcome of social processes. Ultimately, it is these social processes which are the hard part, and often the stumbling block, for improving resource stewardship and sustainable land management.

Part I, the introductory chapters, provides an overview of the region in an analysis of the geography and environmental history. This we judge to be important in providing the physical setting of the region and in exploring some of the distinctiveness

[2] A body of literature exists in Russian. Part of the rationale of this book is to "unlock" some of this information and make it more readily available.

of the Central Asian region *sens lat.* We provide background to the evolution of the management of the resources which particularly emphasizes the historical evolution of ethnic and national management systems.

In the next two sections (2 and 3), attention is paid to the concept of land stewardship and the practice of land stewardship as it applies in Tajikistan as a special case study. It is recognized that protecting rural livelihoods is the key to adoption of land stewardship by the impoverished population.

Part IV deals with the important matter of creating the enabling environment. National and lower-level governments must take responsibility as policy makers and policy enforcers to provide an enabling environment for sustainably using the natural resources for the benefit of their own citizens, discouraging short-term profit seeking.

Part V, the last section, poses the following questions: Where are we now? Where do we want to be in the coming decades? And How do we get there?

The general conclusions reached are indeed somewhat pessimistic. There are no quick or easy solutions to the complex of Central Asian land management problems. But based on a better understanding of the situation and of the management systems (including the legal and policy environment) involved in the different parts of Central Asia, there are clear paths forward. It is vitally important that the direction be set and the first steps taken in that direction. If this book can help in the process of advancing better land stewardship in Central Asia, we will feel truly rewarded.

Lanzhou, China Victor Squires
June 2012

About the Editor

Victor Squires is an Australian with decades of experience in research, academia, and as a consultant in Africa, Asia, and the Middle East. Dr. Squires is a visiting adjunct professor at Gansu Agricultural University in Gansu, China, and at the University of Arizona, Tucson, USA. He is the recipient of two gold medals from the Chinese government: in 2008, the award for International Science and Technology Cooperation and in 2011 the Friendship Award, China's highest honor for foreigners.

Contents

Contributors

Mohammad Hossein Emadi International Consultant and Senior Advisor to the Minister of Agriculture, Irrigation and Livestock, Kabul, Afghanistan

Nargis Halimova GITEC Consulting GMBH, Cologne, Germany

Ian Hannam Centre for Agriculture and Law, University of New England, Armidale, Australia

Inam-ur-Rahim Mountain Societies Research Centre (MSRC), University of Central Asia, Bishkek, Kyrgyzstan

Shoaib Ismail International Center for Biosaline Agriculture (ICBA), Tashkent, Uzbekistan

Rodney Jackson Snow Leopard Conservancy, Sonoma, CA, USA

Michael J. Jacobs Department of Ecosystem Sciences and Management, Texas A & M University, Texas, USA

Osamu Kozan University of Kyoto, Kyoto, Japan

Tobias Kraudzun Centre for Development Studies, Institute of Geographical Sciences, Freie Universität Berlin, Berlin, Germany

Barno Kurbanova Donaev Management Consulting, Dushanbe, Tajikistan

Hua Limin College of Grassland Science, Gansu Agricultural University, Lanzhou, China

John E. Leake The Institute for International Development, Adelaide, Australia

Zvi Lerman Department of Agricultural Economics and Management, The Hebrew University of Jerusalem, Jerusalem, Israel

Daniel Maselli Mountain Societies Research Centre (MSRC), University of Central Asia, Bishkek, Kyrgyzstan

Stefan Michel Freelance consultant

T.F. Radjabov Samarkand State University, Samarkand, Uzbekistan

R.M. Rahimon American University of Central Asia, Bishkek, Kyrgyz Kyrgyzstan

Sarah Robinson Division of Ecology and Evolution, Imperial College, London, UK

Cyrus Samimi Department of Geography and Regional Research, University of Vienna, Vienna, Austria

Catherine A. Schloeder Department of Ecosystem Sciences and Management, Fortine, MT, USA

David Sedik FAO, Food and Agriculture Organization of the UN, Rome, Italy

M. Shaumarov Institute of Agricultural Economics and Social Sciences in the Tropics and Subtropics, University of Hohenheim, Stuttgart, Germany

E.V. Shuyskaya K.A. Timiryazev Institute of Plant Physiology Russian Academy of Sciences, Moscow, Russia

Navinder Singh Department of Wildlife, Fish, and Environmental Studies, Faculty of Forest Sciences, Swedish University of Agricultural Sciences, Uppsala, Sweden

Victor Squires College of Grassland Science, Gansu Agricultural University, Lanzhou, China

GITEC Consulting GMBH, Dusseldorf, Germany

Paul J.H. Strong GITEC Consulting GMBH, Cologne, Germany

Kristina N. Toderich PFU-CGIAR, ICARDA Sub-office, Tashkent, Uzbekistan

Kim Andre Vanselow Institute of Geography, Friedrich-Alexander-Universität Erlangen-Nürnberg, Erlangen, Germany

Christoph Wiedemann College of Grassland Science, Gansu Agricultural University, Lanzhou, China

Degang Zhang Freelance consultant

Yerlan Zhumabayev Central Asian Countries Initiative for land management (CACLIM)

Abbreviations and Acronyms*

ACTED	Agence d'Aide a la Cooperation Technique et Developpement
AKDN	Aga Khan Development Network
AALS	In Uzbekistan, Aydarkul-Arnasay Lakes systems
APPA	Appreciative Participatory Planning and Action
ARP	Agriculture Reform Program
Azal	A national NGO in Tajikistan
CAC	Central Asian countries
CAG	Collective Action Groups
CACILM	Central Asian Countries Initiative for Land Management
CBO	Community-based Organizations
CAMP	Kuhiston A national NGO in Tajikistan
CAR	Central Asian Republics, especially a group that are in the USAIDs program
CCA	UN Common Country Assessment
CIDA	Canadian International Development Agency
CLS	In Kazakhstan, Conditional Land Shares
CMPF-SP	Support Project of CACILM Multicountry Partnership Framework
dS/m	A measure of electrical conductivity that reflects soil or water salinity
DLDD	Dryland Degradation and Desertification
EFA	Ecosystem Function Analysis
FAO	Food and Agriculture Organization of the UN
GIS	Geographical Information Systems
GM	Global Mechanism of UNCCD
GRM	An International consulting company based in Australia
Helvetas	A Swiss-based NGO (Schweizer Gesellschaft für internationale Zusammenarbeit)
HBS	Household Budget Survey
ILP/ILUP	Integrated Land Use Planning

*See Glossary for terms in italics.

IAMO	Leibniz Institute of Agricultural Development in Central and Eastern Europe
IDP	In Afghanistan, internally displaced persons
IGDK	In Afghanistan, Independent General Directorate of Kuchi. IGDK operates as a quasi-ministry as a mechanism for facilitating government representation and for solving problems encountered by Kuchi.
IPCC	Intergovernmental Panel on Climate Change
KPD	In Afghanistan, Kuchi Provincial Directors (KPDs) manage business by *shura*
LADA	Land Degradation Assessment in Drylands
LFA	Landscape Function Analysis
MAIL	In Afghanistan, Ministry of Agriculture, Irrigation and Livestock
MAP	Mean annual rainfall
MBTI	In Tajikistan, Inter-Districts Bureau of Technical Inventory
MDG	Millennium Development Goals
MTND	In Afghanistan, Meat Trade News Daily
MWRLR	Ministry of Water Resources and Land Reclamation
MOA	Ministry of Agriculture
MOI	Ministry of Justice
NPF	National Programming Framework
NRM	Natural Resource Management
OTA	In Afghanistan Office of Tribal Affairs
Oxfam	A UK-based international NGO
PAR	Participatory Action Research
PEACE	is a USAID-funded Project in Afghanistan for reducing risk for Kuchi people
ppm	Parts per million
PRSP	Poverty Reduction Strategy Program
PUA	In Kyrgyzstan Pasture Users Associations formed at the level of each *aiyl okmotu*.
PUG	Pasture Users' Group a generally informal group of farmers with common needs and problems
RS	Remote Sensing
SCLC	In Tajikistan the Cadastre Department
SDO	Sanayee Development Organization a registered Afghan development NGO
SIDA	Swedish International Development Cooperation Agency
SNC	In Uzbekistan, a consulting firm
SPA	Strategic Partnership Application of the UNCCD
TA	Technical Assistance e.g., from Asian Development Bank or other donor agency
TajStat	A set of statistics published by the government of Tajikistan
UNECE	United Nations Economic Commission for Europe
UNEP	United Nations Environment Programme

UNFCCC	United Nations Framework Convention on Climate Change
UNFPA	United Nations Population Fund
UNDP	United Nations Development Programme
UNCCD	United Nations Convention to Combat Desertification
UNMAP	United Nations Mine Action Program (mine field clearance)
WB	World Bank
WOCAT	World Overview of Conservation Approaches and Technologies

List of Boxes

List of Figures

List of Tables

Part I
The Context

The three chapters in this part provide the context for what follows.

Chapter 1 deals with the physical setting and definition of the Central Asian region for the purpose of this book. It is broader than that favored by the former Soviet Union that only included the five "stans": Kazakhstan, Kyrgyzstan, Tajikistan, Turkmenistan, and Uzbekistan. Western China and Afghanistan are part of the Central Asian region. This definition follows UNESCO practice.

Chapter 2 introduces the rationale for fostering and promoting sustainable development and explains the concepts of land stewardship.

Chapter 3 traces the history of ideas and perceptions about land and land use from the beginning of the nineteenth century through the colonization of much of the region by Czarist Russia and later by the Soviet Union through to the post-Soviet era of independence.

Chapter 1
Setting the Stage: Key Features of the Present-Day Central Asian Region: An Introduction to the Wider Historical, Social, Political, Economic, Cultural, and Ecologic Contexts of the Region in a Nutshell

Daniel Maselli and Inam-ur-Rahim

Abstract The Central Asian region is a vast area encompassing six countries, Afghanistan, Kazakhstan, Kyrgyzstan, Tajikistan, Turkmenistan, Uzbekistan, and part of western China. This definition follows UNESCO's classification. The chapter focuses on describing Central Asia's biophysical context. It outlines the key features, challenges, responses, and opportunities. It is meant as an introduction to the wider historical, social, political, economic, cultural, and ecologic contexts of the region and sets the stage for what follows in the remaining chapters.

Keywords Economy • Energy • Agriculture • Communication • Political system • Landscapes • Migration

Key Points

- Central Asia is presently home to over 64 million people with considerable differences regarding urban-rural distribution (Table 1.4). The average population growth varies from 0.9 % for Kyrgyzstan to 1.8 % for Tajikistan. Especially within the poorer countries, out-migration rates of people moving to other countries are remarkably high, with currently Kyrgyzstan topping the list followed by Tajikistan. In both countries, the generated remittances play a key role within the overall national economy and are an important factor of investment and innovation/changes in people's livelihoods.
- The five nation states that evolved from the Soviet republics and the contiguous territories of western China and Afghanistan that collectively, according to UNESCO, make up the Central Asian region (CAR) are in many respects very different from each other both with regard to their biophysical and their socio-economic features.

D. Maselli (✉) • Inam-ur-Rahim
Mountain Societies Research Centre (MSRC),
University of Central Asia, Bishkek, Kyrgyzstan
e-mail: daniel.maselli@ucentralasia.org; inam.rahim@ucentralasia.org

V. Squires (ed.), *Rangeland Stewardship in Central Asia: Balancing Improved Livelihoods, Biodiversity Conservation and Land Protection*,
DOI 10.1007/978-94-007-5367-9_1, © Springer Science+Business Media Dordrecht 2012

1 Overview

The territory of present-day Central Asia encompasses a vast area of more than 4 million km^2 comprising the surface of the five former Soviet republics (now independent states) – Kazakhstan, Kyrgyzstan, Tajikistan, Turkmenistan, and Uzbekistan – that make up the bulk of the Central Asian region (CAR). However, the term "Central Asia" also includes adjacent territories such as the southern outskirts of the Russian Federation, western Mongolia, Afghanistan, and western China with the Xinjiang Province as well as parts of the Tibetan Plateau. The borders (Fig. 1.1) between countries in the CAR are of rather recent making and do not run along historical boundaries of former kingdoms or *khanats* that dominated large portions of the area over centuries.

The territory of Central Asia is landlocked and situated at the crossroad of most of the historical civilizations including China to the east, India and Persia to the south, Russia with Siberia to the north, and Turkey with the remainder of Middle East to the west. The famous silk route passed through the Central Asian territory and connected people from diverse cultures to interact in its few urban centers often

Fig. 1.1 Map of the political entities that comprise the five former Soviet republics and their borders with neighboring countries

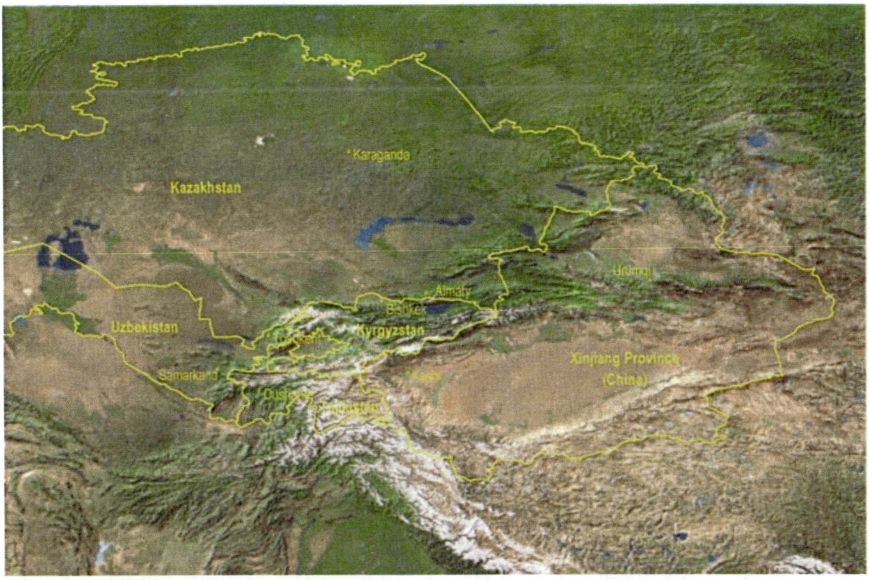

Fig. 1.2 Satellite imagery of the Central Asian region and borders of the former Soviet republics

placed in desert oasis. This led to the creation of centers of learning like Samarkand and Bukhara that have contributed enormously to our current knowledge.

Besides the few urban centers, large portions of the Central Asian territory consist of grassland steppes with a lot of commonalities due to interactive nomadic cultures. Moreover, the vast steppes used by nomads through their mobile lifestyle made rigid boundaries rather an unnecessary and even critical obstacle (Jacobs and Schloeder, Chap. 6). Mobility allowed for regular interaction among people of different ethnic background. However, after the forced sedentarization and mass collectivization induced by the Marxism-Leninism doctrine, these borders are nowadays continuing to limit movements. They hamper adaptation to both climate change and global change. In a general perspective, the traditional nomadic practices developed over centuries and even millennia in Central Asia are more than ever at risk despite the "political freedom" gained by the dissolution of the former Soviet Union in the early 1990s. After independence, the new nation states adapted different strategies to move forward. In a synoptic perspective, Central Asia embraces thus an impressive wide spectrum of nation states and ethnic territories that are very different from one another both with regard to their biophysical and their socioeconomic features. Despite this diversity, long-distance horizontal and vertical movements – i.e., transhumance – were and for the time being remain a common feature characterizing land use in Central Asia and shaping people's life as well as their environment. Similarly, as for the vast plains, the huge mountain systems are a key characteristic of Central Asia. They have conserved more traditional features as compared to many flat and rather uniform lowland areas (Fig. 1.2).

2 Biophysical Characteristics

The relief of Central Asia is characterized by an impressive diversity of large-scale landscapes (Figs. 1.3 and 1.4). The northern part of the territory is dominated by vast grassland steppes while relatively low-altitude flat deserts and semideserts prevail in the central and western areas. Among the most prominent deserts are the "Kara-Kum" or "black sand" in Turkmenistan, the "Kyzylkum" or "red sand" in western Uzbekistan, and the "Taklamakan" desert ("land of poplars" or "gardens of the desert") in western China.

In the south and east, large mountains raise up to altitudes of 7,495 m a.s.l. (Peak Communism or Lenin). The major mountain ranges are the high Pamirs in southwestern Tajikistan adjacent to the Hindu Kush range in Afghanistan and Pakistan, the Tien Shan in Kyrgyzstan with its prolongation into western China's Xinjiang Province, and the Pamir-Alai in southeast Kazakhstan connecting to Mongolia.

These mountain ranges have a direct influence on the climate, land cover, and land use in Central Asia. They host a high number of glaciers and act as water towers to the entire region. These high mountains toward the south also retain the northern Siberian cool winds which remain in the territory keeping winter temperatures well below freezing point while at the same time preventing moist clouds of the monsoon rain system approaching mainly from the Indian Ocean in the south to reach Central Asia. The mountains thus act as barriers keeping the territory relatively arid. The dominant climate in Central Asia is thus continental, characterized by

Fig. 1.3 Yurts are a feature of traditional way of living (photo D. Maselli, Kyrgyzstan 2008)

Fig. 1.4 Vast steppes and snow-capped mountains dominate large areas

cold winters and hot summers with arid and semiarid regimes in the vast lowland plains and semiarid to semihumid regimes at higher altitudes.

2.1 *Kazakhstan*

This country covers more than two-thirds of the entire surface of the five Central Asian republics. It is thus by far the largest country in the region (Table 1.1) and the 9th largest country worldwide. Nevertheless, only little more than 8 % of its territory is categorized as arable land. With altitudes ranging from −132 (*Vpadina Kaundy*) to 6,995 m a.s.l. (*Khan Tangiri Shyngy* or former *Pik Khan-Tengri*), the Kazakh landscapes encompass a huge variety of ecosystems that translate into a variety of traditional and well-adapted land use systems including in particular mobile/nomadic pastoralism. The territory is dominated by vast flat steppes that stretch from the Volga in the west to the Altai mountains in the east and from the plains of western Siberia in the north to the deserts and oases in the south.

2.2 *Kyrgyzstan*

Historically, *Kyrgyzstan* was formally annexed to Russia under the Czar's authority from 1876 (Rahimon, Chap. 3) and later became a Soviet Republic in 1936. More than 93 % of its territory or close to 200,000 km^2 is mountainous with the lowest

Table 1.1 Key geographic features of five former Soviet republics in Central Asia

Attribute	KZ	KG	TJ	TM	UZ
Surface in km²	2,724,900 (19×)	199,951 (1.4)	143,100 (1×)	488,100 (3.4×)	447,400 (3.1×)
Land	2,699,700 (6.3×)	191,801 (2.2×)	141,510 (1×)	469,930 (3.3×)	425,400 (3×)
Water	25,200	8,150	2,590	18,170	22,000
Border line in km	12,185	3,051	3,651	3,736	6,221
Lowest point in m a.s.l.	−132	132	300	−81	−12
Highest point in m a.s.l.	6,995	7,439	7,495	3,139	4,301

Source: CIA World Factbook, accessed March 2012
Total figures: surface, 4,003,451 km^2 (about 40 % of the total European surface); land, 3,928,341 km^2; irrigated land, 113,206 km^2; arable land, 311,228 km^2

point at 132 m a.s.l. (*Kara-Daryya*) and the highest peak *Jengish Chokusu* (or former *Pik Pobedy*) at 7,439 m a.s.l. (Table 1.1). The country thus hosts a large number of glaciers and high-altitude lakes of which the *Son Kul* is the most famous one where traditionally different tribes used to graze their animals on the rich summer pastures (*jailoos*). The climate varies from dry continental to polar in the *Tien Shan* mountains to a subtropical climate in the *Ferghana* valley and a temperate one in the northern foothill zone along the *Chuy* valley. While large parts of the country's hill and mountain slopes are poorly covered by forest, Kyrgyzstan is home to the world's largest natural-grown walnut forest (*Arslanbob* near *Jalalabad* in the south). The *Issyk-Kul* ("hot lake") is the second largest saline lake of the world after the Caspian Sea and is a declared biosphere reserve (Fig. 1.5).

2.3 Tajikistan

Like Kyrgyzstan, *Tajikistan* is dominated by mountain ranges such as in particular the high Pamirs which are often referred to as "roof of the world." Tajikistan hosts the highest peak in Central Asia, while its waters leave the country via the Syr Darya river toward Uzbekistan at the lowest point of about 300 m a.s.l. in the southeast. With the Alai mountain range along the northern border, the ethnic Kyrgyz nomads populating the eastern Pamirs, and the outreach to the edges of the Ferghana valley, Tajikistan shares many commonalities with Kyrgyzstan. Its role as regional water tower is underpinned by the fact that the glaciers and lakes represent the world's second largest mountain hydropower potential. The Tajik National Park is the largest protected area in Central Asia. It consists of an impressive mountain system surrounded by the Hindu Kush to the south, the Karakorum and Himalayas to the southeast, the Kunlun Shan to the east, the Pamir-Alai to the north, and the Parapamir to the west. The park includes a diversity

Fig. 1.5 Revenue from tourism at Issyk-Kul provides economic diversity (photo D. Maselli, Tajikistan 2008)

of natural landscapes with rare and endangered flora and fauna of regional and global importance such as in particular the snow leopard (*Uncia uncia*) and the red wolf (*Cuon alpinus*) listed in the red book of IUCN (Jackson, Chap. 15). Some of the most spectacular features are the two high-altitude lakes of *Karakul* and *Sarez*, the two highest peaks *Somoni* (7,495 m a.s.l.) and *Lenin* (7,134 m a.s.l.), the two monumental glaciers *Fedchenko* and *Medveji,* and numerous hot and mineral springs (Figs. 1.6 and 1.7).

2.4 Turkmenistan

The territory of *Turkmenistan* is characterized by a subtropical desert climate manifesting as sandy landscape (*Kara-Kum* desert) that forms dunes and rises to mountains in the south at the border with Iran. The highest elevation point in Turkmenistan is at 3,139 m a.s.l. (*Gora Ayrybaba*), while the lowest point is at −81 m below sea level (Vpadina Akchanaya). The Great Balkhan Range in the west (Balkhan Province) and the Köýtendag Range along the southeast border with Uzbekistan (Lebap Province) are the only other significant elevations in the country. The Great Balkhan Range rises up to 1,880 m a.s.l. (Mount Arlan), while the highest summit in Turkmenistan is the Ayrybaba (3,137 m a.s.l.) belonging to

Fig. 1.6 Alpine summer pastures provide seasonal grazing for the flocks and herds of pastoral nomads (photo D. Maselli, Kyrgyzstan 2011)

Fig. 1.7 Tajikistan's Zerafshan valley is utilized by a sedentary population whose livelihood is derived from small plots of irrigated land on the terraced hill slopes and in the fertile valley bottoms (photo D. Maselli, Tajikistan 2009)

the Kugitangtau Range. The major rivers include the Amu Darya, the Murghab, and the Tejen. The climate is mostly arid and subtropical with little rainfall. Winters are mild and dry, while most precipitation falls from January to May with the heaviest precipitation in the Kopet Dag mountain range. Based on the prevailing natural conditions, the country mainly depends on intensive agriculture located in irrigated oases as well as on exploiting substantial oil and gas resources. Its major agricultural products though are cotton as a cash crop for export and wheat for domestic consumption. With more than 300,000 km^2 of permanent meadows and pastures, Turkmenistan is the third largest producer of fresh milk and meat in the region (Table 1.3).

2.5 *Uzbekistan*

Today, *Uzbekistan* controls most of the *Ferghana* valley which is the largest fertile area of Central Asia and home to about half of the entire region's population. Uzbekistan's highest point reaches 4,301 m a.s.l. (*Adelunga Toghi*), while the lowest point is slightly below sea level (*Sariqamish Kuli*, −12 m a.s.l.). Intensive agriculture aimed at producing the "white gold" (cotton) and grain caused considerable environmental damage due to the overutilization of agrochemicals and of scarce water resources leading to soil degradation in particular salinization. Eventually this led to the drying out of the Aral Sea and to environmental pollution through wind erosion transporting toxic materials over large distances. The landscape is mostly dominated by sandy deserts and dunes interrupted by flat and intensely irrigated river valleys along the two major rivers of Central Asia, the *Amu Darya* in the south and the *Syr Darya* in the north. The overall climate is dry and hot in summer with rather mild winters (Fig. 1.8).

2.6 *Xinjiang*

This large region in western China is split by the *Tien Shan* mountain range into two large basins: the *Dzungarian* basin in the north and the *Tarim* basin in the south. Much of the *Tarim* basin is dominated by the Taklamakan desert. The lowest point in *Xinjiang* is the *Turpan* depression, −155 m a.s.l. Its highest point is the famous K2 mountain, which reaches 8,611 m a.s.l. and is located on the border with Pakistan. Other important mountain ranges include the *Pamirs* in the south-east, the *Karakorum* in the south, and the *Altai* in the north. The *Tien Shan* mountain range marks the *Xinjiang*-Kyrgyzstan border at the *Torugart* pass (3,752 m a.s.l.). The Karakorum Highway links Islamabad (Pakistan) with *Kashgar* (China) over the *Khunjerab* pass.

Fig. 1.8 Modern mechanized agriculture is expanding in the Ferghana (photo D. Maselli, Uzbekistan 2007)

2.7 Afghanistan

This country is also regularly considered as being part of Central Asia. The country's highest point is the *Noshaq* at 7,492 m a.s.l. By and large, Afghanistan is characterized by a continental climate with very harsh winters in the central highlands, the glaciated northeast (i.e., around Nuristan), and the *Wakhan* corridor, where the average temperature in January falls below −15 °C, and hot summers in the low-lying areas of the *Sistan* basin of the southwest, the *Jalalabad* basin in the east, and the Turkestan plains along the *Amu Darya* river in the north, where temperatures average over 35 °C in July. Despite having numerous rivers and reservoirs, large parts of the country are dry. The endorheic *Sistan* basin is one of the driest regions in the world. Aside from the usual rain falls, Afghanistan receives snow during winter in the *Hindu Kush* and *Pamir* mountains, and the melting snow in the spring season enters the rivers, lakes, and streams. However, two-thirds of the country's water flows into neighboring countries of Iran, Pakistan, and Turkmenistan.

3 Land Use

Despite the fact that all CAR countries are still heavily involved in and dependent on agricultural production, the overall output (and partially also the productivity) has declined tremendously after independence and remains generally at a very poor level despite governmental efforts supported by bi- and multilateral partners. The stagnation

Table 1.2 Key land use figures in the five former Soviet republics

Attribute	KZ	KG	TJ	TM	UZ
Arable land in % (in km^2)	8.28 (223,535)	6.55 (12,563)	6.52 (9,226)	4.51 (21,194)	10.51 (44,710)
Irrigated land in km^2 (2008)	35,560	10,196	7,220	18,000	42,230
Permanent crops in km^2 (2009)	–	750	1,330	611	3,500
Permanent meadows and pastures in km^2 (2009)	1,850,000	92,660	38,750	307,000	220,000
Forest in km^2 (2009)	33,146	9,370	4,100	41,270	32,794
Fresh cow milk in tons (2010)	5,347,540	1,317,300	602,000	2,150,000	6,120,000
Indigenous cattle meat in tons (2010)	546,817	100,217	–	147,900	665,300

Source: FAOSTAT, data for 2010, and CIA World Factbook; accessed June 2012

is due to a number of factors such as in particular the lack of investment opportunities needed, e.g., to replace aged agricultural machinery and improve technology, overutilized and degraded soils and vegetation cover, as well as inappropriate water management. Generally, the current agricultural sector is suffering from a deep economic, social, and ecological crisis that requires new orientation. However, the lack of appropriate policy and land reforms including their ineffective implementation hamper the required structural adjustment of land use in the entire region. Western China (Xinjiang and Tibet) has benefited by higher levels of investment, inward migration, and more effective land reforms (Hua, Chap. 14; Hannam, Chap. 17). Afghanistan is caught up in seemingly interminable war (Emadi, Chap. 5; Jacobs and Schloeder, Chap. 6).

Permanent meadows cover the prevailing portion of the land, and hence pasturing is the dominant land use and animal husbandry the dominant occupation. The overall forest cover in Central Asia is extremely poor: 131,660 km^2 or just about 3 % of the entire land surface (Table 1.2). It would require considerable efforts to increase forest cover in a substantial manner within all Central Asian countries. The forced reversion to self-sustenance and specific incidences, such as in particular the civil war in Tajikistan in the 1990s, have put forest cover under heavy stress and have further exacerbated its deterioration (Table 1.3).

With regard to agricultural production, Central Asia is generating an impressive amount of goods with a total value of nearly 22 billion international dollars in 2010. Despite its vast territory, Kazakhstan is only the number 2 in the region since Uzbekistan with the fertile and irrigated Ferghana valley produces about 1.6 times as much as its large neighbor. Live animals and livestock products bear huge potentials for proper processing and export. However, this potential has not so far been properly exploited.

4 Environmental Issues and Natural Hazards

Most CAR countries also suffer from a variety of environmental threats and mostly human-induced degradation processes (Fig. 1.9). Kazakhstan, e.g., has a number of radioactive or toxic chemical sites stemming from the former Soviet Union defense

Table 1.3 Net agriculture production value (in $1,000) for the five former Soviet republics

Attribute	KZ	KG	TJ	TM	UZ
Total	6,486,959	1,520,594	1,213,099	2,470,080	10,270,284
Camel milk	–	–	–	–	170
Cow milk	1,471,536	349,039	104,064	231,627	1,211,421
Goat milk	3,759	2,215	19,732	–	10,235
Sheep milk	10,514	14,019	–	–	–
Camel meat	1,142	–	–	–	1,257
Cattle meat	1,098,547	270,724	77,372	399,520	1,794,035
Goat meat	48,908	13,255	–	23,505	–
Horse meat	113,849	33,852	–	–	3,424
Sheep meat	334,752	111,935	106,843	353,027	272,137
Wool	71,933	20,853	11,041	72,698	50,717
Livestock	3,756,942	867,670	342,608	1,158,664	3,659,022
Food	–	–	–	–	8,491,591
Crop	–	–	–	–	7,064,968

Source: FAOSTAT, data for 2010

Fig. 1.9 Cutting of trees and other woody plants for fuel is a daily chore, often done by children and women (photos by D. Maselli, Tajikistan 2008)

industries and test ranges. They represent a considerable health risk both for humans and animals. The diversion of water for irrigation from the two main rivers – the Amu Darya and the Syr Darya – that used to flow into the Aral Sea has dried up its shore and made its surface shrink in an alarming manner. As a consequence, a harmful layer of chemical pesticides and natural salts covers the soil that is picked up by the wind and blown away as noxious dust storms. Nowadays, the concerned area is one of the heaviest polluted ones in Central Asia.

Mining along with inappropriate precautionary environmental measures is threatening considerable areas in Central Asia such as the high-altitude gold mines in Kyrgyzstan that threaten glaciers and their melting waters as well as traditional summer pastures (*jailoos*). Furthermore, uranium and mercury tailing sites are of concern to large parts of lowland population including in particular the Ferghana valley.

In most cases, both technical know-how and funding are required to take appropriate measures that will help avoid further spoiling vast areas needed as living ground for future generations.

Fig. 1.10 Accelerated soil erosion is a common result of deforestation, a process that was accelerated as a consequence of the civil war in Tajikistan (photo D. Maselli, Tajikistan 2006)

The sudden and severe interruption of energy supply to the region has also triggered serious environmental damages in particular related to the collection of firewood for heating and cooking in many rural places (Vanselow, Chap. 4). Cold winter temperatures coupled with traditional cooking habits using simple but ineffective stoves and oven have contributed to the degradation of many remaining (riparian) forests and shrub vegetation. Given the general scarcity of wooden fire resources, dried dung cakes remain a widely used substitute. However, the burning of dung interrupts the natural cycle of nutrients, thus negatively impacting on soil fertility which in turn leads to reduced agricultural productivity and smaller harvests causing a vicious circle.

Many parts of Central Asia are exposed to a variety of potential natural hazards and related disasters. The biggest threat emanates from the rather frequent and partially strong seismic activities that generate regular earthquakes of different magnitudes. Landslides, floods, droughts, avalanches, or mud flows are causing damages to the infrastructure and require considerable reconstruction and maintenance work particularly in remote mountain areas. More recently with the rapid acceleration of glacier melting, the risk for incidences caused by glacial lake outburst flows (GLOFs) has increased and requires due attention and adequate preventive, mitigation, and awareness, rising efforts at all levels including in particular local authorities and communities. Along with the expected and partially already felt climate change, the potential for more frequent and severe pest calamities, heat waves, fires, or long periods of deep cold increases, too, calls for adequate response mechanisms to increase people's resilience (Fig. 1.10).

5 Demography

Central Asia is presently home to over 64 million people with considerable differences regarding urban-rural distribution (Table 1.4). The average population growth varies from 0.9 % for Kyrgyzstan to 1.8 % for Tajikistan. Especially within the poorer countries, out-migration rates of people moving to other countries are remarkably high, with currently Kyrgyzstan topping the list followed by Tajikistan. In both countries, the generated remittances play a key role within the overall national economy and are an important factor of investment and innovation/changes in people's livelihoods.

The global phenomenon of moving toward urban centers applies to Central Asia as well. This development is strongly correlated with the level and increase of the (peri-)urban population. As a consequence, the future generations expected to manage rural livelihoods and their respective natural resources appear to be at risk. The key question related to the exodus from rural areas to central political, economic, and cultural locations is whether there will be sufficient young people left ready to become herders of large flocks of small ruminants especially in the high-altitudinal belts of mountain ranges such as the Tien Shan in Kyrgyzstan or the Pamirs in Tajikistan but also the wide steppes of Kazakhstan. Who will be ready to live a seminomadic lifestyle and refuse modern standard commodities? Herein, the provision of basic services such as health, education, and communication plays certainly a key role as incentive or disincentive for long-term/permanent migration or settlement of the younger generation.

With regard to the composition of the population, Central Asia represents an impressive mix of ethnicities and related cultures partially as a result of the settlement policy of the Soviet Union and partly as the result of long-term historical movements. Herein, the Uralic-Altaic and Indo-European represent the two main ethnic groups under which most of the others can be categorized. While this multiethnicity

Table 1.4 Key demographic features of the five former Soviet republics

Attribute	KZ	KG	TJ	TM	UZ
Population (2012 est.)	17,522,010	5,496,737	7,768,385	5,054,828	28,394,180
Urban population in % (2010)	59	35	26	50	36
Population growth rate in % (2012 est.)	1.235	0.88	1.823	1.143	0.94
Migrants/1,000 population (2012 est.)	0.43	−8.1	−1.21	−1.9	−2.65
Literacy rate in %	99.5 (1999 est.)	98.7 (1999 cens.)	99.5 (2000 cens.)	98.8 (1999 est.)	99.3 (2003 est.)

Source: CIA World Factbook, accessed June 2012

Total population for featured CAS countries: 67,416,137 (about 9 % of the total European population)

Fig. 1.11 People of Central Asia (photos D. Maselli, Tajikistan and Kyrgyzstan, 2006–2009)

could potentially be an asset for the entire region, there is an imminent danger of ethnical conflicts particularly due to phenomena such as exclusion and/or political under-/misrepresentation.

Like in other rapidly growing regions, the age composition in Central Asia is characterized by an impressive youth bulge (Figs. 1.11 and 1.12).

The relative population density among the five former Soviet republics is highest in Uzbekistan followed by Tajikistan with the lowest in Kazakhstan. On one hand, Kazakhstan has the highest proportion of population residing in urban areas as compared to the other states, while on the other hand having the lowest density of population per unit of pastoral and agricultural land. This shows that in Kazakhstan, more and more population is moving to the urban centers due to increasing job opportunities, while, e.g., the people in Tajikistan still depend on subsistence farming. This disparity is confirmed by the proportion of population living below poverty line which is highest in Tajikistan and lowest in Kazakhstan. Due to the better economic conditions, Kazakhstan acts as pull area for the working labor of the other Central Asian nations, in particular Kyrgyzstan and Tajikistan (see Table 1.4). The high literacy rate is comparable among the former Soviet republics, thanks to the Soviet system that offered education to all while, e.g., the literacy rates in neighboring Afghanistan and Pakistan are less than 60 % (Fig. 1.13).

Fig. 1.12 Women in Tajikistan producing traditional handicrafts (photo D. Maselli, Tajikistan 2004)

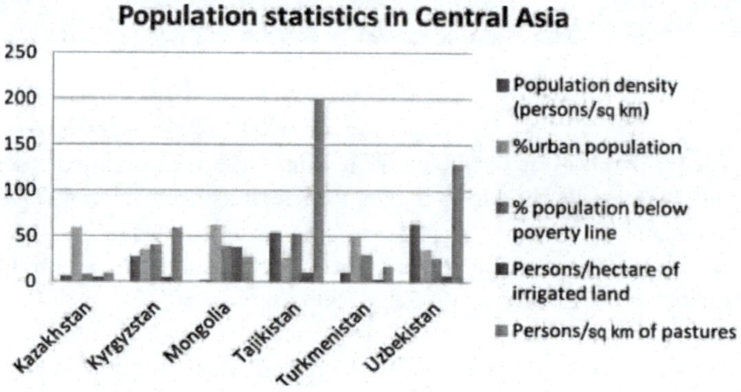

Fig. 1.13 Population statistics for Central Asia

6 Social Features

Since independence in 1990, a sharp increase in the inequality of income distribution has hurt the lower ranks of society in most of the Central Asian states. Many of the former facilities and services provided to the entire population through the Soviet system have collapsed. This increased the vulnerability of many households, in particular single-headed women households (Kurbanova, CIA World Factbook, FAOSTAT) (Chap. 7) (Table 1.5).

Table 1.5 Key social features of the five former Soviet republics

Attribute	KZ	KG	TJ	TM	UZ
Maternal mortality rate in deaths/100,000 live births (2008)	45	81	64	77	30
Infant mortality rate in deaths/1,000 live births (2012 est.)	23.06	30.78	37.33	40.89	21.2
Health expenditures in % of GDP (2009)	4.3	4.1	5.3	2.3	5.2
Life expectancy at birth in years (2012 est.)	69.63	69.45	66.38	68.84	72.77
Fertility rate in children born/woman (2012 est.)	2.41	2.73	2.85	2.14	1.86
Children under the age of 5 years underweight in %	4.9 (2006)	2.7 (2006)	14.9 (2005)	–	4.4 (2006)
Access to improved sanitation facilities in % (2008)	97	93	94	98	100
Urban areas	97	94	95	99	100
Rural areas	98	32	94	99	100
Access to improved drinking water source in %	95 (2008)	76	70 (2008)	83 (2000)	87 (2008)
Urban areas	99	97	94	97	98
Rural areas	90	49	61	72	81

Source: CIA World Factbook, accessed June 2012

7 Economy

The dramatic shift from planned to market economy triggered by the disruption of the Soviet Union caused a sharp drop in the production of the Central Asian economies immediately after becoming independent in the early 1990s. This economic collapse was accompanied by an abrupt fall in living conditions for the majority of the population and by an out-migration of ethnic Russian specialists returning to their native country along with their families. As a result – and despite many efforts trying to address the economic breakdown – poverty is still widespread, especially in rural areas but meanwhile also in peri-urban settlements. Many governmental services have ceased to exist or have been considerably reduced in scope at all levels and in many domains such as in particular health and education. This state failure is continuing to strain the relation between authorities and the population.

With the disappearance of the socialist redistributive economy, the new Central Asian states are forced to find their own position in the global market. Due to their geographical double landlocked position between China and Russia as two economic and political giants, this task is difficult, especially for the poorer and weaker countries such as in particular Tajikistan and Kyrgyzstan. Most of the Central Asian economies remain predominantly agricultural relying mainly on primary exports that expose them to fluctuating world prices and financial turmoil.

Despite its huge territory and natural livestock grazing potential, Kazakhstan with its growing industrial development has become a major pull area for agricultural products such as in particular livestock-related products like milk and meat.

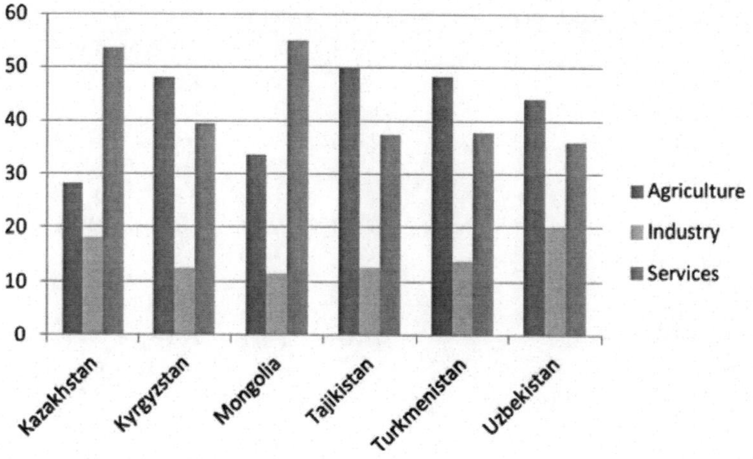

Fig. 1.14 Percent distribution of labor force according to occupation sectors

This provides certain market opportunities for the smaller Central Asian countries to access regional markets. However, competition from China is rapidly increasing as many products such as apples are being imported into Kyrgyzstan. This compromises the market opportunities for certain agricultural and livestock products of the Central Asian states. Nevertheless, the meat market remains a good avenue for income, provided meat quality can be improved and serious issues of animal health are addressed properly and effectively such as in particular brucellosis in Kyrgyzstan or anthrax in Tajikistan (Fig. 1.14).

8 Current Status and Future Prospects of the Five Former Soviet Republics

8.1 Kazakhstan

Kazakhstan is the "economic giant" among the newly created Central Asian countries. Its overall GDP is nearly one-third higher than the total GDP of the four other countries (Table 1.6). Its per capita GDP is considerably higher than in Kyrgyzstan, Tajikistan, and Uzbekistan. The primary/agricultural sector is of much less importance in comparison to the other four countries. Compared to Uzbekistan, the labor force is only about half with a relatively high concentration in the tertiary sector as compared to the other four countries. With regard to unemployment and to inflation rate, Kazakhstan is performing rather well. This translates into a rather moderate portion of its population living below poverty line, i.e., less than 10 %. With regard to the occupation of the labor force, Kazakhstan is currently the only one with less than 30 % in the primary sector (agriculture) and more than 50 % in the tertiary sector (services) as compared to the other four CAS states.

Table 1.6 Key economic features of the five former Soviet republics

Attribute	KZ	KG	TJ	TM	UZ
GDP in billion $	214.5	13.16	16.01	41.51	94.04
Purchasing power parity official exchange rate (est. 2011)	180.1	5.4	6.8	24.1	43.7
GDP real growth rate in % (2011 est.)	6.5	7	6	9.9	7.1
GDP per capita in $ (2011 est.)	13,000	2,400	2,000	7,500	3,300
GDP sectors in % (2011 est.)					
Agriculture	4.6	20.2	18.1	7.8	21.9
Industry	34.2	27.8	22.4	24.1	37.8
Services	55.7	51.9	59.4	68.1	40.2
External debt in billion $ (2011 est.)	95.95	3.74	2.20	0.53	4.44
Health expenditures in % of GDP (2009)	4.3	4.1	5.3	2.3	5.2
Labor force in million	8.7 (2011 est.)	2.3 (2007)	2.1 (2009)	2.3 (2008 est.)	16.3 (2011 est.)
Unemployment rate in %	5.2 (2011 est.)	18 (2004 est.)	2.2 (2009 est.)	60 (2004 est.)	1 (2011 est.)
Distribution of family income/ Gini Index	26.7 (2009)	33.4 (2007)	32.6 (2006)	40.8 (1998)	36.8 (2003)
Budget in billion $ (2011 est.)					
Revenues	38.65	1.32	1.74	4.18	14.97
Expenditures	42.74	1.64	1.78	4.07	15.07
Taxes and other revenues in % of GDP (2011 est.)	21.5	24.4	25.6	17.3	34.3
Budget surplus (+) or deficit (−) in % of GDP (2011 est.)	−2.3	−6.1	−0.6	+0.4	−0.2
Public debt in % of GDP	16 (2011 est.)	28 (2010 est.)	–	–	7.7 (2011 est.)
Inflation rate (consumer prices) in % (2011 est.)	8.3	18.6	14.3	15.0	16.0

Source: CIA World Factbook, accessed June 2012

Total figures for featured countries: labor force 31.8 million; annual oil production 1,914,166 bbl/day; annual gas production 137.2 billion cu m in 2009/2010

Kazakhstan is also both the largest consumer and producer of energy in Central Asia. With a consumption of nearly 78 billion kWh of electricity in 2009, it consumed more electric power than the remaining other four countries (Table 1.7). The regional supremacy of Kazakhstan related to energy production manifests via its impressive and rapidly evolving oil production that boosts its economy and secures the country's international reputation as a new key oil provider on the global market. Kazakhstan's other major export products are ferrous metals, chemicals, machinery, grain, wool, meat, and coal. It has large deposits of petroleum, natural gas, coal, iron ore, manganese, chrome ore, nickel, cobalt, copper, molybdenum, lead, zinc, bauxite, gold, and uranium (Tables 1.8).

Table 1.7 Key energy features of the five former Soviet republics

Attribute	KZ	KG	MN	TJ	TM	UZ
Electricity in billion kWh	(2009 est.)	(2008 est.)	(2010 est.)	(2009 est.)	(2009 est.)	(2009 est.)
Production	75.6	11.7	4.313	16.1	15.5	47.4
Consumption	77.9	7.5	3.375	16.7	13.0	40.1
Oil production in bbl/day (2010 est.)	1,610,000	946	–	220	216,000	87,000
Natural gas production in billion cu m	35.61 (2009 est.)	0.015 (2009 est.)	–	0.038 (2009 est.)	42.4 (2010 est.)	59.1 (2010 est.)
Natural gas proved reserves in trillion cu m (2011 est.)	2.407	0.006	–	0.006	7.504	1.841

Source: CIA World Factbook, accessed June 2012

8.2 *Kyrgyzstan*

Due to its topography, *Kyrgyzstan* is a predominantly agricultural country with nearly 50 % of the labor force engaged in agriculture and contributing to about 20 % of the GDP (Table 1.6). It has the region's second lowest per capita GDP (2,400 USD in 2011) right after Tajikistan (2,000 USD in 2011).

Kyrgyzstan has abundant hydropower, minerals such as in particular gold and rare earth metals, coal, oil, natural gas, and some other marketable deposits like naphthalene, mercury, bismuth, lead, and zinc.

8.3 *Tajikistan*

The civil war 1992–1997 that hit Tajikistan has severely damaged its already weak economic infrastructure causing a sharp decline in industrial and agricultural production. As a consequence, the country today features among the poorest of the former 15 Soviet republics that become independent in 1990. Its per capita GDP in 2011 reached approximately 2,000 USD, and widespread corruption along with weak governance, seasonal power shortages, as well as a high external debt are persisting obstacles toward improving the overall fragile socio-economic situation. Due to the lack of employment opportunities, about 1 million Tajik work as migrants outside the country (40–50 % of the total labor force), and the remittances generated account for about 25 % of total GDP. Tajikistan has the second biggest hydropower generation potentials in the world. In addition to this hydropower, Tajikistan has modest quantities of petroleum, uranium, mercury, brown coal, lead, zinc, antimony, tungsten, silver, and gold as additional natural resources available for exploitation and export.

Table 1.8 Key import and export features of the five former Soviet republics

Aspect	KZ	KG	TJ	TM	UZ
Exports in billion $ (2011 est.)	65.9	2.3	1.8	14.4	13.8
Main export partners 2010 in %	China 20.2 Germany 9.1 Russia 8.5 France 7.1 Turkey 4.5 Canada 4.5 Italy 4.1	Russia 35.7 Uzbekistan 21.9 Kazakhstan 17.3 China 5.4 UAE 4.6 Afghanistan 4.3	Turkey 28.4 Russia 14.4 Uzbekistan 10 Iran 6.2 China 5.6 Norway 4.5	China 28.6 Turkey 10.6 UAE 7.2 Afghanistan 6.5 Iran 6 Italy 5.4 Kazakhstan 4.5	China 21.8 Russia 18.1 Turkey 14.5 Kazakhstan 8.5 Bangladesh 8.5
Imports in billion $ (2011 est.)	32.1	3.7	3.9	9.0	8.7
3 main import partners 2010 with %	Russia 34.3 China 27.7 Germany 5.2	China 61 Russia 17.2 Kazakhstan 5.7	China 35.3 Russia 23.5 Kazakhstan 8.3	Russia 21.6 Turkey 20 China 9.2	Russia 25.4 South Korea 17.3 China 13.9
Major export products	Oil and oil products, ferrous metals, chemicals, machinery, grain, wool, meat, coal	Gold, cotton, wool, garments, meat, tobacco, mercury, uranium, hydropower, machinery, shoes	Aluminum, electricity, cotton, fruits, vegetable oil, textiles	Gas, crude oil, petrochemicals, textiles, cotton fiber	Hydrocarbons (natural gas), gold, cotton
Major import products	Machinery and equipment, metal products, foodstuffs	Oil and gas, machinery and equipment, chemicals, foodstuffs	Petroleum products, aluminum oxide, machinery and equipment, foodstuffs	Machinery and equipment, chemicals, foodstuffs	Machinery and equipment, foodstuffs, chemicals, ferrous and nonferrous metals

Source: CIA World Factbook, accessed June 2012

8.4 Turkmenistan

Based on the prevailing natural conditions, the country mainly lives on intensive agriculture located in irrigated oases as well as on exploiting substantial oil and gas resources and the third producer of fresh milk and meat in the region (Table 1.3).

8.5 Uzbekistan

Besides agriculture and in particular the cotton monoculture inherited from Soviet times, Uzbekistan is trying to diversify its export portfolio by developing in particular the infrastructure to exploit the available oil and natural gas reserves as well as other natural resources. Like most of its neighbors, Uzbekistan has deposits of coal, gold, uranium, silver, copper, lead, zinc, tungsten, and molybdenum that are in high demand on the global market and may help boost the economic development of the country (Tables 1.8).

Twenty years after independence, some of the economies have started recovering, showing signs of steady improvement and growth, mainly thanks to the availability of high-demand natural resources such as energy carriers in particular. Unfortunately the GDP growth rates have not yet helped to reduce poverty and social inequality – partially due to inefficient governmental service delivery structures as well as persisting political instability in many instances.

Unemployment rates are generally high, and the job markets are far too small to absorb the entire workforce – in particular youth. Considering that in many of the countries nearly half the population is under the age of 20 and thus about to join the workforce. This is a key issue that needs to be addressed in the context of regional sustainable development as it might again contribute to a drastic increase of unemployment and trigger more migration flows toward foreign destinations.

9 Politics

Since their independence in the early 1990s, the political changes within the Central Asian republics have had a critical impact on their respective political, social, and economic development as well as their integration into the worldwide market economy. Changes have been dramatic and will continue to shape the future of society in the region. At present, Table 1.9 gives an overview of the main systemic features of the political systems in the five Central Asian republics.

10 Communication, Infrastructure, and Mobility

Connectivity in all its facets is a key aspect of modernization in the present-day Central Asia. The vast often thinly populated flat territories and the huge mountain ranges represent natural obstacles and challenges for securing an adequate communication and transportation for many rural communities (Figs. 1.15, 1.16 and Table 1.10).

Table 1.9 Key political features of the five former Soviet republics and Mongolia

Attribute	KZ	KG	MN	TJ	TM	UZ
Capital	Astana	Bishkek	Ulaanbaatar	Dushanbe	Ashgabat	Tashkent
Government type	Republic with authoritarian presidential rule and little power outside the executive branch	Republic	Republic	Republic	Declared secular democracy and presidential republic; de facto authoritarian presidential rule	Republic with authoritarian presidential rule
Juridical system	Civil law system	Civil law system	Civil law system	Civil law system	Civil law system with Islamic influences	Civil law system
Executive power	President (elected by popular vote for 5-year term), Chairman of Supreme Soviet; Prime Minister (appointed by President) Head of government; Cabinet with Council of Ministers appointed by President	President (elected by popular vote for 6-year term); Cabinet of Ministers proposed by Prime Minister, appointed by President; Prime Minister nominated by strongest party (coalition)	Presidential candidates nominated by political parties represented in State Great Hural and elected by popular vote for a 4-year term	President (elected for 7-year term, renewable 1×), Head of State and Chairman of Supreme Assembly; Cabinet with Council of Ministers appointed by President, approved by Supreme Assembly; Prime Minister Head of Government, appointed by President	President (elected for 5-year term; Chief of State and Head of Government); Cabinet of Ministers appointed by President	President (elected for 7-year term); Cabinet of Ministers with Prime Minister

(continued)

Table 1.9 (continued)

Attribute	KZ	KG	MN	TJ	TM	UZ
Legislative power	Bicameral parliament (Senate and Mazhilis)	Unicameral Supreme Council (members elected by popular vote to serve 5-year terms)	Unicameral State Great Hural (76 seats; 48 members directly elected from 26 electoral districts; 28 members proportionally elected based on party's share of total votes; all 4-year term)	Bicameral Supreme Assembly with National Assembly (upper chamber) and Assembly of Representatives (lower chamber)	Unicameral National Assembly (second chamber "People's Council" abolished in 2008)	Bicameral Supreme Assembly with Senate and Legislative Chamber
Judicial power	Supreme Court and Constitutional Council	Supreme Court and Constitutional Court (judges appointed for 10-year terms by Supreme Council on recommenda-tion of President)	Supreme Court, appeals' court for people and provincial courts; judges nominated by General Council of Courts, approved by President	Supreme Court (appointed by President)	Supreme Court (appointed by President)	Supreme Court (nominated by President, confirmed by Supreme Assembly)

Political parties	12 parties with various additional pressure groups	6 parties and several pressure groups	4 main parties	7 parties with splinter parties and unregistered parties	1 party without opposition (political parties allowed to register since Jan 2012)	5 parties without major opposition or pressure group
Administrative division	14 provinces	7 provinces	21 provinces	2 provinces and 1 autonomous province	5 provinces	12 provinces

Source: CIA World Factbook, accessed June 2012

Fig. 1.15 Mechanized conveyance is used to transport equipment and portable housing to and from the summer grazing – generally over unmade roads and crossing rivers over hazardous bridges (photo D. Maselli, Kyrgyzstan 2008)

Fig. 1.16 Mountainous terrain and remote locations contribute to the problems of trade, lack of public transport services, and poor communications in many Central Asian countries (photo D. Maselli, Tajikistan 2008)

Table 1.10 Key communication features in five former Soviet republics and Mongolia

Attribute	KZ	KG	MN	TJ	TM	UZ
Internet hosts	53,984 (2010)	97,976 (2010)	20,865 (2011)	1,504 (2010)	794 (2010)	47,718 (2010)
Internet users in million (2009) in % of population	5.3	2.2	10.4	0.7	0.08	4.7
Railways in km 2010	15,079	470	1,908	680	2,980	3,645
Roadways in km	93,612 (2008)	34,000 (2007)	49,249 (2010)	27,767 (2000)	58,592 (2002)	86,496 (2000)
Waterways in km	4,000 (2010)	600 (2010)	580 (2010)	200 (2010)	1,300 (2008)	1,100 (2009)

Source: CIA World Factbook, accessed June 2012

References

Emadi MH (2012) Better land stewardship to avert poverty and land degradation in Afghanistan. In: Squires VR (ed) Rangeland stewardship in Central Asia. Springer, Dordrecht, pp 91–108 (Chapter 5, this volume)

Hannam I (2012) International perspectives on legislative and administrative reforms as an aid to better land stewardship in Central Asia. In: Squires VR (ed) Rangeland stewardship in Central Asia. Springer, Dordrecht, pp 407–430 (Chapter 17, this volume)

Hua L, Zhang D (2012) Engaging with land users; the first steps on a long road. In: Squires VR (ed) Rangeland stewardship in Central Asia. Springer, Dordrecht, pp 333–356 (Chapter 14, this volume)

Jackson R (2012) Fostering community-based stewardship of wildlife in Central Asia: transforming snow leopards from pests into valued assets. In: Squires VR (ed) Rangeland stewardship in Central Asia. Springer, Dordrecht, pp 357–380 (Chapter 15, this volume)

Jacobs M, Schloeder C (2012) Extensive livestock production: Afghanistan's Kuchi herders, risks to and strategies for their survival. In: Squires VR (ed) Rangeland stewardship in Central Asia. Springer, Dordrecht, pp 109–128 (Chapter this 6, this volume)

Kurbanova B (2012) Constraints and barriers to better land stewardship: analysis of PRAs in Tajikistan. In: Squires VR (ed) Rangeland stewardship in Central Asia. Springer, Dordrecht, pp 129–163 (this volume)

Rahimon RM (2012) Evolution of land use in nomadic culture in Central Asia with special reference to Kyrgyzstan and Kazakhstan. In: Squires VR (ed) Rangeland stewardship in Central Asia. Springer, Dordrecht, pp 51–70 (Chapter 3, this volume)

Vanselow KA, Kraudzun T, Cyrus Samimi C (2012) Land stewardship in practice – an example from the Eastern Pamirs of Tajikistan. In: Squires VR (ed) Rangeland stewardship in Central Asia. Springer, Dordrecht, pp 71–90 (Chapter 4, this volume)

Chapter 2
Better Land Stewardship: An Economic and Environmental Imperative, If There Is to Be Sustainable Development

Victor Squires

Abstract This chapter attempts to get to grips with the concept *of land stewardship* and its links with sustainable development (SD) in the context of the Central Asian region (CAR as defined in Chap. 1). The idea is to convey clarity to the concept by elucidating the principles and practices which can make it work, particularly in rangeland-based production systems and the respective local rural populations. However, the diversity of rangeland-based agriculture and livestock raising systems throughout the CAR, especially in the five Central Asian "stans," Kazakhstan, Kyrgyzstan, Tajikistan, Turkmenistan, and Uzbekistan, makes it rather difficult to characterize the consequences of a successful transition from Soviet style to more sustainable farming systems and the adoption of an attitude of land stewardship among the rural community and those who attempt to regulate its use (policy-makers and other government officials).

SD is a concept that people know about, but opinions differ as to what it means in concrete term actions. The key question is: "What do we want to maintain, for what purpose and for whom?" Many possibilities exist. These are elaborated in this chapter and elsewhere in the book as is some examination of the relationship between SD and land stewardship and what being a land steward really means.

Keywords Sociocultural • Intergenerational equity • Ecology • Economics • Rangeland-based industries • Ecological goods and services • Stewardship priorities • Interconnectedness • Land tenure systems • Farming systems • Sustainable land management • Biosophysical approach • Landlessness • Geobotanic survey • Land degradation • WOCAT • Land users' views

---

V. Squires (✉)
GITEC Consulting GMBH, Cologne, Germany

College of Grassland Science, Gansu Agricultural University, Lanzhou, China
e-mail: dryland1812@internode.on.net

V. Squires (ed.), *Rangeland Stewardship in Central Asia: Balancing Improved Livelihoods, Biodiversity Conservation and Land Protection*,
DOI 10.1007/978-94-007-5367-9_2,

Key Points

- In its broadest sense, land stewardship is the recognition of the collective responsibility to retain the quality and abundance of land, air, water, and biodiversity and to manage this natural capital in a way that conserves all of its values, be they ecologic, economic, or sociocultural. Stewardship is a journey, and the commitment that comes with being good land stewards takes time and effort.
- There is nothing new in the quest by humankind for permanent ways of using the land entrusted to them. The intent of this book is to identify which players, policies, and procedures can best contribute to a more sustainable way of living of the CAR (rural) societies and ensure that land is used in a way that will allow future generations to enjoy the benefit stream that can flow from land that is used in a sustainable way. This gives rise to the concept of *intergenerational equity.*
- Sustainability of land use systems should be a high priority in all countries of the CAR because agriculture (including animal husbandry) is the cornerstone of the economic health and well-being of all them. The meaning of sustainability presents problems, its meaning "in theory" is commonly intuited, but "in practice" it is seldom really explained or understood. Likewise, there are the ongoing responses to periodic changes in population density, weather patterns, competing land uses, alternative economic uses, and natural condition of resource base for each rangeland site. Perhaps what is more appropriate is "more adaptable and sustainable ways of living for the CAR societies."
- Maintaining the productive quality of and its resources for continued future use is one of the most important challenges directly confronting the rural population in CAR, but the problems also impact on the wider society of all these countries.
- The first priority in all CAR rangelands is to maintain and restore the ecological sustainability of the watersheds and rangelands for present and future generations. To achieve and maintain the ecological sustainability of the rangelands while balancing the diverse economic and social needs of rangeland inhabitants along with conserving rangeland biodiversity and watershed values is a major challenge. The wider societies in CAR countries—both rural and urban—have an enormous stake in fostering progress toward profitable, environmentally friendly farming systems.
- Regardless of how we define sustainable use of land, it is ultimately the land user who must establish practical, sound practices. So the land users' point of view of what constitutes sustainability is the most important perception of all. There is a clear need to bridge the gap between production and income objectives of the land users on the one hand and the long-term objective of preserving natural resources on the other.

1 Introduction

The maintenance of land and its resources for continued future use are the most important environmental problems directly confronting the rural population in Central Asia, but the problems also impact on the wider society of all these countries (see below). Even defining the problem and deciding which outcomes are best is fraught with difficulties, confusion, and conflict (Brown et al. 2008; Kreutzmann 2012). In this chapter, there is an attempt to get to grips with the concept *of land stewardship* (and of sustainability) and convey it with clarity, elucidating the principles and practices which can make it work, particularly in rangeland-based industries and related rural populations. Stewardship is defined by Worrell and Appleby (2000) as the responsible use (including conservation) of natural resources in a way that takes a full and balanced account of the interests of society, future generations, and other species, as well as of private needs, and accepts significant answerability to society. In a farming context, stewardship refers to the notion that farmers are stewards of the land and that farming is a way of life that places implicit responsibility on farmers to look after the land for future generations. Stewardship has relevance to aspects of land tenure and property rights, which makes it applicable across a wide range of fields of resource use. Stewardship is important in addressing land tenure, which is one of the major contributors to land degradation, mainly in developing countries, where land tenure systems limit the possibility people have to take full responsibility for the land (see below).

In its broadest sense, stewardship is the recognition of the collective responsibility to retain the quality and abundance of our land, air, water, and biodiversity and to manage this natural capital[1] in a way that conserves all of its values, be they environmental, economic, social, or cultural. Although the definitions are almost limitless, there really are two clear and fundamental elements of "stewardship"—*awareness* and *action*. That means:

1. Recognizing the collective responsibility to retain the quality and abundance of our natural resources.
2. Putting that awareness into action by making the appropriate decisions for how to best use and manage these resources not only for today but for future generations as well.
3. But implementation of decisions and monitoring outcomes are also required as the steward needs to periodically adapt improvements or changes to emerging conditions in response to past and ongoing interventions/actions.

Being or becoming a good land steward implies:

- Understanding the concept of giving *value to ecological goods and services*. We depend on ecological goods and services everyday for our health, social, cultural, and economic needs. Ecological functions are the base resources that sustain our lives. The sustainability of communities and economies depends upon an ability to maintain or restore the ecological functions of both urban and rural landscapes. *Ecological goods* are the products of the processes and interactions of natural systems. The natural world provides us with the essential services we require for life. These services are called *ecological services*.

[1] They are the resources and benefits provided by the ecosystem that are essential for human survival and economic activity. See discussion on ecological goods and services.

- Recognizing important *stewardship priorities*. CAR is a region of diversity—socially, culturally, and ecologically. Ecologically, this vast region is blessed with unique and distinct landscapes (Maselli, Chap. 1). Each has its own unique natural features: of which climate is one important shaping factor, and each supports its own distinct species of animals and plants. Over the past century, these landscapes have been intensely impacted by human development. Long-time residents of an area can pinpoint the visible and dramatic changes that have occurred in their local natural landscapes (Kurbanova, Chap.7). In highly populated areas of the country, native grasslands, aspen-dominated parkland, wetlands, and other natural features have been virtually eliminated. Natural resource exploitation (including mining and urban and infrastructure development) is increasing exponentially and continues to significantly impact remaining natural landscapes.
- Learning from those who provide good stewardship examples. This is where stewardship begins.
- Applying key *stewardship principles* in all our land and resource use decisions. Effective stewardship helps maintain and restore the function of the natural resources (air, land, water, biodiversity); we rely on to produce the goods and services we depend. There are four guiding principles of environmental sustainability:

1. *Caring for the system as a whole*—Adopting an ecosystems' holistic resource management approach includes understanding the fundamental roles and values of natural systems, building up biological fertility in the soil, incorporating an understanding of the ecological cycles on the landscape (water, energy, nutrients), and how land-use practices can either benefit, be in harmony, or negatively impact these cycles and other land-users' flora and fauna.
2. *Conserving resources*—Maximizing efficiency and striving to reduce the consumption of renewable and nonrenewable resources and long-term optimization versus short-term maximization of production.
3. *Maintaining and enhancing stability in nature*—Sustaining and encouraging natural biological diversity and complexity and maintaining natural areas and functions on the land (i.e., wildlife habitat conservation).
4. *Applying cultural values*—Caring for the health of the land for future generations and long-term economic stability, the link between civilization (urbanization) and the land-base and ecosystems that are vital to survival, and the intrinsic value and right of all life on Earth to exist.

The resource base for agriculture, including animal husbandry on rangelands, unless husbanded carefully and replenished continually, will dwindle in its capacity to produce at levels required to meet the demands of burgeoning population and changed market demands from an increasingly urbanized society. Farming systems collapse or are forced to change when they become unprofitable to the farmer or when they impose on farm families, neighbors, or rural communities (or perhaps whole nations) excessive indirect costs or burdens. These indirect costs arise, for example, from increased frequency and severity of natural disasters such as floods or landslides that are attributable to poor land management.

There is nothing new in the quest by humankind for permanent ways of using the land entrusted to them. The challenge of this book is to identify which players, policies, and procedures could contribute to a more sustainable society and ensure that land is used in way that will allow future generations to enjoy the benefit stream that can flow from land that is used in a sustainable way. This gives rise to the concept of *intergenerational equity*.[2]

The topic of sustainability of agricultural land *sens lat.* (including rangelands) should be a high priority in all countries of the CAR, especially the higher altitude areas (Kreutzmann 2012) because agriculture (including animal husbandry) is the cornerstone of the economic health and well-being of all them. There is no more important question than that of the sustainability of agricultural ecosystems. Desertification, deforestation, and accumulation of chemicals in soils and waters are of increasing concern in many ecosystems in most countries in the CAR. Many definitions of sustainability have been presented, and this is as it should be. The word sustainable may imply a steady state.[3] If one sees a steady-state situation, one must look over horizons to some distant goal. A careful reading of the development literature reveals as many ideas about direction as there are authors, so consensus on an equilibrium point would be impossible. A workable definition is *"an agriculture that can evolve indefinitely toward greater human utility, greater efficiency of resource use, and balance with the environment that is favorable both to humans and to most other species"* (Harwood 1990).

This definition is heavily value-laden but is consistent with the parameters of an emerging social and political agenda for agricultural development. It is also very generic. To understand the process by which it is translated into substance in any national setting, some sense is needed of political agendas, the translation of these agendas into policy, and the roles, the agenda, and the policy in national development. Quite clearly, there are differences between the various countries that make up the CAR.

Several scenarios for sustainable land use have been articulated, and most stress the following:

- The interconnectedness of all parts of a farming system, including the farmer and his family
- The importance of the many biological balances in the system
- The need to maximize desired biological relationships in the system and to minimize use of material and practices that disrupt these relationships

Sustainability involves the complex interactions of biological, physical, and socioeconomic factors and requires a comprehensive approach in order to improve existing systems and develop new ones that are more sustainable.

[2] It means that we inherit the Earth from previous generations and have an obligation to pass it on in reasonable condition to future generations.

[3] Sustainable development (SD) means not a steady state as such but can/should imply the increase of ecologic, economic, and sociocultural capital.

The notion of sustainability presents problems; its meaning "in theory" is commonly intuited but "in practice" is seldom really explained or understood, especially whenever the prevailing condition of a given rangeland is not inherently productive and whenever an alternative form of land use is far superior in a given case. Likewise, there are the ongoing responses to periodic changes in population density, weather patterns, competing land uses, alternative economic uses, and natural condition of resource base on each rangeland site. Perhaps what could be agreed upon in the context of this book is "more adaptable and sustainable ways of living for the Central Asian societies."

Sustainable land use should involve the successful management of resources to satisfy changing human needs while maintaining or enhancing the natural resource base while avoiding environmental degradation. The emphasis in CAR countries is on rural income, and employment and trying to handle environmental problems, or instilling a land ethic (land stewardship) by the passing of laws against land degradation simply will not work:

- Key constraints hampering the sustainable use of pastures/rangelands in CAR. Weak infrastructure in many of the countries is a major constraint to a higher productivity/rentability and transporting and marketing crop and livestock products.
- Financial and administrative systems are often biased toward urban consumers.
- Land tenure systems can discourage land users from conserving natural resources and investing in future productivity.
- Most countries in the CAR lack laws to protect forests and rangelands from indiscriminate exploitation, and the implementation is also a problem.

2 Society's Stake in Better Land Stewardship

What are the expectable/desired benefits of more widespread adoption of responsible land stewardship in CAR countries? To answer this question requires some sense of where rangeland-based farming and animal husbandry is now, relative to more sustainable land use, and how it might change as progress is made in adopting cropping patterns and grazing management systems more consistent with the principal features of sustainability previously identified.

First, here are some general points regarding the sustainability of agriculture at the present time. The economic scorecard is well known and clearly not good in several key respects. The CIS countries[4] generally have low GDP and low scores on human development index (HDI). Many rural people are on or below the poverty line, and the continuation of the overall contribution of rangeland-based production systems (including livestock) to national economic activity and long-term prospects is problematic (Kurbanova, Chap. 7; Lerman, Chap. 8; Sedik, Chap. 9).

[4] Commonwealth of Independent States – former member countries of the Soviet union

Second, the problems faced by rangeland-based industries (Strong, Chap. 10) that could undermine sustainable land use differ greatly by region, both in degree and character. Many, if not most, farming systems currently practiced in CAR are unsustainable. Adjustments in livestock husbandry and cropping practices and technologies surely will be needed for most farming systems to remain viable, even for subsistence level outputs. The transition to sustainable production systems leads to development of diversified (and more specialized) producers whose enterprises capture the comparative advantage of the rangeland resource and location in relation to markets for their outputs.

The cost of these practices to land users to move toward more sustainable systems will not be great and generally can be spread over a variety of agronomic and ecological benefits such as moisture and nutrient retention *in situ*, improved soil, and higher forage yields. An exception could be on the more steeply sloping land that will require terracing or other costly structural practices to keep soil erosion in check.

Third, sustainable farming systems and practices will be adopted when, and only if, they offer farmers a convincing opportunity to earn higher profits than from any other systems.

Finally, we must remember that current systems often are used because they were used the year before, are proven, reduce short-term risk, and require modest investment and a low level of management skill and equipment.

The diversity of rangeland-based agriculture and livestock rising throughout CAR makes it difficult to characterize the consequences of a successful transition to more sustainable farming systems and the adoption of a sense of land stewardship among the rural community. Nonetheless I will try. If we assume that sustainable agricultural and animal husbandry practices (including adoption of better grazing management) will be incorporated into specialized farming systems, this trend would greatly facilitate the arrest and reversal of land degradation. By fostering spring deferment of grazing, rest rotation grazing, use of crop rotation, minimum tillage, early weaning, better winter housing, improved management of hayfields, and so on so productivity rises and incomes increase.

The principal benefits to society from arresting and reversing land degradation will be higher household incomes, increased productivity of farming systems, conservation of biodiversity, increased carbon sequestration, and fewer floods and landslides. Wildlife habitat will be improved (Jackson, Chap. 15), and over time, new recreational opportunities and tourism opportunities will arise. Over time, the economic value of these benefits, while difficult to quantify, surely will exceed several billion dollars each year.

The wider society in CAR countries (both rural and urban) has an enormous stake in fostering progress toward continuously productive farming systems. The two most dramatic near-term benefits from such progress will involve, first, improved economic performance and increased household incomes, made possible by increased offtake from livestock. Higher benefits that will flow from development of more specialized farming system. For example, highlands can focus on breeding and lowland farmers can take the younger livestock for fattening and finishing to meet the market demand for meat, milk, and other livestock products, and from the burgeoning urban populations (Leake, Chap. 18).

3 Sustainable Resource Management, for Whom?

The "ecosystem stewardship" approach proposed by Chapin et al. (2010) integrates three strategies for sustainable development: reducing vulnerability to expected changes, fostering resilience to sustain desirable conditions in the face of perturbations and uncertainty, and transforming from undesirable trajectories when opportunities emerge. Each of these is applicable to sustainable land management (SLM) in the pastures and rangelands.

Sustainability is a concept that most people know about, but opinions differ as to what it means. *Sustainable development* means "development to meet the needs of the present, without compromising the ability of future generations to meet their own needs." Sustainable development assumes the alignment of development decisions with environmental considerations. The key question is just "what do we want to maintain?" Many possibilities exist. Do we want to maintain:

(a) The rural population and community structure at existing levels?
(b) The biological and ecological integrity of the region?
(c) The financial viability of farmers and herders?
(d) The culture and traditions of the farmers and herders?

Once a decision is made as to which of these (singularly or in combination) is the main aim, then the action taken to achieve this aim can be specified.

The impact of people on the land resources depends on a number of "pillars":

- Values and beliefs
- Cultural norms
- Knowledge generation and transfer
- Research and development
- Business and financial institutions' social and other service systems
- Legal and justice systems
- Civic and political institutions

A program for the sustainable development of the rangelands can be seen through three strategic approaches, each of which is supported by selected programs of activity, summarized in Table 2.1.

Approach 1 covers essential institutional support. The most important being land tenure initiatives. *Approach* 2 addresses livestock and rangeland development. Livestock production is the key economic driver that will underpin any sustainable support needed for the rangelands. The rangelands require appropriate support for

Table 2.1 Three strategic approaches in a program to develop sustainable land use

Approach 1	Approach 2	Approach 3	
Institutional/policy development	Livestock and rangeland development	Diversification	
		3.1 Nonlivestock agricultural production	3.2 Nonagricultural production
1.1: Land tenure	2.1: Fodder production	3.1.1: Agroforestry	3.2.1: Review employment opportunities/ regional and district plans
1.2: Policy	2.2: Veterinary services		3.2.2: Other investment/ business opportunities
1.3: Farmer/community organizations	2.3: Capacity building FFS[a] for improved husbandry		3.2.3: Capacity building—vocational training
1.4: Capacity building	2.4: Pasture management		
1.5: Legislation	2.5: Feeding/nutrition		
1.6: Rural finance	2.6: Land degradation		

[a]FFS Farmer Field Schools—farmer-led, farmer-organized training in the field

land degradation remedial measures to ensure continued viability of the land that the livestock sector depends on. *Approach* 3 covers nonlivestock production activities. These are highly relevant if livestock economic pressures on the rangelands are to be relieved, to aid recovery.

Central Asian societies, especially rural societies, are deficient in a crucially important ingredient. The missing ingredient is an understanding by the majority of the population of:

- Sensitivity and interdependence of living systems
- The place of humans in nature
- Relations between cultural and natural processes

Correcting this deficiency is essential for attaining an ecologically sustainable society which is at the same time satisfying in terms of quality of life. Such an attainment will require a *biosophysical*[5] approach to priority-setting and decision making throughout society. The challenge to today's generation is how to devise land use systems that will maintain the productive capacity of the land for the benefit of both present and future generations.

[5] *Biosophy* is the science and art of intelligent living based on the awareness and practice of spiritual values, ethical-social principles, and character qualities essential to individual freedom and social harmony.

4 Sustainability and Change in the Natural Environment

There is a tendency to consider the earth's ecosystems as constant and unchanging when left to natural processes, but clearly this is not so. Time series photographs and the results of geobotanic surveys document the changes in ecosystems as evidenced by loss of woody plants that were cut for fuelwood and loss of forage species as a result of overutilization. As the changes in the pattern and intensity of rainfall and slowly rising temperatures over the past 20 years or so have shown, the biophysical environment is changing (Fig. 2.1).

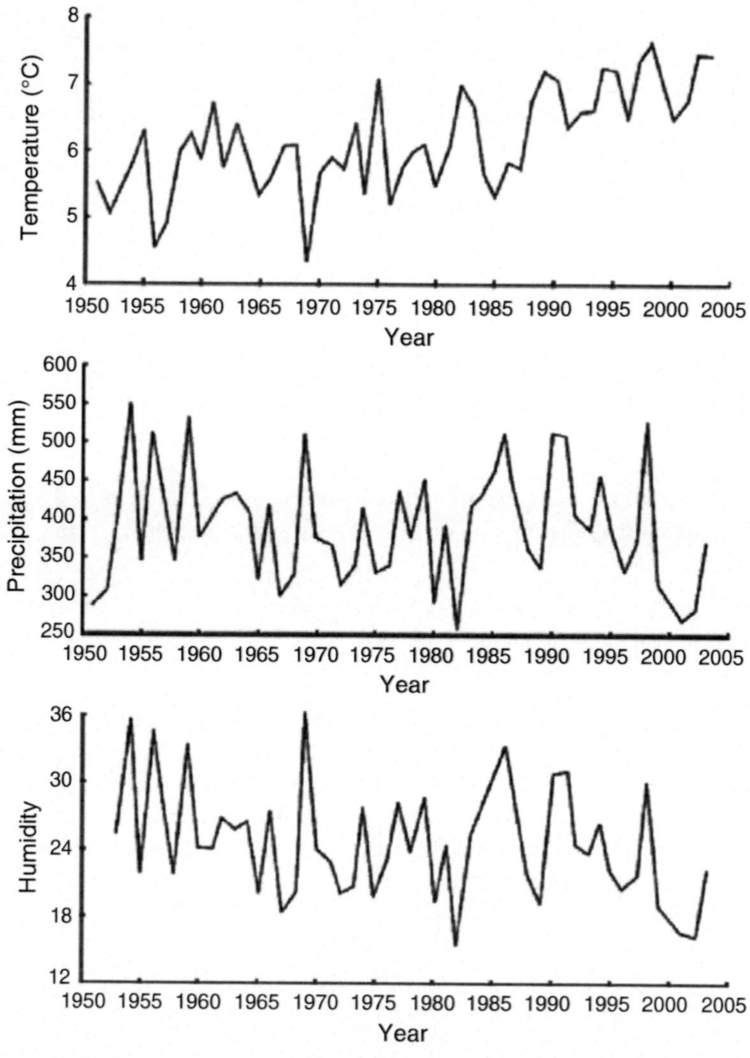

Fig. 2.1 Time series data from Junggar Basin, Xinjiang in western China over a 50-year period showing variability in temperature, rainfall, and humidity

As people strive for a better life, there has been a tendency to see development as synonymous with economic growth and wealth in monetary terms. Both capitalist and socialist systems are based on the quest for economic growth, which in turn is recognized as the mark of progress—the measure by which a nation's well-being has been gaged. The long-term implications of this approach has come to the fore as ecological disasters such as the Aral Sea crisis in western Kazakhstan have become more widely known.

The ideal now is sustainable development (SD). This ideal is a concept that combines two conflicting visions of social change. The ideal originates from environmental concerns rather than financial considerations and is largely based on the conviction that present economic practices, especially those applied in the countries of Central Asia, cannot preserved the Earth's productive potential for future citizens. Thus, policies aimed at satisfying short-term economic and social needs have conflicted with the need to account for what has become known as "*intergenerational equity.*" It has been argued by several commentators that sustainable land use can be achieved without destroying the environment while improving the livelihood of low-income people, but CAR presents special challenges.

In the great majority of cases, the rural poor in the CAR countries have only two options: remain as landless laborers suffering chronic poverty and malnutrition (Kurbanova, Chap. 7) or take the opportunity to occupy marginal grazing and croplands. With most of the good agricultural land concentrated in the hands of an elite class of landlords, the redistribution of land becomes critical both to social justice and to the sustainability of societies (Kurbanova, Chap. 7; Halimova, Chap. 13).

The importance of human–environment interactions to the condition of land compels attention to adaptive management. In order to reconcile concerns and agendas at a higher strategic level, identification of synergies, conflicts, trade-offs, interconnections, feedbacks, and spillover effects among multiple objectives, drivers, actions, policies, and time horizons is crucial. Once these issues are transparent, coordinated action can be put into place (Cowie et al. 2011).

But what does sustainable development mean in practice? It means inter alia using the land in ways that do not degrade it. A consensus definition of sustainable land management (SLM) has proven elusive. Adapting the well-known definition of sustainable development devised by the World Commission on Environment and Development (UN 1987) to the case of land, we consider SLM to be "the management of land to meet present needs without compromising the ability of future generations to meet their own needs." There are now in use a number of different labels for sustainable land use, but the term regenerative agriculture is the term I prefer. In my opinion, enhanced regeneration of natural resources is essential to the achievement of a sustainable form of land use. Other aspects will contribute to sustainability, but regeneration is the component of sustainability that is most important. The term "regenerative" can become part of the language of renewal, reconstruction, and permanence of rural people (and ultimately urban people as well).

There is a need to develop production systems that maximize positive synergies between the various elements of a specific system. SD is commonly defined through the three "pillars" consisting of ecological, economical, and social components. Essentially, sustainable development (SD) comprises a set of strategies and tools to:

- Integrate biodiversity conservation and ecosystem structure and function with economic development
- Maintain ecological integrity of the rangeland system so as to conserve biodiversity and carbon sequestration capacity
- Ensure satisfaction of basic human needs such as food, shelter, and security
- Achieve equity and social justice
- Provide for social self-determination and cultural diversity

Land degradation (LD) is a major problem for most of CAR's rangelands (Gintzburger et al. 2003). Accelerated soil erosion in all of its forms (gullies, sheet and rill erosion, slumping, and in some places wind erosion) is widespread. Hence, there is a clear need to tackle the root *causes* of LD and not just to deal with its *consequences*. Most efforts in the past have been aimed at "solving" minor problems such as "how to get more forage per hectare" rather than at dealing with the underlying causes of low productivity such as insecure land tenure, unclear boundaries for the assigned grazing user rights (Halimova, Chap. 13), or lack of clear policy on how to balance livestock numbers and feed supplies (Michalk et al. 2010). There is need to raise awareness among both land users (herders and farmers) and the technical staff at all levels (from national to village [*kishlak*] level) of the keys to sustainability and the realities of the market economy. Many LD problems have their origin outside of the agriculture sector. They arise from legislation and policies developed in the cities and from pressures exerted by market forces—including international ones that affect trade and world commodity prices.

The policy approach of the government of each of the CAR countries toward rangeland degradation and farmer/herder livelihoods cannot be viewed in isolation of policy developments and operation in other parts of society. In the transition from a centrally planned to a more market-oriented economy, the types of policies, the mix of policies, and the broader policy and institutional setting have all undergone dramatic changes. The rural sector has not been immune from the general reforms and changes occurring elsewhere in society.

5 Challenges Faced in Reversing Land Degradation

- There have been successful examples of reversal of land degradation as detailed in the book "Where the Land is Greener" (WOCAT 2007). Similarly, useful work on soil erosion control has been done by the Soil Science Research Institute of Tajikistan and doubtless by other research agencies in other CAR countries. More needs to be done to replicate these proven practices and approaches and both replicate them and facilitate scaling-up. A deeper insight into the mechanisms and processes of recovery and restoration of degraded lands has been provided by Tongway and Ludwig (2010).

- Economies in CAR, especially in the five "stans," are in transition toward a market economy at a time when GDP is still the lowest of the CIS countries. This imposes considerable constraints on government spending and presents a number of serious challenges. Low relative incomes and the incidence of poverty are pervasive in rural areas and a major challenge for all levels of government.
- There has been a massive increase in livestock numbers and a more intensive use of the forage resource. This intensive use has led to severe degradation and lower productivity as well as massive reductions in carbon sequestration potential and in biodiversity.
- Some of the key issues and challenges confronting the pastoral lands and their users include increasing human population, excess grazing pressure, increasing land degradation, more intensive use of the rangelands, and the link between livelihoods, ecological services, and degradation.
- Growth of livestock industries is constrained by the availability and cost and availability of feed inputs. Another key input is labor. Population growth in pastoral areas has exceeded that in other parts of each country. There is a high proportion of the male population involved in work as migrant labor. For example, many of the able-bodied men from Tajikistan between the ages of 18 and 50 years have gone to work in Russia. At the same time, the relative scarcity of capital in the poor areas has constrained growth in the pastoral region.

There are six major focus areas in rural development in CAR that impact on any attempt to adopt and promote SD. These are:

Managing structures (fragmented and chaotic structures, overlapping mandates, and economies of scale)
Managing policies (lack of comprehensive rangeland management policy framework; inconsistency in planning, legislation, and programs; coordination with other policies)
Managing institutions (powers and responsibilities, capacity to carry out tasks, coordination, facilitative vs. interventionist approaches)
Managing people (structural adjustment of pastoral and agricultural industries, settlement policies, population policies)
Managing livestock (livestock industry development, technology [feed, breeding, grazing management])
Managing markets (price determination and macrolevel management of markets, microlevel management of markets)

The major challenges to be addressed in the rangelands include:

- The need to improve information on extent and state of the rangelands, and how they are changing over time.
- The need for rangeland technicians to refine existing models of rangeland ecology and to work with economists, livestock specialists, and pastoral development experts to design appropriate management systems for livestock production.

- Linkages between the ecological aspects of conserving the biodiversity and watershed values of the rangelands and the economic benefits and goals of sustainable development. The problems of the region's pastoral areas need to be more clearly articulated.
- Need to better integrate the mainstreaming of biodiversity conservation programs for the rangelands with other development and environment activities.

For the rangelands, the first priority in all CAR rangelands is to maintain and restore the ecological sustainability of the watersheds and rangelands for present and future generations. Restoring and maintaining the ecological sustainability of the rangelands while balancing other needs is a challenge. Diverse economic and social needs of rangeland inhabitants, along with pressure to conserve rangeland biodiversity and watershed values, add to the difficulties faced. Meeting the challenge requires that there is a move away from a focus of sustaining livestock outputs from the rangelands to one of sustaining ecological processes and a wide variety of goods, services, conditions, and values. Many rural people face a downward spiral of decreased grazing land, of increased crop encroachment, and spiraling firewood requirements. These forces contribute to the impoverishment of the rural population and to accelerated land degradation. The trend is being exacerbated by recurring drought, and vulnerability to drought is one of the main indicators of long-term environmental and social sustainability of these farming systems (Squires 2011).

Within rangelands,[6] ecological sustainability requires maintaining the composition, structure, and processes of the rangeland ecosystem. The concept of *ecological sustainability* provides a foundation upon which the management of rangelands can contribute to the goals of economic and social sustainability. Implementation of ecological sustainability into development plans for rangeland areas is not a precise process; there are many unknowns and risks that cannot be controlled. Therefore, planning for rangeland sustainability should acknowledge the following features of rangeland systems:

- The dynamic nature of ecological systems
- The significance of natural processes
- The uncertainty and inherent variability of ecological systems
- The impact of cumulative effects (including climate change)

When developing actions to reverse LD and improve livelihoods and conserve biodiversity, there is a need to:

- Leave options open by not preempting future actions
- Conserve habitat for native species of plants and animals
- Raise productivity of ecological systems
- Reduce uncertainty through adaptive management and continuous learning

[6] See Squires 2011, for a fuller discussion of rangeland, including widely accepted definitions and the goods and services that derive from it.

Since rangelands vary considerably across the CAR, development programs need to be focused at local community levels. This requires improved community participation and the development of sustainable participatory mechanisms for community-based natural resource management (Kurbanova, Chap. 7). There is a clear need to bridge the gap between production and income objectives of the land users (Michalk et al. 2010) on the one hand and the long-term objective of preserving natural resources on the other.

There is recognition now too of the fact that there are few management options available to the land users. Those that do exist fall into two categories:

Reduce total grazing pressure (from livestock, from mammalian competitors such as rodents and wildlife, and from grasshoppers and other invertebrate pests) by reducing herd/flock sizes through heavier culling and through adoption of precision management to cull unproductive animals. Breed improvement also falls into this category as a longer term strategy, but it is not a panacea. Improved breeds will not perform well unless they get better feed.

Increase feed supply and/or utilization efficiency (by planting sown pastures and fodder crops; by utilizing crop residues in a better way, e.g., urea treatment; by conserving fodder as hay or silage). Better ration formulation for penned animals helps to make better use of the available feed and allows the tailoring of the ration to the specific animal's need. Of course reducing the competition from rangeland pests like rodents and grasshoppers should be part of the strategy to reduce grazing pressure.

Increased attention to livestock–environment interactions is therefore of critical importance in maintaining the CAR rural resource base. An integrated approach is required to reverse the present downward trend in rangeland productivity. The objective is not simply to revegetate the degraded areas by means of reseeding or imposition of grazing bans, but rather, it involves the management of livestock (the majority of which are owned and controlled by village-based households).

This management involves more than adjusting the grazing pressure. It calls for adjustment of the animal husbandry system and greater understanding of the linkages between grazing livestock and the rangeland on which they depend, including the impact of abiotic elements such as climate. This last point is particularly important in terms of the impending impacts of global climate change (Oxfam 2009). Many CAR countries, especially Tajikistan, are ill-prepared for climate change (Fig. 2.2).

6 Climate Change Impacts Are Complicated by Environmental Management Weaknesses

Environmental problems, independent of climate change, have presented serious challenges to most of the CAR countries although not every country has the exact same suite of problems. Many countries lack management practices needed to protect the natural resource base on which economic activity depends. Shortcomings

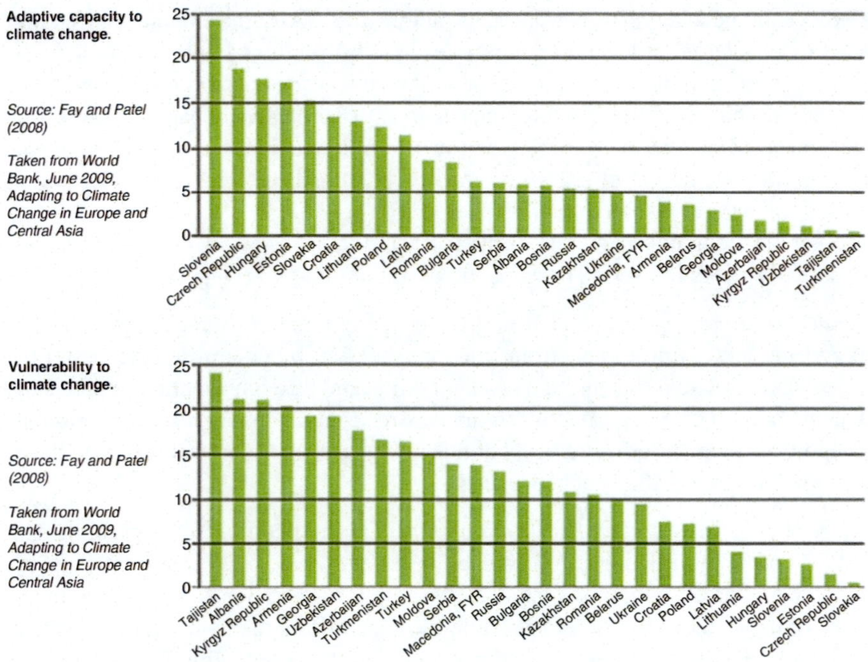

Fig. 2.2 Tajikistan is the country most vulnerable to climate change in the eastern Europe and central Asian bloc, but Kyrgyzstan, Uzbekistan, and Turkmenistan are also vulnerable. Many central Asian nations have low adaptive capacity to cope with climate change

are evident in management of soil fertility, water use, forest management, and illegal logging. Projecting current management practices into an era of accelerating climate change raises concerns. Worries about not only about social and economic setbacks in farming and forestry but also impacts on ecosystem stresses, including biodiversity loss and damage to watersheds and rural landscapes.

Failure to address land degradation problems is particularly worrisome because climate change could make today's problems worse through a pattern of alternating droughts and intense rainfall. Institutional and management weaknesses stem mainly from the complex transition from centrally planned communist-era governance models (Rahimon, Chap. 3). Though the most difficult decades have passed, a legacy of distorted specialization and rigid, poorly-funded institutions remain (Squires, Chap. 12).

Vulnerability to climate change will be dominated by socioeconomic and legacy issues (Fay et al. 2009). Resilience to a changing climate—whether to a climatic shock or to changing averages—depends heavily in the state of the system that it impacts, whether human, physical, or ecological. Thus, a short drought may be manageable for a farmer coming out of a prosperous year but ruinous if it follows another dry year that drained household savings or reduced herd/flock size.

Decades of mismanagement and neglect have diminished CAR countries' natural resistance. Under the socialist system, economic growth was pursued in blatant disregard for prevailing natural conditions.

Land users' ability to adapt to a changing climate depends on the elements of well-functioning farming systems. Such elements include:

- Locally relevant agricultural research in techniques and crop varieties and animal genotypes
- Training in new technologies and knowledge-based farming approaches
- Private enterprises, or public or cooperative organizations, for inputs such as seeds and machinery, and access to affordable rural finance for such inputs
- Physical infrastructure and logistical support for strong, transporting, and distributing farm products
- Strong links with local, national, and international markets for agricultural products
- Timely access to climatic and forecasting information and the skills needed for their interpretation and application

Rangelands grazed by livestock support a forage crop capable of intercepting and storing large amounts of solar energy and, consequently, support livestock production at low cost, if managed properly. There has been slow realization that livestock are *tools* for managing the rangeland vegetation resource and marketing its forage and that livestock are *not an end in themselves.*

There is still a lot to learn about how to manage rangelands for higher energy interception. Vegetation is the central variable in the rangeland system which is externally affected by the amount and timing of precipitation and other weather factors such as wind, freezing conditions, and drought. Severe LD also prevents vegetation from regenerating (Squires et al. 2009).

The grazing subsystem is driven by external factors—*vegetation productivity* which in turn is dependent on other factors (see above). Livestock's domination in grazing systems limits the opportunity for rangeland ecosystems to recover and is the driving force in rangeland degradation. Another important driving force in the rangeland ecosystem is the human dependence on livestock as the main source of income.

7 The Land User's View of Land Stewardship

Regardless of how we define sustainable use of land, it is ultimately the land user who must establish practical, sound practices. So the land users' point of view of what constitutes sustainability is the most important perception of all. Farmers and herders seek first a reasonable financial return on their capital and labor. They seek efficient production techniques that do not demand too much physical effort (labor), personal time, or capital. Financially they aim to operate farming and livestock raising methods that can adapt to risk and changes in markets and weather. Personally and socially, they prefer a farming method that keeps their customs alive, meets their peer groups' expectations in land stewardship, and gives their offspring a sound resource base for future family prosperity. In working toward these goals for the family farmers, we must be mindful of the fact that land users cherish the freedom to act independently, to operate their farming/livestock enterprise, and to work with minimum regulation and interference. They value the right to seek information and assistance when and from whom they choose and to decide for themselves how best to manage the land.

One of the forces affecting the way land users manage their resources (soil, water, vegetation) is their security of land tenure. Clearly, people with insecure tenure and annual (may be renewable) leases will be less concerned about the long term than someone who has lifetime land use tenure (Kurbanova, Chap. 7; Halimova, Chap. 13). The attitude of the land user has a critical bearing on the significance of "sustainability." One of the aims of good land stewardship is to use the resources in a way that leaves something of value to the coming generations. The idea of intergenerational equity would have no meaning if government policymakers avoid all responsibility to ensure secure tenure and equitable division of land resources.

The idea of stewardship is based on landholders regarding themselves as temporary custodians of the nation's resources or as end users of the land. This long-term unselfish view brings with it a respect for the landscape and the humility associated with frugal living and an appreciation of nature. It is the essence of a personal worldview based on sustainability.

8 Summary and Conclusions

It is clear that for CAR countries, with all of their complexity and their dwindling resource base, widespread poverty, and burgeoning populations, better land stewardship is an ecological and economic imperative.

A national action plan for the rangeland should be developed by each country to systematically address the real problems of rangeland degradation in ways that have a long-lasting impact on halting and reversing the trend of rangeland degradation. A long-range, logically structured action plan needs to be developed to ensure that efforts being made to stabilize and improve rangeland ecosystems are focused and coordinated. The government needs to provide the necessary policies and institutional support to ensure that technical solutions have an enduring and positive impact on the rangelands. The action plan must impact a wide range of institutional, regulatory, financial, educational, and physical forces if it is to be effective in halting and reversing rangeland degradation. A big part of this is to change the mind-set of the primary land users (farmers and herders) and foster the development of a land ethic based on land stewardship. An ethic of stewardship for rangeland aims to promote sustainable land use and to develop sustainable communities. The importance of working with local communities is elaborated elsewhere in this book (Kurbanova, Chap. 7; Hua, Chap. 14).

To promote ecological sustainability of CAR's rangelands, the ministry concerned with nature protection should begin to play a more active role in rangeland research, monitoring of the rangeland environment, and engaging in policy dialogue on environmental sustainability of the rangelands. Such ministries do not have the livestock production orientation of the Ministry of Agriculture (or local equivalent) and should help shape the research, policy, and development agenda to ensure that ecological sustainability of rangeland ecosystems is pursued.

Acknowledgements The work reported here was generated from experience gained from the ADB/GEF Rural Development project, and the author was engaged as a consultant by a consortium of GITEC Consult GmbH/Donaev Management Consulting under the direction and guidance of the PMU in Dushanbe.

References and Further Reading

Brown GC, Waldron SA, Longworth JW (2008) Sustainable development in Western China: managing people, livestock and grasslands in pastoral areas. Edward Elgar, Cheltenham, 294 p

Chapin FS III, Carpenter R, Kofinas GP, Folke C, Abel N, Clark WC, Olsson P, Stafford Smith DM, Walker BH, Young OR, Berkes F, Biggs R, Grove JM, Naylor RL, Pinkerton E, Steffen W, Swanson FJ (2010) Ecosystem stewardship: sustainability strategies for a rapidly changing planet. Trends Ecol Evol 25:241–249

Cowie AL, Penman TD, Gorissen L, Winslow MD, Lehmann J, Tyrrell TD, Twomlow S, Wilkes A, Lal R, Jones JW, Paulsch A, Kellner K, Akhtar-Schuster M (2011) Towards sustainable land management in the drylands: scientific connections in monitoring and assessing dryland degradation, climate change and biodiversity. Land Degrad Dev 22:248–260

Fay M, Block RI, Ebinger J (2009) Adapting to climate change in Eastern Europe and Central Asia. World Bank, Washington, DC

Gintzburger G, Toderich KN, Mardonov BK, Mahmudov MM (2003) Rangelands of the arid and semi-arid zones of Uzbekistan. CIRAD/ICARDA, Montpellier, 426p

Halimova N (2012) Land tenure reform in Tajikistan: implications for land stewardship and social sustainability: a case study. In: Squires VR (ed) Rangeland stewardship in Central Asia. Springer, Dordrecht, pp 305–332 (Chapter 13, this volume)

Harwood RR (1990) A history of sustainable agriculture. In: Edwards CA, Lal R, Madden P, Miller RH, House G (eds) Sustainable agricultural systems. Soil and Water Conservation Society, Ankeny, pp 3–19, 696p

Hua L, Zhang D (2012) Engaging with land users; the first steps on a long road. In: Squires VR (ed) Rangeland stewardship in Central Asia. Springer, Dordrecht, pp 333–356 (Chapter 14, this volume)

Jackson R (2012) Fostering community-based stewardship of wildlife in Central Asia: transforming snow leopards from pests into valued assets. In: Squires VR (ed) Rangeland stewardship in Central Asia. Springer, Dordrecht, pp 357–380 (Chapter 15, this volume)

Kreutzmann H (ed) (2012) Pastoral practices in high Asia: agency of 'development' effected by modernisation, resettlement and transformation. Springer, Dordrecht, 341p

Kurbanova B (2012) Constraints and barriers to better land stewardship: analysis of PRAs in Tajikistan. In: Squires VR (ed) Rangeland stewardship in Central Asia. Springer, Dordrecht, pp 129–162 (Chapter 7, this volume)

Michalk D, Hua L, Kemp D, Jones R, Takahashi T, Wu J, Nan Z, Xu Z, Han G (2010) Re-designing livestock systems to improve household income and reduce stocking rates in China's Western rangelands. In: Squires V, Hua L, Zhang D, Li G (eds) Towards sustainable use of rangelands in North West China. Springer, Dordrecht, pp 301–324, 354 p

Oxfam (2009) Reaching tipping point: climate change and poverty in Tajikistan. Oxfam International, Dushanbe, 22p

Squires VR (ed) (2011) Range and animal sciences and management, Encyclopedia of Life Support Systems, vol 1.UNESCO/EOLSS, Oxford

Squires VR (2012) Governance and the role of institutions in sustainable development (Chapter 12, this volume). In: Squires VR (ed) Rangeland stewardship in Central Asia. Springer, Dordrecht, pp 275–304, 22p.

Squires VR, Lu X, Lu Q, Wang T, Yang Y (2009) Degradation and recovery in China's pastoral lands. CABI, Wallingford, 284 pp

Tongway DJ, Ludwig JA (2010) Restoring disturbed landscapes: turning principles into practice. Island Press, Washington, DC, 210 p

UN (1987) Our common future/Brundtland report. United Nations World Commission on Environment and Develop, Oxford University Press, 400 p

WOCAT (2007) Where the land is greener: case studies and analysis of soil and water conservation initiatives worldwide. Co-published by CTA, UNEP, FAO and CDE

Worrell R, Appleby MC (2000) Stewardship of natural resources: definition, ethical and practical aspects. J Agric Environ Ethics 12:263–277

Chapter 3
Evolution of Land Use in Pastoral Culture in Central Asia with Special Reference to Kyrgyzstan and Kazakhstan

R.M. Rahimon

Abstract The nomadic culture is deeply ingrained in Central Asia and has been part of the lifestyle of its peoples for millennia. This chapter traces the transformation of the land relations, evolution of the perception of pasturelands, and the various forms of the property rights from the time before colonization by Russia in the nineteenth century to the present day. The effects of socio-ecological transformations such as climate change, the collapse of the Soviet empire, and civil war are examined. In recent years, the transition to sedentarization, the privatization of land, and the conversion of rangeland for cropping has become important.

This chapter considers changes in land relations during the following historic periods: precolonial period (from the beginning of nineteenth century to the end of the nomadic influence), colonial epoch of the Russian empire (the end of the nineteenth and beginning of the twentieth centuries), the socialist period (1917–1991), and the post-Soviet period. The current situation in the context of post-Soviet restructuring and the transition to a market economy has been analyzed.

Keywords Land tenure Sharia law • Russia • Soviet • Collectivization • Land reform • Nomads • Transhumance • Intergenerational equity • Pastoral culture

Key Points

- Throughout Central Asia, there were three basic agricultural groups: the sedentary, nomadic, and seminomadic, whose livelihood was characterized by, respectively, crop growing, livestock, and a combination of both. In order to understand the essential features of land/people relations at present, it is necessary to analyze the transformation over time, from the period when they were year-round nomads to the period of colonization and to the present-day heterogeneous economy.

R.M. Rahimon (✉)
American University of Central Asia, Bishkek, Kyrgyz Kyrgyzstan
e-mail: ruslanlyon@gmail.com

V. Squires (ed.), *Rangeland Stewardship in Central Asia: Balancing Improved Livelihoods, Biodiversity Conservation and Land Protection*,
DOI 10.1007/978-94-007-5367-9_3, © Springer Science+Business Media Dordrecht 2012

- In the past, the ownership of land was entirely divorced from ownership of whatever happened to exist upon it, including water, crops, or livestock, so having user rights to arable land alone did not always provide the preconditions for independent agricultural production.
- From the point of view of preservation of ecological balance in the preindustrial period, nomadic communities, certainly, possessing various skills and knowledge about environment, tried not to create an imbalance because cattle survival and their prosperity depended on nature. Originally all of the land was once held under communal rules in which land was primarily seen as a resource for survival of everyone belonging to the clan or group dwelling in a certain area.
- The reforms of collectivization during the Soviet era were more to do with sedentarization and with formation of new settlements in the form of collective farms and state farms. The changes were made in the pasture organization and the economic status of a particular land resource.
- The acquisition of independence is characterized by a new configuration of social and economic system of values, development of market relations, disintegration of collective farms and state farms, and land reforms. It is obvious that the land, and pastures as a whole, became the core strategic resource, in the conditions of a new geopolitical situation.

1 Evolution of Land Use in Pastoral Nomadic Culture of Central Asia

1.1 The Setting

Much of Central Asia is desert surrounded by the Pamir-Altai, Tien Shan, and the Altai mountain ranges located on the southeastern, southern, and eastern borders. In the north, deserts merge into semidesert and then into steppe zones (Maselli, Chap. 1). The steppes never maintained stable borders because they moved meridionally in both humid and dry periods. This instability had a direct effect upon the economic systems of the populations who occupied these border regions. In the mountainous regions, uncountable ridges define foothill and hill valleys.

Small rivers originate in and nourish the valleys; these rivers are also the headwaters of the two great waterways of the Central Asia, the Amu Darya River and the Syr Darya Rivers, which carry their waters to the Aral Sea. The fertile Ferghana valley is richly supplied with water from the upper Syr Darya River. In addition to the Syr Darya River, there are other great rivers in the eastern Central Asia. There are the Ili, Chu, Talas, Kizilsu, Kaskadaria, and Sokh, among them, flowing from the mountains to the fertile territory of the southern Kazakhstan known as the Semirechye (seven rivers). The largest lakes in this area are the Balkhash and the Issyk Kul. Many other rivers, however, do not reach the larger reservoirs but disappear into the sands of the plains (e.g., the Zeravshan, the Sanzar, the Tedjen, the Murgab, the Sarysu, and the Chu). Large rivers such as the Amu and Syr Darya also periodically changed their course, creating new river channels and deltas. These rivers irrigated various zones of the desert plain and helped to develop agriculture and semisedentary cattle-breeding. It is worth saying that the Syr Darya River and the Amu Darya River definitively shaped the landscape of the west Central Asia. Their deltas were homeland for the original cattle-breeding societies, which contributed to the development of the earliest and powerful Central Asian Chorasmian state.[1]

1.2 Pastoralism Has a Long History in Central Asia

The origins, spread, and development of pastoral economies on the Eurasian steppe have been the subject of significant research and debate (Bendrey 2011). Any particular pastoral system may be envisaged as a product of a number of interacting variables: the characteristics of the animals, the environment, and the human culture. Several economic-cultural types developed within the limits of the Central Asian historical-cultural area. They were settled plow farmers (who used irrigation systems), semi-settled stock breeders and farmers, and nomads and seminomads—cattle breeders and stock breeders of the steppes. But different subtypes that existed within each economic-cultural type of cattle breeders and nomads reflected local specificity of the landscape zones. So, in the north of Kazakhstan, severity of winters (with low temperatures and strong winds) forced cattle breeders to drive cattle to the south, where herds were concealed inside of ridges or in sand dunes. The cattle breeders of the southern warm deserts tended their herds locally during the year or drove them to the mountain meadows.

The mountain steppes and meadows were used as summer pastures in the east. In the winter, cattle were hidden in the deep valleys, and alpine pastures were used in the summer. As a rule, seasonal winter sites were used from year to year for many centuries. There were warm steppe zones in central Kazakhstan characterized by a pattern of sheep-breeding and horse-breeding with long migrations. In the southwest of Turkmenistan, camels, sheep, and goats predominated in the herd. The particular routes of movement were caused by the location of water sources and wells. In contrast, in the southern and eastern regions of Turkmenistan, where mountains are found, various systems of stock-breeding (distant pasture, nomadic pasture, pastoral pasture, barnyard-distant pasture, and mountain pasture) were developed. The seasonal migrations of the stock breeders mainly involved altitudinal migration from the lowlands to the uplands. But cattle-breeding went with agriculture, to some extent, almost everywhere.

[1] Chorasmian state abolished by the Achaemenid king Cyrus II the Great in the mid-sixth century BC.

Historically, Kazak livestock have been shepherded between seasonal pastures under several types of production unit, each considerably larger than the individual family unit. Under the pre-Soviet traditional system, Kazak nomadism was organized around subclans, the *aul* group of up to 100 families who traveled together to share herding and defense, while higher tribal and judicial authorities coordinated interclan movements.

In the last two centuries, the extent of flock mobility has expanded and contracted in response to fundamental changes in national, political, and economic systems (Alimaev and Behnke 2008; Alimaev and Temirbekov 2003). These changes are first recorded in the period of Russian imperial expansion into Kazak tribal lands during the nineteenth century.

Seasonal livestock mobility by Kazak pastoralists once allowed the efficient exploitation of natural pasture variability (Alimaev and Temirbekov 2003). Contraction of mobility leads to ecosystem fragmentation, which was summarized by Hobbs and Galvin (2003) as:

> a diminished ability of large herbivores to access natural heterogeneity in vegetation and topography. As fragmentation occurs, ecosystems are simplified by breaking up interdependent spatial units into separate entities… the result of this simplification is a reduction in the scale over which complex interactions among environment, large herbivores and human management takes place. (Hobbs and Galvin 2003)

This chapter considers changes in land relations during the following historic periods:

- Precolonial period (from the beginning of nineteenth century to the end, in case of the Kyrgyz nomadic influence of Qoqand Khanate)
- Colonial epoch of the Russian empire (the end of the nineteenth and beginning of the twentieth centuries)
- The socialist period (1917–1991)
- The post-Soviet period

2 Changes in Land Relations and Control of Access to Pastures and Water Under Different Regimes

2.1 *The Precolonial Era (Nineteenth Century)*

While it is clear that pastoralism has existed in Central Asia for centuries, it is outside the scope of this chapter to delve too deeply into the past. Excellent accounts of the earlier eras are to be found in Yablonsky (2003), a publication in English that for the first time gives to the English-speaking reader access to the little known and generally inaccessible Russian literature on the archeology of the earliest Eurasian nomads.

On the eve of the Russian conquest in 1876, the Kyrgyz were organized in leagues (*el-jurt*) (cf. Valihanov c.1860: 83–4, Soltonoev c.1934: II, 176). A league had a common territory composed of various types of seasonal pastures and recognized the authority of a single ruler: the *chon manap*.[2] Access to water and to pastures (according to the season) was dispensed on the authority of one man. The clan's supervision system in Kyrgyzstan was also aimed at provision of well-being and preserving of originality, independence, and integrity of an ethnic formation (Grodekov 1889). Generally, grazing pressure was much lower at that time as the livestock inventory and human populations were low. The situation in other CA countries varied, but Jacquesson (2011) touches on some aspects of precolonial land relations.

2.2 The Colonial Epoch of the Russian Empire

In 1876, czarist Russian troops defeated the Kokand Khanate[3] and occupied northern Kyrgyzstan. The late nineteenth century saw extensive Russian and Ukrainian immigration into northern Kyrgyzstan. A 1916 rebellion against Slavic domination was met with harsh reprisals; a third of the Kyrgyz population fled to China. Soviet domination followed, and in 1924 present-day Kyrgyzstan was designated the Kara-Kyrgyz ['black Kyrgyz'] Autonomous Region. Soviet pressure to collectivize and settle the nomadic Kyrgyz changed the cultural landscape of their mountainous land as Stalin offered its official republic status in 1936.

Il'yasov (2002) gives an interesting overview of land relations on the eve of the Soviet invasion and defines the following patterns of ownership: *amlyak* (state), *mulk*, and *vakuf*. According to Sharia[4] law, *amlyak* land belongs to the state in the name of the supreme governor Khan (*chon manap*); all the others were users and not owners. In this case, the state ownership was a prerogative power of the nomadic nobility. At the beginning of the colonial epoch, imperial Russia took away from the khans the right to dispose of the land. A new political superstructure was created, with new statuses and powers of the nomadic nobility and a change in traditional system of mutual relations:

> A cursory illustration of the state of native agriculture and animal husbandry in Central Asia as it existed during the turn of the 19th century would depict a feudalistic relationship between sedentary society and the soil, the applied technology in a biblical stage of development and a fiercely conservative Islamic sentiment found in all levels of society, which combine to formulate an environment inconducive to change. (Michael Kurasz, pers, comm., 2010)

---

[2] Chon manap is a chieftain. Manap is like a Russian duke. Kyrgyz manaps were simply the best people, holding no more honors, among other authorities of the clan.

[3] A state in Central Asia that existed from 1709 to 1876 within the territory of modern eastern Uzbekistan.

[4] The jurisdiction over land depended on the "Shariat"—the law of Islam, as interpreted by the local clergy. The majority of land in the region being regarded as "useless" belonged to no one man but to society at large.

Clan organization (see above) was mostly destroyed under collectivization in the early 1930s, as pastoralists were forced into state-run farms. By the 1940s, large groups of animals were once again being herded to remote pastures, this time supplied with technical inputs by the state collective farms. By the end of the Soviet period, state farms with up to 60,000 head of small stock orchestrated a regimented system of seasonal movement, dividing labor into brigades separately responsible for shepherding and other tasks such as harvesting hay, veterinary inputs, transport, marketing, and social infrastructure for the state farm workers.

The ownership of land was entirely divorced from ownership of whatever happened to exist upon it, including water, crops, or livestock, so affluence in arable land alone did not always provide the preconditions for independent agricultural production. This appears to ignore the ancient Quranic passage which commands that "water cannot be the subject of ownership but of usage." However, there were exceptions to this in some areas of Turkmenistan and Kyrgyzstan, where the population was predominantly nomadic. Here, personal attachment to the land was not that intense, and therefore, some resources were used communally.

Throughout Central Asia, there were three basic agricultural groups: the sedentary, nomadic, and seminomadic, whose livelihood was characterized by, respectively, crop growing, livestock, and a combination of both. A number of key concepts at this point deserve explanation: The "*metayage* system"—Islamic law—provides that the decision made in establishing the lease of a piece of land must be derived from the decision to what use the land shall be put. This principle supposedly serves as the basic foundation of the *metayage* system of land use in Central Asia. It is a series of negotiations and contracts made pertaining to the combination and application of resources toward agricultural production.

Some researchers accepted *metayage* as a way to ensure property preservation. From the economic anthropology point of view, this concept is a way to allow redistribution. Redistribution, as well as reciprocity, is the phenomenon in more traditional societies, when the right to redistribute these or other resources is for the purpose of maintenance of prestige and system of the political power. The *Mulk* property assumed the privately owned lands concentrated in hands of the secular and spiritual nobility, not bearing service to "sovereign" for using the earth. Besides, it is interesting to note the inheritance situation upon the death of the so-called private lands. "If *bai (rich men in the village, chiefdom in the nomadic unit)* used the same site of the communal land under pastures over a long period, gradually the right for use of it turned to the right of private possession. But the right of possession did not mean the property right to this land: the land was not given to a plot, and was not on sale neither in estate, nor out of it. "After the death of the feudal lord the site was passed on to his sons and relatives who used this land as communal, like the lord did" (Shahmatov 1964; p.103). During the colonial period, the Russian empire started to interfere to the right of private usage by the right of inheritance (Box 3.1).

The following type of the property, *wakuf* lands, passes in time to eternal possession of clergy on the basis of the donation certificate. If according to Sheriyat, *wakuf* land was really transferred to spiritual establishment as usual *wakuf,* it remained in the hands of the donator and passed then to its successors, and establishment, in which "*wakuf*" was endowed and received the certain income ostensibly for realization of that purpose into which name "*wakuf*" was entered. The "*wakuf*" institute

Box 3.1 Inheritance of Land in Kyrgyzstan in Pre-Soviet Times

For example, Shabdan Dzhantaev asks to fix a site to his children, but after his death in 1912, the Turkestani general governor refuses, ruling that children of Shabdan Baatyr have no merits that the given possession could not be transferred to them. It is necessary to note, certainly, that the land was a stumbling block for the nomadic nobility. "With the transition to settled way of life, the communal form of land tenure which, in essence, allowed owning a considerable quantity of cattle would be broken as well as the power that came with it." "Infringement of the communal form of land tenure and establishment of the homestead precinct form of land tenure deprived Manaps' and Bais' right to use sites of relatives." Certainly, this was included into a complex of the reasons for the Basmachi revolts in the south of the population where Sharia patterns of ownership were practiced.

Based on Il'yasov S.I. Zemelnye otnosheniya v Kirgizii v konce 19-nachale 20 vekov, Frunze, izd-vo Akad. Nauk, 1963 p.164 and p.275

was an important means of strengthening the economic basis of Muslim clergy (Il'yasov 1963 p.101). As a result, the whole Shariat patterns of ownership have been poorly developed among Kyrgyz nomads, except for some cases in the south. In this case, it was the system of reference points in various disputes and disassembling, both between the tribes, and in affairs with new colonial administration. It is possible to draw a conclusion that borders for tribally controlled pasturelands defined before and during the imperial epoch have been defined more or to less in natural ways. Hazanov (2002), generalizing approaches to property in different nomadic communities, comes to the following conclusions:

- Existence of some tendencies to a private property establishment of some kinds of pastures, with reference to separate societies where nomads politically appear in subordinate position (even in nomadic societies, private property is limited and is less significant than in settled societies).
- Absence of a clear boundary on the territory belonging to nomadic formation.
- The property on pastures belongs to all nomadic community, but in practice, they are distributed as possession or using between its various divisions.
- Realization of function of distribution (redistribution) pastures of nomadic aristocracy between the relatives and a choice of the best pastures as compensation, it was perceived by the Soviet ethnographers as *de facto* private property fact (Hazanov 2002). The close interrelation with such factors, as environment, specific structure of herd, the social organization, and the whole complex of mutual relations is obvious.

From the point of view of preservation of ecological balance in the preindustrial period, pastoral communities, certainly, possessing various skills and knowledge about environment, tried not to create an imbalance because cattle survival and their prosperity depended on nature. Originally all of the land was once held under

Fig. 3.1 Pastoral nomadic herders on the move to summer pastures. Their gurs/yurts are transported by camels

communal rules in which land was primarily seen as a resource for survival of everyone belonging to the clan or group dwelling in a certain area. Belonging to the clan, the group, the kinship, or the *kishlak* was expressed as submitting oneself to the customs of the clan, the group, the kinship, or the village.

Respect for the most essential of resources often had a religious connotation and still does in many places in CA. A strict set of rules was in place, manifesting the appropriate and respectful way to use and exploit the resource. Land was seen as too important a means of existence to give private persons any individual power of control, allocation, and exploitation over it in perpetuity. The individual was tied to the explicit consent of the group or the clan, the kinship, or the kishlak. By social convention, the individuals also were tied to the land. Then land imposed cultural and religious obligations upon the users, for example, taking into account some duties to ancestors and to a yet unborn offspring.[5] These lands were in fact common pool resources.[6] Under customary rules, use and exploitation of land is bound with a social mortgage on every tenure right.

The pastoral nomad migrates twice per year in Central Asia—uphill in spring for greener pastures and downhill in fall to warmer altitudes (Fig. 3.1). His livestock is his fortune, lifestyle, and transportation. He generally owns no land and despises the

[5] The concept of intergenerational equity.

[6] The term underlines the openness of the resource but not in the sense of open unconditional access!

art of tilling the soil. Surprisingly, he operates within a closed environment. His semiannual movements are between the same summer and winter pastures via the same basic routing which his ancestors used. To wander haphazardly would be to encroach upon the territories of others; however, this traditional macro-system of land use is not as complex or as intensive as the metayage system.

The Russian intrusion into CA's agriculture had from the outset little effect upon the metayage system, other than constricting it somewhat as the Russians gradually occupied the better lands. The Tsar's policy was to divide the land among the bigger native landowners and the European farmers (Pierce 1960; Allworth 1989). However, this does not imply a malevolent disposition on the part of the Russians. Some commentators maintain that the Tsarist administrators often sided with the natives' claims to land, and they passed laws as to how much land Russians can buy. Kyrgyzstan laws forbade the local Kyrgyz to lease out land to individuals outside of their own community and that new settlers could lease land only at the public treasurer's office. However, this proved to be poorly enforceable. Many of the Soviet works cited tend to refer to poor Russians and poor native peasants with the same title "byednyak," unlike the references to proletarians, which may imply that a degree of camaraderie existed between the two groups.

Under the Soviets, Kyrgyzstan was to become the breadbasket of CA. Its climate and land are largely unsuitable for cotton (Allworth 1989). The settlement and education of the nomads here were given primary importance; the presence and tradition of the Kyrgyz nomads decided the prominence of cattle-breeding in Kyrgyzstan's economy.

The division of animal husbandry from agriculture had originally established the nature of an economy based upon barter trade, and all trade within the nomadic communities in CA, up until the Russian penetration, could be translated into sheep and goats. This included labor contracts, marriage agreements, and an array of services which reached from field fertilizing to bankers' gold deliveries.

In order to understand the essential features of land/people relations at present, it is necessary to analyze the transformation over time, from the period when they were year-round nomads to the period of colonization and to the present-day heterogeneous economy.

Jacquesson (2010) examines pastoralist and descent systems in Kyrgyzstan from their integration in the Tsarist colonial system to the present day. Based on extensive archival work as well as over a decade of ethnographic fieldwork in northern Kyrgyzstan, Jacquesson gives a chronological and detailed account of the impact of Tsarist rule; the formation of the Kyrgyz Republic, collectivization, and sedentarization; the Soviet livestock economy; and the dismantling of this system.

Kassam (2009) considered the effects of socio-ecological transformations such as climate change, the collapse of the Soviet empire, and civil war in the valleys of the Pamir mountains in the historical Badakhshan region, now divided between Afghanistan and Tajikistan. Preliminary findings indicate concern for food sovereignty, evidence of biocultural impacts of climate change, increasing burden on women, debilitating opium addiction, ecological importance of sacred sites, and other priorities related to sustainable livelihoods, such as energy needs (for fuel and lighting) and physical and social infrastructure in the form of roads and schools.

3 The Soviet Period

Changes in the twentieth century were initiated with the communist revolution and subsequent collectivization of Kazak livestock into state-managed farms (1920–1935). Further changes encompassed the post-Stalinist state farm expansion and emphasis on high livestock output (up to the end of the USSR in 1991). In the last decade, the pastoral Kazaks had to cope with the immediate post-Soviet economic crisis, dissolution of state farms, and loss of 70% of the national flock (1992–1999). The current period is one of the rebuilding private livestock holdings starting in 2000 (see below).

Some Soviet sources considered nomads to be an underprivileged group whose deprivation drove them to nomadic life. Nomads were classified as "the backward people" by the Soviet ideologists and researchers. They were forcibly settled down during industrialization, and efforts were made to integrate them into the mainstream.

It is necessary to note that with demarcation and delimitation of borders in 1924 in Central Asia, also impacted (but insignificantly) on management of pastures (Pierce 1960), the reforms of collectivization were more to do with sedentarization and with formation of new settlements in the form of collective farms and state farms. The changes were made in the pasture organization and the economic status of a particular land resource. Sheep yards are constructed, herds are grouped, and also large-scale support of cattle-breeding with the rational organization of agro-economic activities is carried out.

With transformation and formation of new settlements, the system of values, and also social status, changed. Now, the simple nomad shepherd, the Kazakh, or Kyrgyz became part of a working proletarian class, "the engine of social progress," and the founder of a new order. It is interesting to observe a huge role of the Soviet state in maintenance of life of shepherds in their new social space. In connection with the process of adaptation to settled life, formation of collective farms and state farms, and resettlement of the basic part of the population in valleys, pastures are converted to "specialized places," no longer the usual communal land but part of new economic system with rational organization. Pastures were structured from the point of view of profitability and concentration of agricultural resources. Despite various geographical, economic factors and animal husbandry scales, a state policy role in Kyrgyzstan and Kazakhstan was identical.

The Soviet state mode supported the life on pastures, thereby, stimulating extensive livestock production, especially from cattle (Box 3.2).

3.1 Pasture Use in Kazakhstan: An Example from the Eskeldinsky Area

Before giving a short description of pasture use, it is necessary to note that Kazakhstani pasture management system differs from that of Kyrgyzstan. Unlike the Kyrgyzstani practice where the cattle are contained in a stall in the winter, the cattle in Kazakhstan

Box 3.2 Description of the Life of Shepherd/Cattle Herders in Collectives and Support Given by the State

The field research was carried out by the author on summer pastures, concerning village Aldabergenov, in the Eskeldinsky area near Almaty, Kazakhstan. The traditional pasture areas are located in the foothills of Dzungar Ala Tau, about 30 km from the Kazakh-Chinese border. In the Eskeldinsky area, summer pastures are at an altitude of about 1,800 m and approximately in 80 km from village Aldabergenov. The users of pastures such as Teris Akan Zhazygy, Tyolyoktu a Torah have settled down.

In addition to veterinary, zootechnical, and infrastructural support, there was support to shepherds/cattle herders as "the advanced working class of a socialist society" for their mobile way of life. Vehicles were sent periodically to the remote pastures with supplies and amenities. Essential commodities were offered for sale at low prices, and from time to time, traveling hairdressers and troupes of amateur performers put on plays and musicians performed. Film screenings were offered at nearby frontier posts. All of this was done to provide a life and leisure of workers who were absent from the village. Doctors were delivered to the hard-to-reach spots in mountain by the helicopter if required. The shepherds/cattle herders who met production targets regularly received awards and discounts for acquisition of home appliances. On account of their trade-union affiliation, the Kazakh and Kyrgyz shepherds/cattle herders were eligible to rest in the best resorts of the Soviet Union. Mr. Kanagat Serikbaev, a wool cutter of the village Aldabergenov, received his reward of a tourist trip to the east Europe countries in 1985—an excursion he remembered nostalgically when interviewed by the author 20 years later. He also underlined the impossibility of realization of this trip in modern conditions. Certainly, many shepherds remembered with pleasure about the Soviet period, but the majority of them did not feel sorry about the current situation.

during the winter time was driven away to the steppe to what are called rundown[7] lands. The basic winter rundown lands are located in sandy steppes of Southern Pribalhashja. The pasture of Ortaozek is 160 km from the main homestead, in a southwest strip of Sary-Ishyk-Otrau. Winter pastures on the vegetative structure are suitable for all kinds of agricultural animals which are on rundowns. In the Soviet period, collective farms also had additional winter pastures for temporary usage from the state fund Akdala which is 130 km northwest of Ortaozek. It was not possible to use this land completely during the whole of winter because of a lack of water. Therefore, only in years with ample snow herds of horses and a flock of sheep are driven here, and then for only a short time. Spring-autumn pastures of Moinkum are located to the south of

[7] A local term, also called distant pastures.

Box 3.3 Life on a Collective Farm "Rundown Center" in Kazakhstan

A farm after the name of Aldabergenov of the village Aldabergenov is one of few collective farms in Kazakhstan. It had one of the most advanced infrastructures in area. So, for example, the rundown center of village Aldabergenov in the end of 1970s, so-called administrative headquarters, included library, veterinary clinic, apartment houses, medical center, trading post, and other facilities for serving workers such as first-aid post, a shop, club, a warehouse, hotel, and office. The rundown centers had a constant radio communication with collective farm. Movies were periodically shown in rundowns by means of a mobile film projector. There was also a red corner in the yurt where newspapers and magazines gathered. In an interview with the former manager of a "red yurt" on a pasture of Belkuduk in the end of 1960, he said that he usually relayed information of events occurring in the country, and abroad, and explained this or that to the public. In this yurt, it is possible not only to have a talk and peruse fresh magazines and newspapers, or read an interesting book, but one could listen to music by radio, to play drafts and chess, to look at amateur performances and concerts, and to receive the necessary information. The first-aid post was in a standard small house which consisted from five rooms where there was a drugstore and a hospital with five cots. In the summer, the manager of the medical aid station, together with the hospital attendant, accompanied shepherds/cattle herders to the pastureland, and the medical aid station wandered from a yurt to yurt. It was possible to call helicopter in case of a serious illness or accidents and transfer the patient to the Ortaozeksky first-aid *raion* institute of public health services.

winter rundowns that, by the way, is near to a branch line of the Turkestan-Siberian highway. These pastures possess various water sources, and also the vegetative cover is much more variegated in comparison with winter pastures. Summer pastures are at a distance about 100 km on high-mountainous Alpine meadows of Dzungarian Ala Tau. The distance between summer and basic winter pastures of village Aldabergenov is about 250 km. Thus, as we see the basic structure in the Kazakhstan system was a collective farm with a settlement, occupied rundown center, and some kind of administrative base. The logistic problem consisted of organizing social and economic life of cattlemen throughout the 5–6 months of the year when they were more than 100 km from their household. Box 3.3 gives a summary of the measures taken by the state to alleviate the sense of isolation.

4 Contemporary Stage

The immediate aftermath of de-collectivization in the 1990s left most formerly employed pastoralists atomized into nuclear family units and having obtained very few livestock or capital assets from the privatization of state farms. With neither subclans nor state farms to provide support, individual families with their small flocks could not cope with the scale of investment and effort required for long-distance migration.

The acquisition of independence is characterized by a new configuration of social and economic system of values, development of market relations, disintegration of collective farms and state farms, and land reforms. It is obvious that the land and pastures as a whole became the core strategic resource, in the conditions of a new geopolitical situation. The liberalization of the economy is a necessary precondition for successful land reform and farm restructuring. Decreased state interference in agricultural production has stimulated the private sector toward higher productivity and toward overall economic well-being.

Land reform is an important part of the transition to a market economy in the Central Asian region (CAR). The reform demands solutions to new challenges created by the privatization and transfer of trading assets from the state to the private sector in the absence of an established land tenure framework. In addition to the legal details of reform, there are many and varied political and social factors that influence the reform though changes that are brought about in the relationship between the people and the resources on which they depend. Throughout the nineteenth and twentieth centuries, despite the historical epoch and sociopolitical formations, the pattern of ownership on pastures always remained as a social phenomenon both in practice and in collective perception of the population of Kyrgyz and Kazakh people. All attempts to persuade them about the necessity for privatization failed because of unacceptability of the idea and the world view of land according to nomadic traditions.

The transformation from a centralized command economy to a market-driven economy requires a continuous decentralization of decision making from government authorities to independent farmers and herders. This requires that a set of legally defined property rights for land and water be developed and adopted by the governments of each of the CA countries. The first steps have been taken by allowing long- term leasing (inheritable user rights) and completing the legal framework for land reform (Halimova, Chap. 13; Hannam, Chap. 17). The current status of land reforms and farm restructuring (see below) shows exceptionally big differences among the individual countries (Robinson, Chap. 11).

In Kyrgyzstan, significant progress has been made on land privatization and farm restructuring, although the process is not complete. Land privatization and farm restricting have been accomplished through many decrees issued by the president and the government.

In Kazakhstan, initial privatization of large farms has been completed; however, the restructuring of the farming sector is still in process. The land law passed in 1995 still maintained state ownership of agricultural land (Robinson, Chap. 11). The government is defining new policies toward livestock and rangeland management, as revenues from the booming mineral economy are released for rural reconstruction and agricultural research. Entirely new systems of private livestock and rangeland management have developed, about which there is little information. Urgent and controversial questions about the rangelands have arisen in Kazakhstan including whether to encourage private rangeland ownership, how to prevent overgrazing on

Table 3.1 Land ownership types and farm types in Uzbekistan

Types of land ownership	Types of farms
State	Sovkhozes
Private	Privatized livestock farms (land under buildings)
Lifetime inheritable possession	Small holdings
	Private orchards and vineyards
Perpetuity	Kolkhozes, agricultural cooperatives
Leases	Lease holdings
	Private farms

common land, whether to charge fees for grazing on state land, how to prevent livestock mortality during drought and blizzards, and what scale of private farm operation is viable and should be given state support. There is a need for contemporary and relevant data to guide policies on extensive livestock management.

In Turkmenistan, the land reform process is slow, but some progress is being made. The historical forms of land ownership in Turkmenistan cannot be ignored. The *sanashik* form of land ownership is a right on yearly land use. Turkmenistan still retains this form of land ownership.

In Tajikistan, the government has passed several land laws and decrees and distributed 50,000 ha for creation of household plots. Through operation by 1995, more than 240,000 citizens received from 0.08 to 0.15 ha of irrigated land for lifetime use in agriculture. Over 139,000 ha of mainly pastureland has been privatized (Halimova, Chap. 13).

Uzbekistan has been more cautious than other countries, for example, Kyrgyzstan. In Uzbekistan, the law "on lands" was adopted in 1990 which stipulated that land would be under state ownership. However, land could be allocated for short- or long-term lease and also lifetime inheritable possession (law "on dekhan farms," 1992). Table 3.1 shows the different forms of land tenure in Uzbekistan (Robinson, Chap. 11; Halimova, Chap. 13; Shaumarov, Chap. 16).

5 Kyrgyzstan Under the Spotlight

The tenure situation in Kyrgyzstan is more clear and land tenure reform is further advanced. As a result of more than 70 years of development, the agricultural sector is now the basic sector of the economy of the Kyrgyz Republic (about 40% of GDP and providing gainful employment to more than 50% of the active population of the country). Agriculture, including animal husbandry, acts as a link for development of other areas of economy, in a private sector (private cattle-breeding, farming), in a process of industry, and in a sphere of services. More than 80% of the goods in the consumer market including food stuffs are made from agricultural raw materials.

Now the land is being privatized in Kyrgyzstan. The system has changed after the reform in agrarian relations and structure. More than 90% of the land in former collective farms and state farms was distributed to more than 80,000 citizens (former workers on the state-controlled land), and small private farms has been created. There are about 900,000 ha of arable land (about 69% of the total) that was distributed to farmers and others. The republic's basic direction in agriculture is animal husbandry which is based on cheap forages from natural pastures. The total area of natural pastures of the republic is 9.1 million hectares (Mha), 3.9 Mha of which is summer pasture, 2.8 Mha is spring-autumn pasture, and winter pasture is 2.4 Mha. The land reform campaign promoted creation of a multistructural economy in the agriculture sector; however, it was not supported with corresponding financing (rural credit) to facilitate the carrying out of all necessary agrotechnical operations on allotments. Regrettably, the lack of finance and the small scale of farming operations have contributed widespread unemployment in rural areas and an increase in poverty. The overstocking of pasture has led herders to seek more remote pastures which extends the distances of livestock movement. The speed of movement has accelerated as more and more livestock are on the stock routes. This and the greater exposure to disease en route can negatively affect the health of cattle (Sitnyanskii p. 187).

According to Article 21 of the Land Code of the Kyrgyz Republic (GOK 1999) about the management of the lands for agricultural purposes and the lands for pastures, the state is the sole owner and access to pastures is by rental (lease) agreements. There are various approaches under discussion to change aspects of the situation governing the use of pasturable lands (Robinson, Chap. 11). For example, formation of pasture user groups (PUGs) that would regulate the use of the assigned land, collect grazing fees, and maintain pasture quality.

The rationale for this change is that the users will have a role as land stewards. As custodians of the leased land, they will ensure that the level of use is sustainable in the long term and will do as much as they can to rehabilitate and improve the pasture. The cattle breeder-peasant will be interested in avoidance of degradation and loss of productive potential. Giving guaranteed user rights to specific pastures and by fixing a fair rent for their use coupled with system of monitoring on the rented sites will help foster a better attitude among the users. Over time it is expected that the PUGs will develop a sense of responsibility for rational use and improvement of the condition of the pasture. This will help counter the problems emerging now as, for the first time, the national sheep inventory has passed eight million. The republic now faces a forage shortage, overstocking, and pasture degradation. Measures undertaken by the state departments only partially solved the problem of ecological balance.

6 Conclusions

Kyrgyzstan and Kazakhstan are now agrarian-industrial countries with multistructural economy, and pastoral nomadism is considered only in a context of traditional practice. "Nomad," as an image and even more as a type of seminomadic practice, continues to exist in Kyrgyzstan and Kazakhstan until the present day.

In the new geopolitical and geoeconomic conditions, independent Central Asian republics attach important strategic significance to pastures, as the necessary base in agricultural sphere, and to their preservation as an element of traditional culture and the shepherd's practice. But there is risk of things going seriously wrong if vested interests are allowed to dominate the land tenure reform process (Kurbanova, Chap. 7; Halimova, Chap. 13). Jacquesson (2010) argues that in the post-Soviet period, pastoralism has been marginalized in a number of CAs as a subsistence activity rather than becoming commercialized, although there are some attempts in the development of agricultural commerce from the governmental sides from both governments. It is interesting to note that Kyrgyzstan has been often a commercial base of cattle for a southern Kazakhstan. But still, pastoral wealth has been concentrated in the hands of those who were already in influential positions in the late Soviet period (see Kerven et al. 2008).[8]

Acknowledgment The author is deeply indebted to Dr. Victor Squires, University of Adelaide, Australia, for his assistance with this chapter.

References and Further reading

Alimaev II, Behnke R (2008) The shifting balance between migratory and settled pastoralism; Land tenure and livestock mobility in Kazakhstan. In: Galvin K, Reid R, Behnke R, Hobbs T (eds) Fragmentation in semiarid and arid landscapes: consequences for human and natural systems. Springer, Dordrecht, pp 225–44

Alimaev II, Temirbekov S (2003) Policy-driven livestock dynamics and rangeland fragmentation in Kazakhstan. In: Rangelands in the new millennium. Proceedings of the VII international rangeland congress, Durban, South Africa

Allworth E (ed) (1989) Central Asia: 130 years of Russian dominance – a historical overview, 3rd edn. Duke University Press, Durham

Bendrey R (2011) Some like it hot: environmental determinism and the pastoral economies of the later prehistoric Eurasian steppe. Pastor Res Policy Pract 1:8–15

Fedorovich BA (1973) Environmental conditions in arid zones of the USSR and livestock husbandry development. In: Abramzon SM, Orazov A (eds) Studies in economic history of peoples of middle Asia and Kazakhstan (trans: Temirbekov S, Almaty, 2001). Leningrad, Nauka, pp 70–74 [in Russian]

Galvin KA, Reid RS, Behnke Jr RH, Hobbs NT (eds) (2009) Fragmentation in semi-arid and arid landscapes: Consequences for Human and Natural Systems. Springer, Dordrecht, p 411

GOK (1999) Zemel'nyj kodeks KR, g. Govt. of Kyrgyzstan, Bishkek, 2 June 1999

Grodekov NI (1889) The Kyrgyz and Kra-Kyrgyz of Sydaria oblast. Moscow [in Russian]

Halimova N (2012) Land tenure reform in Tajikistan: implications for land stewardship and social sustainability: a case study. In: Squires VR (ed) Rangeland stewardship in Central Asia. Springer, Dordrecht, pp 305–329 (Chapter 13, this volume)

Hannam I (2012) International perspectives on legislative and administrative reforms as an aid to better land stewardship in Central Asia. In: Squires VR (ed) Rangeland stewardship in Central Asia. Springer, Dordrecht, pp 407–429 (Chapter 17, this volume)

Hazanov AM (2002) Kochevniki i vneshnii mir.- Almaty: Daik-press, [In Russian]

[8] Website Gos.Agenstva po okruzhaewej srede i lesnogo hozjajstva KR.

Hobbs T, Galvin G (2003) Preface to Session on fragmentation of rangelands: Ecological and economic implications. In: Rangelands in the New Millennium, Proceedings of VII International Rangelands Congress, Durban South Africa, 26 July–1 August 2003

Il'yasov SI (1963) Zemelnye otnosheniya v Kirgizii v konce 19-nachale 20 vekov, Frunze, izd-vo Akad. Nauk

Jacquesson S (2010) Pastoréalismes: anthropologie historique des processus d'intégration chez les Kirghiz du Tian Shan intérieur. Ludwig Reichert Verlag, Wiesbaden. xi + 281 pp. [In French]

Jacquesson S (2011) Power play among the Kyrgyz: state versus descent. In: Charleux I (ed) Representing power in modern inner Asia: conventions, alternatives and oppositions. Western Washington University, Centre for East Asian Studies, Bellingham, pp 221–244

Kassam K-A (2009) Viewing change through the prism of indigenous human ecology: findings from the Afghan and Tajik Pamirs. Hum Ecol 37(6):677–690

Kerven C, Alimaev II, Behnke R, Davidson D, Franchois L, Malmakov N, Mathijs E, Smailov A, Temirbekov S, Wright I (2004) Retraction and expansion of flock mobility in Central Asia: costs and consequences. Afr J Range Forage Sci 21(3):159–169

Kerven C, Alimaev II, Behnke R, Davidson G, Malmakov N, Smailov A, Wright I (2006) Fragmenting pastoral mobility: changing grazing patterns in post-Soviet Kazakhstan. In: Bedunah DJ, McArthur ED, Fernandez-Gimenez M (comps) Rangelands of Central Asia: proceedings of the conference on transformations, issues, and future challenges, Salt Lake City, UT, 27 January 2004. Proceeding RMRS-P-39. U.S. Department of Agriculture, Forest Service, Rocky Mountain Research Station, Fort Collins

Kreutzmann H (2009) Transformations of High Mountain pastoral strategies in the Pamirian Knot. Nomadic Peoples 13(2):102–123

Kreutzmann H (ed) (2012) Pastoral practices in high Asia: agency of 'development' effected by modernisation, resettlement and transformation, Advances in Asian human-environmental research. Springer, New York, 350 p

Kurbanova B (2012) Constraints and barriers to better land stewardship: analysis of PRAs in Tajikistan. In: Squires VR (ed) Rangeland stewardship in Central Asia. Springer, Dordrecht, pp 129–163 (Chapter 7, this volume)

Pierce RA (1960) Russian Central Asia: a study in colonial rule. University of California Press, Berkeley

Robinson S et al (2012) Pastoral tenure in Central Asia: theme and variation in the five former Soviet Republics. In: Squires VR (ed) Rangeland stewardship in Central Asia. Springer, Dordrecht, pp 239–274 (Chapter 11, this volume)

Shahmatov VF (1964) Kazahskaya pastbiwno-kochevaja obwina, Almata, izd-vo Akad. Nauk Kaz. SSR

Shaumarov M et al (2012) Participatory management of desert rangelands to improve food security and sustain the natural resource base in Uzbekistan. In: Squires VR (ed) Rangeland stewardship in Central Asia. Springer, Dordrecht, pp 381–406 (Chapter 16, this volume)

Sitnyanskii Y (1998) Sel'skoe hozyaistvo kirigzov: tradisii i sovremennost, M., izd-vo RAN

Sneath D (1999) Spatial mobility and Inner Asian pastoralism. In: Humphrey C, Sneath D (eds) The end of Nomadism?: society, state, and the environment in Inner Asia. Duke University Press, Durham, pp 218–277

Tolybekov SE (1963) Kochevoe obwestvo kazahov v 17-nachale 20 veka, Almata, izd-vo Kazgosizdat

Yablonsky LT (2003) The archaeology of Eurasian nomads. In Archaeology (Hardesty DL (ed)) in Encyclopedia of Life Support Systems (EOLSS), Developed under the Auspices of the UNESCO. Eolss Publishers, Oxford

Part II
Land Stewardship: From Concept to Practice

The four chapters in this section provide a summary of experience with land stewardship in the Central Asian region as well as detailed accounts of the situation in Afghanistan and Tajikistan.

Chapter 4 deals with the outcomes of extensive field work in several "stans" in the post-Soviet era and assesses the extent to which land stewardship has been advanced or retarded by events, policies, and legislative actions.

Chapter 5 turns the focus to Afghanistan, an important part of the Central Asian region, and examines past, present, and planned measures to deal with the critical land management\land tenure issues.

Chapter 6 turns attention to the livestock subsector in Afghanistan and to the plight of the nomadic peoples who struggle to survive in this ravaged and war-torn country.

Chapter 7 reports the outcomes of a comprehensive program of PRA in Tajikistan and assesses the barriers to adoption of more sustainable land use practices against a background of rural poverty, uncertain land tenure, and a discriminatory legislative and policy environment.

Chapter 4
Land Stewardship in Practice: An Example from the Eastern Pamirs of Tajikistan

Kim André Vanselow, Tobias Kraudzun, and Cyrus Samimi

Abstract This chapter reviews previous research and summarizes our own findings that are based on extensive fieldwork conducted between 2007 and 2010 in the Eastern Pamirs region that covers part of eastern Tajikistan, western China, and northern Afghanistan. The entry point is an overview of the post-Soviet transformation processes in Central Asia that are related to pastures as common property. This is followed by a description of the politico-historical development in the Eastern Pamirs as it characterizes the basic conditions of economic and social life over time.

Keywords Commons syndrome • Mountains • Orographic environment • Socioeconomic status • Land user rights • Tenure • Fuel wood • "*Teresken* syndrome"

Key Points

- The region is characterized by hostile living conditions resulting from a high elevation and distinct aridity. The altitude ranges from around 3,500 m a.s.l. in the wide valleys to more than 7,000 m in the northern and western mountain ranges. This orographic environment causes low temperatures year-round, with an annual average of only minus 1°C in the valleys. The extreme aridity, with precipitation values partly below 100 mm per year, is also primarily determined orographically.

K.A. Vanselow (✉)
Institute of Geography, Friedrich-Alexander-Universität Erlangen-Nürnberg, Kochstraße 4/4, 91054 Erlangen, Germany
e-mail: Kim.Vanselow@geographie.uni-erlangen.de

T. Kraudzun
Centre for Development Studies, Institute of Geographical Sciences, Freie Universität Berlin, Malteserstraße 74-100, 12249 Berlin, Germany

C. Samimi
Department of Geography and Regional Research, University of Vienna, Althanstraße 14, 1090 Vienna, Austria

V. Squires (ed.), *Rangeland Stewardship in Central Asia: Balancing Improved Livelihoods, Biodiversity Conservation and Land Protection*,
DOI 10.1007/978-94-007-5367-9_4, © Springer Science+Business Media Dordrecht 2012

- The economic and social dimensions also strongly influence land use practices, which are substantially connected to the environmental dimension of "unsustainability" in the Eastern Pamirs. An improved land stewardship must therefore include the sustainable management of the essential natural resources – in this case pastures and fuelwood.
- A key element of land stewardship is property rights. They have a strong influence on the use and access to natural resources and thus on their environmental sustainability. The dissolution of the state farms and the end of the external inputs of the Soviet economy led to small-scale pastoralism primarily on a household basis. The present-day utilization patterns, however, are strongly determined by conflicting claims related to legal pluralism. Most notably, they differ according to the socioeconomic status of the livestock owners.

1 Introduction

Mountains play an important role in the world's sustainable development (United Nations 1992; Jansky and Pachova 2006; Maselli 2012). Despite the acquired knowledge about their importance, the Global Environment Outlook 3 states that most "mountain commons are ecologically under-managed and suffer from the classic 'commons syndrome': while all seek to benefit, stakeholders lack coordination, incentives and instruments for joint care" (UNEP 2002:68). This is particularly true for the region in the outermost east of Tajikistan, known as the Eastern Pamirs, where rangeland stewardship is highly affected by difficult physical environmental prerequisites and politico-historical implications of land tenure and pasture practice.

2 Geographical Setting of the Eastern Pamirs

The Eastern Pamirs, predominantly a high-mountain plateau in Central Asia, are located in the east of Tajikistan and partly in China and Afghanistan (Fig. 4.1). The region is characterized by hostile living conditions resulting from a high elevation and distinct aridity. The altitude ranges from around 3,500 m a.s.l. in the wide valleys (Fig. 4.2) to more than 7,000 m in the northern and western mountain ranges. This orographic environment causes low temperatures year-round, with an annual average of only minus 1 °C in the valleys (Miehe et al. 2001). The extreme aridity, with precipitation values partly below 100 mm per year, is also primarily determined orographically. This means the high ranges shield the plateau from the precipitation of the westerlies and the Indian summer monsoon (Walter and Breckle 1991). Hence, the Eastern Pamirs can be described as a cold high-mountain desert that is characterized mostly by scattered, slow-growing vegetation (Agakhanjanz 1979; Walter and Breckle 1991; Breu and Hurni 2003; Abdullaev and Akbarzadeh 2010).

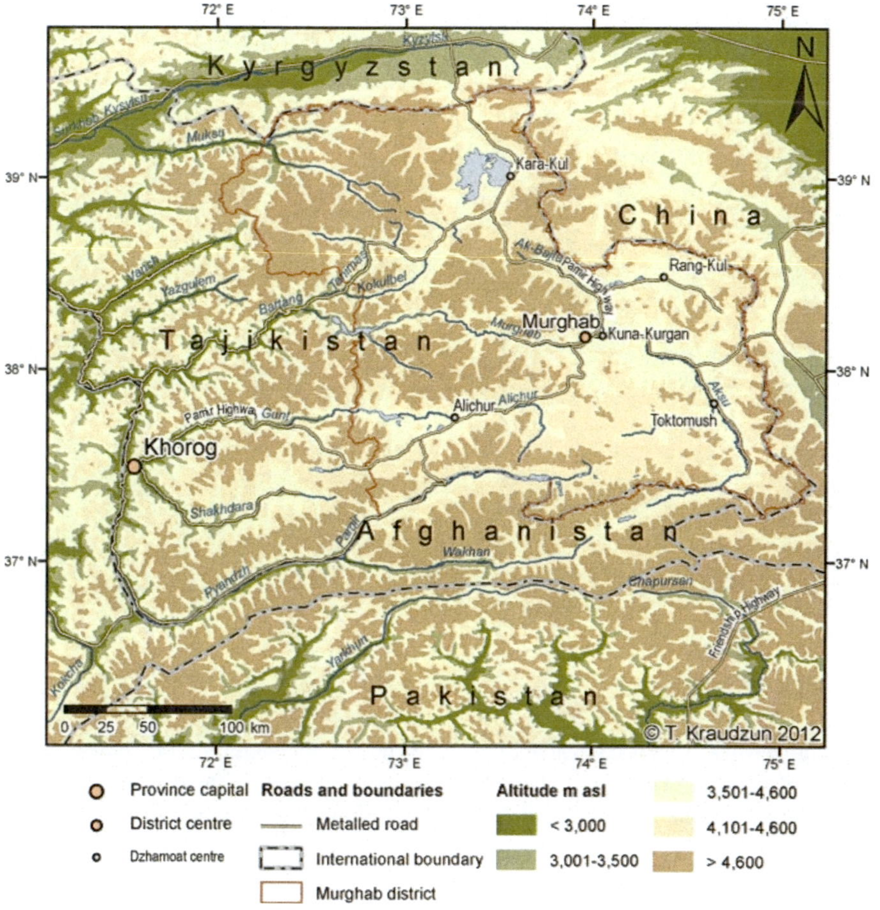

Fig. 4.1 Overview on the Eastern Pamirs

2.1 *Politico-Historical Setting of the Eastern Pamirs*

Due to the harsh climate, this area was virtually uninhabited before the nineteenth century. Only some Kyrgyz nomads moved to this high plateau – partly during the summer season, partly also permanently – to graze their livestock. Their pasture camps were, however, spread over a huge area resulting in a very marginal land use (Maanaev and Ploskikh 1983; Kraudzun 2012). This situation changed dramatically with the Russian and British expansion in Central Asia in the second half of the nineteenth century, today known as the "Great Game" (Kreutzmann 2009). This conflict has resulted in the permanent colonization of this peripheral high-mountain region in particular by sedentarizing Kyrgyz pastoral nomads (Maanaev and Ploskikh 1983; Antonenko 1985). Since pastoralism with extensive livestock

Fig. 4.2 Photo of typical *pamer* (plain and wide valley in the Eastern Pamirs)

herding of yaks, sheep, and goats has been the primary agricultural option over time, during the Soviet Union, the production of meat was assigned as the region's main economic task (Kreutzmann 2011b). Therefore, highly subsidized collective and state farms were set up (Rahimon, Chap. 3).

The allocation of pastureland was subject to an elaborate management plan, usually with four seasonal pasture camps and additional imports of fuel and forage, to ensure a well-balanced utilization of all pastures (Kraudzun 2012; Vanselow et al. 2012). However, the dissolution of the Soviet Union and the subsequent Tajik independence in 1991 brought an end to that system. This in turn impeded the provisioning of the region and led to grave supply problems, particularly with regard to forage, fuel, and transportation resources. Recently, indications of extensive environmental degradation have been ascribed to the post-Soviet land use changes and pastoral practices, including the extraction of firewood and overuse of pastures near villages (Breu et al. 2005). Due to the lack of subsidies, bridging long distances between the seasonal pastures poses a major problem to the majority of today's smallholders (Domeisen 2002; Vanselow et al. 2012). Furthermore, the limited supply and high costs of imported fossil fuels led to an increased use of dwarf shrubs as energy resource (Breu et al. 2005; Förster et al. 2011). However, these plants are also important forage plants, particularly in winter (Breckle and Wucherer 2006; Breu 2006; Kraudzun et al. 2012).

The described developments pose major threats to a sustainable land stewardship in the Eastern Pamirs. Since they are determined by the physical and biological environmental preconditions as well as the political and socioeconomic legacies of the recent history, an interdisciplinary approach was realized which combines data and methods from both human and natural sciences. As such, this chapter reviews previous research and summarizes our own findings that are based on extensive fieldwork conducted between 2007 and 2010 within the scope of the project "Transformation Processes in the Eastern Pamirs of Tajikistan – Changing Land Use Practices, Possible Ecological Degradation and Sustainable Development" (funded by the Volkswagen Foundation).

The entry point is an overview of the post-Soviet transformation processes in Central Asia that are related to pastures as common property. This is followed by a description of the politico-historical development in the Eastern Pamirs as it characterizes the basic conditions of economic and social life over time. The economic and social dimensions strongly influence land use practices, which are substantially connected to the environmental dimension of "unsustainability" in the Eastern Pamirs. An improved land stewardship must therefore include the sustainable management of the essential natural resources – in this case pastures and fuelwood.

3 Property Rights of Common Pool Resources

A key element of land stewardship is property rights. They have a strong influence on the use and access to natural resources and thus on their environmental sustainability (Demsetz 1967; Bromley 1992; Robinson and Whitton 2010). For common pool resources, such as pastures, Schlager and Ostrom (1992) defined a property rights schema with four levels that are associated with particular rights. The first level is the authorized user, who holds the right of access and withdrawal; this means he has the right to enter a specific property and obtain the "products" of a resource. The claimant additionally has the right to regulate internal use patterns of the property. The next higher level is the proprietor, who can exclude the use by others for his property. The highest level is the owner who can sell or lease the property and associated rights (Halimova, Chap. 13).

Based on the assumption of open-access to common resources, Hardin (1968) postulated the "tragedy of the commons." He assumed that each user of an open-access common good ignores costs imposed on other users. In consequence, individual decisions would accumulate to an overuse and the possible depletion of the specific resource. This means that people only follow their own short-term preferences that lead to effects that are against anyone's long-term interest. According to Ostrom et al. (1999), however, the open-access regime is only one out of four types of property rights that are related to common pool resources. The second and third types are group and individual property. Here, individuals are at least proprietors in the meaning of Schlager and Ostrom (1992). This means they hold rights of access and use of a resource and can exclude the use of others. This is essential as defined users and a clear delimitation between the resources are important attributes for the sustainable use of common pool resources. The difference between these two types is related to the ease with which the individuals can buy or sell a share of the resource. The last type they call government property. This means that the resource is owned by the state. Therefore, national, regional, or local authorities have the right to regulate use patterns and access rights, and they can decide to sell or lease the resource.

This concept of property rights is based on static definitions such as statutory law. From a legal anthropological perspective, however, the claim of property rights is superimposed by the coexistence and interdependence between multiple legal orders such as state, customary, and local laws (Meinzen-Dick and Pradhan 2001; von Benda-Beckmann 2002). The result is a legal pluralism that enables people to rely on different legal bases and therefore act flexibly regarding their use of natural resources (see Sect. 4 below).

During the Soviet era, rangeland in Central Asia was owned by the state and managed by kolkhozes (collective farms) and sovkhozes (state farms), representing Ostrom et al. (1999) fourth property type. Since these production systems revealed to be very ineffective, land reform processes were introduced in the new Central Asian states soon after their independence (Mearns 1996; Wolfgramm et al. 2010). In order to increase efficiency and to avoid degradation (referring to the proposed negative impacts by open-access regimes), common pool resources were extensively privatized leading to different approaches of pasture stewardship and management. In most cases the collective and state farms were dissolved, and their livestock assets were equally distributed to the former employees' households. Pasture land, however, remained state-owned in most cases (Peyrouse 2009; Bichsel et al. 2010; Schoch et al. 2010; Steimann 2011).

This is the case, for example, in Kyrgyzstan, where soon after independence in 1991, market economy reforms were introduced, which resulted in the privatization of the former collective and state farms in 1993. The end of subsidies after the Soviet dissolution and the split of livestock assets into small units, however, led to a severe market break of the agricultural economy in the mid-1990s (Bichsel et al. 2010; Steimann 2011). A second reason for the failure of early privatization is the lack of husbandry knowledge as the new livestock owners were most often highly specialized workers in the state farms. As a result of the crisis, increasing poverty led many people to revert to an agricultural subsistence strategy (Wilson 1997; Ludi 2003; Schoch et al. 2010; Wolfgramm et al. 2010). This situation is not specific to Kyrgyzstan but can be observed in other Central Asian states as well (Herbers 2006; Kreutzmann 2012; Rahimon 2012). In Kyrgyzstan, until lately, three categories of pastures were differentiated.

Pastures in the vicinity of villages were managed by the subdistrict administration. Spring and autumn pastures, usually located at some distance to the village, were under the control of the district administration. Finally, remote summer pastures were managed by the provincial administration. Since 2002, individuals and groups could lease pastureland according to a tenancy system. With the privatization, however, the socioeconomic disparities were aggravated (Ludi 2003;

Bichsel et al. 2010; Steimann 2011). Today, there are many poor households with very few or no livestock. For these herders it is not possible to practice the traditional, ideal three-pasture annual cycle, instead there is an intensified use of pastures close to villages (Schoch et al. 2010). This situation is also reported from Kazakhstan (Kerven et al. 2004) and Tajikistan (see below). By contrast, a few affluent households own above average-sized herds and secure use rights for the best pastures (Steimann 2011). This system also led to an increasing allocation of village pastures to outsiders, resulting in a loss of control over these pastures (Undeland 2005). This formal regulation of pastureland, however, is often disregarded. In practice, the majority of non-tenants ignore the tenancy agreements. They perceive pastures as common property and insist on unlimited access, which leads to high grazing pressure, particularly on common pastures in the vicinity of villages (Wolfgramm et al. 2010).

The majority of herders choose their pastures according to common law rather than state law. Most of them utilize the same pastures their parents used during the Soviet Union, a practice which is often better recognized than a formal tenancy agreement (Steimann 2011). In 2009, a new pasture law was introduced which transfers the right to allocate and manage pastures to so-called grazing committees, democratically elected boards of the local associations of pasture users. However, it is doubtful if the new administrative system will reduce existing inequalities (Jacquesson 2010).

In Tajikistan, state farm assets of livestock were distributed to individual households starting in 1999 (Herbers 2006; Robinson et al. 2008). Pasture land, however, is still owned by the state, but users have access to pasture resources according to a tenancy system based on the Law on Dekhan Farms (Halimova, Chap. 13). Three main types of tenure arrangements can be differentiated: permanent heritable land use, long-term use, and leasing (cf. Robinson and Whitton 2010; Robinson et al. 2010). The first type is based on the establishment of so-called dekhan farms (i.e., peasant farms) by individuals, families, or groups. Here, the land certificate is held by the head of a farm, and shareholding members receive share documents for specific land resources. They can be founded by former sovkhoz or kolkhoz workers by jointly applying for a share of land of the former collective farm where they used to work.

A special case are collective dekhan farms that comprise the entire land of the former collective or state farm. The second type is based on agreements for a long-term use (25 years) between the management of collective dekhan farms and the district land committee. Here, members of the dekhan farms do not receive a share document for the allocated land resources. The third type applies only to remote pastures that remained unallocated. They constitute part of the state fund and may be leased out to any party (individual, collective dekhan farm, or state enterprise) on an annual basis.

Despite many parallels, the transitional conditions in the Eastern Pamirs differed from that of the neighboring regions, which led to specific requirements for a sustainable land stewardship.

4 Development of Land and Livestock Tenure in the Eastern Pamirs

4.1 Land and Livestock Tenure Prior to the Soviet Era

Prior to the nineteenth century, the Eastern Pamirs were nearly uninhabited, with only some Kyrgyz nomads using the area as summer pastures. Modern forms of land tenure, not to mention land stewardship in the current sense, did not exist at that time. The herders were, however, organized in kinship units which acted as political and economic units and jointly moved to specific seasonal pastures (Rahimon, Chap. 3). Generally, these groups were ruled by the head of a wealthy household, who decided on land use issues and owned most of the livestock (Maanaev and Ploskikh 1983; Kreutzmann 2011a; Kraudzun 2012).

In the second half of the nineteenth century, this situation changed in the course of the "Great Game" between the Russian and British empires (Kreutzmann 2009). Borders, in the present sense of the word, did not exist in the Pamirs at that time; however, in 1893 the Russian tsarist army set up the military base "Pamirsky Post." This garrison represents the first nucleus of a permanent settlement in the region and grew in the aftermath to the district's principal town, Murghab. Herdsmen decided to settle within its vicinity as local commodities could be exchanged with products imported by Russian traders. This can be regarded as the first step away from nomadism and extremely scattered settling toward central locations (Hedin 1899; Kraudzun 2011; Vanselow 2011). In 1895, a joint committee of the two empires drew the boundary between the areas of influence along the Amu Darya river system (including Panj and Oxus), and the Eastern Pamirs became incorporated into the Russian Empire. In doing so, existing claims were ignored, as were land use systems and interrelations between peoples, for example, the traditional seasonal routes of the mobile animal husbandry of the Kyrgyz nomads were cut (Kraudzun 2011; Kreutzmann 2011a, b).

4.2 Land and Livestock Tenure During the Soviet Era

The main reason for the substantial provisions applied to the Soviet Pamirs was based on the geopolitical importance of the international border shared with China and Afghanistan (Kreutzmann 2006). In order to control this vast border region, the Russian administration tried to attain the loyalty of the Kyrgyz kinship groups. Later on, this goal was pursued by the Soviet power by means of the establishment of appropriate infrastructure, improved living conditions, and last but not least, of an effective administration (Kraudzun 2012).

In 1924, the Tajik Autonomous Socialist Soviet Republic was founded, at first being only a subdivision of the Uzbek Socialist Soviet Republic. The Gorno-Badakhshan Autonomous Province, however, was created out of the Pamirs area with a nominal autonomy status already in 1925, whereas the Tajik Socialist Soviet Republic became a "full" Soviet Republic only in 1929 (Kreutzmann 1996). In the early stages of the newly founded administrative region, the economic and social organization of pastoralism was still controlled by wealthy livestock owners. However, in the following years, the Soviet power strived to convince the local population of the advantages of collective farms in a state-run economy, sometimes even with force, and the extremely peripheral region became gradually integrated into the Soviet *Gosplan* system (Kraudzun 2012).

Collectivization began around 1940 with the establishment of 11 kolkhozy, which ended up after several rearrangements in one sovkhoz and four kolkhozy. Therefore, Kyrgyz herdsmen were successively sedentarized in permanent villages, where the production infrastructure of the collective and state farms was set up (Kraudzun 2012). With regard to the harsh environment, the primary land use strategy adapted to the region has been pastoralism with extensive livestock herding (Kreutzmann 2011a; Kraudzun 2012). Therefore, the administration allocated the breeding of livestock as the region's main task for the installed collective and state farms, which were then delivered to processing plants in the adjacent lowlands (Kraudzun 2012; Vanselow et al. 2012). Consumable durables such as fuel, staple foods, and additional winter forage were exclusively imported and brought to the region on the Pamirs highway which was advanced from Osh via Murghab to Khorog in 1934 and extended to Dushanbe by 1940 (Kreutzmann 2003, 2004; Kraudzun 2011). The land titles were held by the kolkhozy and the sovkhoz which allocated the pastureland to their members on the basis of an elaborate management plan (see Sect. 4 below).

4.3 Land and Livestock Tenure After the Soviet Dissolution

Despite many parallels between the post-Soviet Central Asian republics (see Sect. 5 below), the transitional conditions in the Eastern Pamirs differed from that of the neighboring regions. Here, the economic problems that occurred during the last years of the Soviet Union led to a search for alternatives to reduce the extremely high subsidies for the state-owned farms in the Eastern Pamirs. Therefore, in 1988, a reform was implemented, which passed the responsibility for economic success of the state-owned farms to an individual basis, that is, to the kolkhoz and sovkhoz members. The farms' animals were leased out to the employees who were made responsible for the economic efficiency of their livestock assets. Each leaseholder was assigned a herd and a set of seasonal pastures and had to acknowledge receipt in the state farm's bookkeeping. All kinds of support by the state farm, such as transportation services or additional forage supply, were summed as the leaseholder's expenses that had to be balanced with the delivered animals as production output. In the former labor-division framework of the Soviet Union, the tenants were employed, for example, as teachers, bureaucrats, or medical doctors. Although many of them had no experience in livestock breeding and herding, this system performed well during the last few years of the Soviet Union but collapsed shortly after the Tajik independence (Kraudzun 2012).

Most significantly the end of the Soviet system stopped the provisioning of the region from outside and led to grave supply problems and consequently to the failure of the tenancy system (Kraudzun 2012). The postindependence Tajik Civil War (1992–1997), which forced many refugees to seek shelter in the Pamirs, aggravated the supply situation (Breu and Hurni 2003; Hurni et al. 2004; Breu et al. 2005) and hampered further reorganization and inhibited steps toward improved land stewardship and livestock management (Kraudzun 2012).

In particular, the temporary lack of law enforcement by the authorities facilitated the herders to act on their own behalf. Although pasture land and most livestock were both still officially owned by the state farms, most leaseholders considered it as their own. They took over control of the pastures and integrated the animals into their private herds. Due to the lack of cash, animals became the main currency. Particularly those entrusted animals were used as medium for exchange and then usually declared by the herders as loss. The increasing "disappearance" of state-owned livestock led to the intervention of the administration. Repeated censuses revealed a massive loss of livestock and therefore large debts of the leaseholders toward the state farms. Meanwhile, however, hyperinflation of the Russian ruble meant that the compensation payment for the missing livestock assets did not pose big problems to most leaseholders. In consequence of the complete failure of the tenancy system, the state farms returned to a variant of the previous Soviet-style management system for the following years (Kraudzun 2012).

Then, in 1999, four of the five still existing collective farms were disintegrated, and new collective structures called *associations of dekhan farms* (ADF) inherited their land titles and remaining livestock. These associations have to be understood as a special case of collective dekhan farms as the leased land corresponds to that of the previous kolkhozes. De jure, however, it is an association of dekhan farms that have to be individually registered – which at present is (still) not the case (Kraudzun 2012; Vanselow et al. 2012).

The livestock was equally distributed among the population of the subdistrict where the former collective farms were located. The allocation procedure was implemented by the district administration and communicated to the population as a distribution of the livestock. This misinformation led to a general perception of a direct privatization of the collective farms' livestock assets. Although the allocated livestock is officially still owned by the ADFs, the people have integrated the animals into their herds and perceive them as their own (Kraudzun 2012).

Concerning pasture land, the ADFs are responsible to equitably distribute pasture use rights in order to ensure sustainable grazing; however, they often lack assertiveness because of the legal pluralism that has developed during the transformation process. This means that the ADFs are challenged by different claims of customary, inherited, and perceived user rights. Particularly wealthy livestock owners took advantage of loopholes to protect their interests by switching between the different types of legislation. Today, as a result many pastures are occupied by the former leaseholders. They have simply continued using the same land that had been allocated to them within the tenant system and perceive these pastures as their own private grazing lands (Kraudzun 2012).

5 Development of Land Use Practices

5.1 Land Use Practices During the Soviet Union

During the Soviet Union, pastureland was allocated according to a strict management plan. Usually, the pastures were utilized as exclusive spring, summer, autumn, or winter pasture. This means the collective and state farms shifted all herds after one season, seeking a well-balanced and sustainable utilization of all pastures in the Eastern Pamirs (Coughenour 2008; Vanselow et al. 2012). Such extensive relocations were only possible because these farms were supported by exceptional subsidies and great organizational efforts, including the transport of the herder families and their livestock from one seasonal pasture to another. In particular, summer pastures were often located far from the permanent settlements, for example, in the subdistrict of Kuna Kurgan, several summer pastures were (and still are) situated in the Chong Pamirs (Great Pamirs) near the Afghan border, more than 100 km away from the central village (Vanselow 2011; Vanselow et al. 2012).

Furthermore, forage resources were supplemented by imports from the Kyrgyz SSR, including 2,000–5,400 t of winter fodder that was grown on an exclave territory in the productive Alai valley and allocated to the state farms of the Eastern Pamirs (Vanselow et al. 2012). In particular, the supply with emergency forage in strong snow-rich winters reduced livestock losses. Additionally, herdsmen received assistance for the time-consuming work during the lambing and shearing period (Kraudzun 2012).

Energy demands that were first satisfied by increasing the harvesting of dwarf shrubs led to pasture degradation and were substituted by an annual share of 18,000 t of Siberian coal since 1961 (Kraudzun et al. 2012). Particularly, during the gestation period, the extra provision of combustibles led to the limitation of weather-related livestock losses (Kraudzun 2012).

Another strong advantage of the collective herds of the state farms is the temporal and spatial flexibility. The integrated management of all animals of a state farm's territory allowed the adjustment of herd sizes to variable forage conditions in space and time. In practice, this means that the pastures were assessed by a visual on-the-spot inspection by farm specialists to determine the particular carrying capacity and hence suitable maximum herd size. Furthermore, even during winter pastures with a relatively high risk of snowfall, the area could be used because in the case of shortcomings additional forage inputs were delivered or the herds were moved to preserved pastures (Vanselow et al. 2012).

In the subdistrict of Kuna Kurgan, for example, such preserved areas were located predominantly in the Kara Suu and Pshart valleys as well as in the northeast of Aksu River (Fig. 4.3). The riparian spring meadows along Murghab River in the Madian

Fig. 4.3 Pasture use during the Soviet Union (*left*) versus present using schemes (*right*)

valley were used exclusively for hay making. The meadows surrounding Murghab town were strictly excluded from use in the pasture management system. The utilization of the remaining pasture land was restricted to only one season. Spring pastures were located in the side valleys of the Madian valley, parts of the area north of Aksu River, and large areas in the central part of the subdistrict. Exclusive summer pastures were limited to the extreme south of the subdistrict in the Chong Pamirs (Great Pamirs) near the Afghan border. Some pastures in the preserved valleys of Kara Suu and Pshart, however, were also partly used as summer pastures. Autumn pastures were located in the northeast of the subdistrict north of Aksu River, whereas winter pastures cover a large area south of Aksu River (Fig. 4.3). In sum, in the subdistrict Kuna Kurgan, 274,000 ha (61%) were accounted for as winter and spring pastures, whereas only 175,000 ha (39%) were allocated as summer and autumn pastures (Vanselow et al. 2012).

5.2 *Land Use Practices After the Soviet Dissolution*

The dissolution of the state farms and the end of the external inputs of the Soviet economy led to small-scale pastoralism primarily on a household basis. According to the statutory law, the allocation of pastureland is incumbent upon the ADFs. They act according to their possibilities to bring about improvements of the land use practices. The present-day utilization patterns, however, are strongly determined by conflicting claims related to legal pluralism. Most notably they differ according to the socioeconomic status of the livestock owners. Three different groups may be distinguished in the Eastern Pamirs. A small number of successful breeders reached certain prosperity and hence were able to seize user rights for a well-defined set of large-scale, productive pastures. A second group with medium-sized herds has to come to agreements within this group and negotiate their demand with the first group of affluent and thus influential livestock owners. To some extent they also look for pasture "niches"; this means they switch to pastures with fewer claims. The majority, however, are smallholders. Particularly this group is not able to practice mobile pastoralism with long-distance rotations between four seasonal pastures.

Today, most of them are organized in herding groups that graze their animals year-round on the pastures in the vicinity of the settlements (Vanselow et al. 2012). For Kuna Kurgan subdistrict, such a permanent use could be approved for the areas in the proximity of Murghab town, the formerly spared pastures in the Kara Suu valley and the former spring pastures north of Aksu River (Fig. 4.3). As a result, some authors observed an overexploitation of these easy-to-reach pastures, whereas extensive but remote areas lie fallow (Domeisen 2002; Breu and Hurni 2003; Robinson et al. 2010). This is generally the case but more differentiated when viewed in detail (see below). Overall, however, only 55–64% of the herders in the Kuna Kurgan subdistrict relocate their livestock to a specific autumn pasture. Concerning the shift to a specific spring pasture, the proportion is even less (26–38%). In consequence, all pasture areas – formerly only used in one particular season – are today utilized in at least two seasons. Apart from the already discussed areas, the pastures in the Pshart valley and northeast of Aksu River – primarily preserved during the Soviet era – are today utilized as summer and autumn pastures.

The formerly exclusive hay meadows in the Madian valley are today spared only during the summer months. Hay is still harvested here at the end of the vegetation period but used as pastureland during the remaining seasons, particularly in winter. The adjacent valleys of the Madian valley – all former spring pastures – suffer today under summer- and springtime grazing. The same is true for the vast areas in the central and southern part of the subdistrict that were allocated as exclusive summer, autumn, or spring pastures and today are grazed in summer and autumn, areas in the southeast additionally in spring. The extensive winter pastures south of Aksu River are still used in winter, but the grazing period was extended until the end of spring. Hence, the grazing period on all pastures is extended, leading to a decrease of the recovery time for the vegetation. Apart from the mentioned groups, some households without own livestock are employed by livestock owners as herders, or they hire out as fuelwood diggers.

During the Soviet state farm system, the herd sizes were flexibly adjusted to the spatial and temporal variability of forage conditions. Today, the herd sizes are determined by the wealth of the owner and are thus not flexibly adapted to the ecological conditions. For example, many of the smallholders and owners with medium-sized flocks use pastures for which they succeeded to negotiate informal use rights with powerful livestock owners or because they chose a vacant pasture. This procedure, however, does not take into account the herd size. Therefore, it often leads to livestock numbers that are inadequate for the pasture potential. Unfortunately, this practice is not challenged because the responsible ADF did not develop mechanisms for pasture administration and control (Vanselow et al. 2012).

Furthermore, the present-day herdsmen have to rely exclusively on their own forage supplies from their limited available hay plots because the external forage supply stopped with the end of the subsidies. In consequence, winter pastures with inconsistent weather conditions cannot be used anymore. This makes winter pastures today the most limited resource. In contrast to the proportion during the Soviet time, at the moment only 193,000 ha, one third of all accessible pastures, can be used without possible weather-related shortages as winter and spring pastures, whereas 358,000 ha, double the area, are available for summer and autumn use. The adapted winter pastures, however, are utilized until the summer pastures are snow free and green enough to feed the livestock and are thus particularly prone to severe degradation (Vanselow et al. 2012).

5.3 Fuelwood Extraction

A second important risk factor for the land resources of the Eastern Pamirs is the extraction of fuelwood. From 1961 onward, *teresken* (*Krascheninnikovia ceratoides)* harvesting was banned by law in order to protect this important natural

Fig. 4.4 Photo of a *teresken* transport. Harvesting of woody plants is common and poses a threat to ecosystem stability

resource. In return, a coal fund was introduced within the budget of the Tajik SSR to purchase 18,000 t per year for the Murghab district. After the Tajik independence, the subsidies for coal ceased, and fuel supplies were not delivered to Murghab district anymore (Kraudzun et al. 2012). The lack of fuel and high costs of imported fossil fuels caused an intensified use of dwarf shrubs, primarily *teresken* but also wormwood (*Artemisia* spp.) as a substitute energy resource (Breu and Hurni 2003; Droux and Hoeck 2004), ignoring the ban that is officially still in force (Fig. 4.4). These plants, however, are also important for the protection of soils from erosion and as forage plants for livestock and wild animals, particularly in winter when fodder resources are scarce (Agachanianc 1966; Jusufbekov and Kasach 1972). Consequently, these dwarf shrubs are especially prone to overexploitation, which would have severe impacts on the ecology of the pastureland (Aknazarov 2003; Akhmadov et al. 2006; Breckle and Wucherer 2006; Breu 2006; Kassam 2009). This circumstance was first discussed in various letters exchanged between S.W. Breckle (Bielefeld) and O.E. Agachanianc (Minsk) and named as the "*teresken* problem."

Today it is discussed in literature even as "*teresken* syndrome" (Breckle and Wucherer 2006). In recent years this topic has attracted much interest, particularly in the work of development programs (BMZ 2005; GEF 2005; Hoeck et al. 2005; BMZ 2007; Doempke 2008). Several generalized statements about the distribution and exploitation of *teresken* in the Eastern Pamirs, however, give the impression that these dwarf shrubs are now a thermal energy resource without important alternatives and will possibly be used until their extinction. For example, Doempke (2008:13) reports that "...already 50% of all *teresken* is believed to have been lost. It has been totally eradicated within a radius of about 100 km around Murghab town."

The results from our own fieldwork indeed confirm the described competitive use of dwarf shrubs as fuel and forage; however, they clearly show that such numbers are inaccurate estimations. Seriously degraded areas of dwarf shrub vegetation are limited to main roads and easily accessible locations around permanent villages. Hence, they do not exceed 6% of the analyzed area of more than 8,000 km^2. In contrast, intact dwarf shrub stands make more than 50% of this area and could be found less than 5 km from Murghab town, however, only on slopes or in small tributary valleys. We furthermore ascertained that herders with pasture rights in the harvesting regions often give their permission to the harvesting. They argue that old *teresken* plants produce only few leaves; consequently, their fodder value is rather limited. Herders often commented that harvesting brigades seek contact with them prior to extraction. Severe conflicts between *teresken* diggers and herders are therefore rare. Finally, the share of dwarf shrubs in the energy mix is recently decreasing in the Eastern Pamirs. Increasing livestock numbers support the replacement of dwarf shrubs with animal manure, and coal imports from mining areas in the Alai valley (Kyrgyzstan) became a regular supply line. Therefore, the use of fuelwood is meanwhile relatively expensive compared to alternative energy resources.

6 Conclusions

In summary, it can be concluded that the particular transformation processes of the Eastern Pamirs prevented the establishment of enforceable institutions for a sustainable land stewardship. After the privatization of the former collective and state farms' herds, the total livestock number in the region has increased and almost reached those of the Soviet era, which has led to a high demand for pastures. Social disparities among different groups of pastoralists and unsettled questions about formal user rights, however, resulted in inappropriate claims. Due to lack of financial resources, today most smallholders do not relocate their livestock to different seasonal pastures; therefore, pastures suffer primarily under a distinct multi-seasonal pasture use. In particular, the livestock numbers grazing on the pastures in the proximity of the settlements are far above the optimal stocking rates based on calculations of the existing pasture potential and show severe signs of overgrazing. Large areas of valuable pastures are claimed by powerful livestock owners who took advantage of the confusion during the transformation process and now are able to exclude the use by others.

The modification of grazing patterns could be identified as caused by a legal pluralism. Officially, the ADFs as holders of the land titles are responsible to distribute the pastures based on the users' herd sizes. In practice, however, this system is not fully achieved. Today, many individual claims are enforced by justifying them as based on customary and inherited law dating back to the tenancy period. Hence, conflicts among the different groups seem to be inevitable. To feed the still growing livestock numbers, however, it is necessary to adapt pasture using schemes and to solve the energy problem. If the stewardship of these two resources is not improved, parts of the fragile high-mountain desert ecosystem of the Eastern Pamirs will become critically endangered which could lead to desertification in the long run.

References and Further Reading

Abdullaev K, Akbarzadeh S (2010) Historical dictionary of Tajikistan. Scarecrow Press, Lanham

Agachanianc OE (1966) Osnovnie problemy fizicheskoy geografii Pamira. Izd. Akademija Nauk Tadshikskoj SSR, Dushanbe

Agakhanjanz OE (1979) Besonderheiten in der Natur der ariden Gebirge der UdSSR. Petermanns Geographische Mitteilungen 123(2):73–77

Akhmadov K, Breckle S, Breckle U (2006) Effects of grazing on biodiversity, productivity, and soil erosion of alpine pastures in Tajik mountains. In: Spehn E, Liberman M, Körner C (eds) Land use change and mountain biodiversity. CRC/Taylor & Francis, Boca Raton, pp 239–247

Aknazarov KA (2003) The present situation of pasture land in Eastern Pamirs. Bielefelder Ökologische Beiträge 18:30–32

Antonenko BA (1985) Socialisticheskie agrarnye preobrazovanija v Gorno-Badachshanskoj avtnomnoj oblasti. In: Masov RM (ed) Ocherki po istorii Sovetskogo Badachshana. Izdat. Donish, Dushanbe, pp 220–253

Bichsel C, Fokou G, Ibraimova A, Kasymov U, Steimann B, Thieme S (2010) Natural resource institutions in transformation: the tragedy and glory of the private. In: Hurni H, Wiesmann U (eds) Global change and sustainable development – a synthesis of regional experiences from research partnerships, vol 5, Perspectives of the Swiss National Centre of Competence in Research (NCCR) North-South. University of Bern/Geographica Bernensia, Bern, pp 255–269

BMZ (2005) Desertifikationsbekämpfung 2005, Bundesministerium für wirtschaftliche Zusammenarbeit und Entwicklung, BMZ-Materialien Nr. 10

BMZ (2007) Desertifikationsbekämpfung, Bundesministerium für wirtschaftliche Zusammenarbeit und Entwicklung, BMZ CD-ROM 012

Breckle S, Wucherer W (2006) Vegetation of the Pamirs (Tajikistan): land use and desertification problems. In: Spehn E, Liberman M, Körner C (eds) Land-use change and mountain biodiversity. CRC/Taylor & Francis, Boca Raton, pp 225–237

Breu T (2006) Sustainable land management in the Tajik Pamirs. The role of knowledge for sustainable development. PhD thesis, University of Berne, Berne

Breu T, Hurni H (2003) The Tajik pamirs: challenges of sustainable development in an isolated mountain region. Centre for Development and Environment (CDE), Berne

Breu T, Maselli D, Hurni H (2005) Knowledge for sustainable development in the Tajik Pamirs Mountains. Mt Res Dev 25(2):139–146

Bromley DW (1992) The commons, common property, and environmental policy. Environmental and Resource Economics 2(1):1–17

Coughenour MB (2008) Causes and consequences of herbivore movement in landscape ecosystems. In: Galvin KA, Reid RS, Behnke RH, Hobbs N (eds) Fragmentation in semi-arid and arid landscapes – consequences for human and natural systems. Springer Netherlands, Dordrecht, pp 45–91

Demsetz H (1967) Toward a theory of property rights. Am Econ Assoc 57(2):347–359

Doempke S (2008) Mission Report, Pamirs-Alai Transboundary Conservation Area (PATCA). Berlin

Domeisen M (2002) Marginalized by the impacts of transformation: a study of post-Soviet. livestock breeding in the high mountains of the Eastern Pamirs. Master's thesis, University of Berne, Berne

Droux R, Hoeck T (2004) Energy for Gorno Badakhshan! Hydropower and firewood cultivation. Analysis of the energy situation in the Tajik Pamirs and its consequences for land use and resource management. Joint diploma thesis, University of Berne, Berne

Förster H, Pachova NI, Renaud FG (2011) Energy and land use in the Pamirs-Alai mountains: examples from five social-ecological regions. Mt Res Dev 31(4):305–314

GEF (2005) Project executive summary GEF council. Intersessional work program submission. Project: sustainable land management in the High Pamirs and Pamirs-Alai mountains – an integrated and trans-boundary initiative in Central Asia

Halimova N (2012) Land tenure reform in Tajikistan: implications for land stewardship and social sustainability: a case study. In: Squires VR (ed) Rangeland stewardship in Central Asia. Springer, Dordrecht, pp 305–332 (Chapter 13, this volume)

Hardin G (1968) The tragedy of the commons. Science 162:1243–1248

Hedin S (1899) Durch Asiens Wüsten: Drei Jahre auf neuen Wegen in Pamirs, Lop-Nor, Tibet und China. Brockhaus, Leipzig

Herbers H (2006) Landreform und Existenzsicherung in Tadschikistan. Die Handlungsmacht der Akteure im Kontext der postsowjetischen Transformation

Hoeck T, Breu T, Stillhardt B, Schwilch G, Herweg K (2005) Synthesis report baseline survey on sustainable land management in the Pamirs – Alai Mountains, Centre for Development and Environment (CDE). Berne

Hurni H, Breu T, Ludi E, Portner B (2004) Der Tadschikische Pamirs. Entwicklungsprobleme und Perspektiven einer Hochgebirgsregion in Zentralasien. Geographische Rundschau 56(10):60–65

Jacquesson S (2010) Reforming pastoral land use in Kyrgyzstan: from clan and custom to self-government and tradition. Cent Asian Surv 29(1):103–118

Jansky L, Pachova NI (2006) Towards sustainable land management in mountain areas in Central Asia. Global Environ Res 10(1):99–115

Jusufbekov CJ, Kasach AE (1972) Teresken na Pamire. Izd. Donish, Dushanbe

Kassam KA (2009) Viewing change through the prism of indigenous human ecology: findings from the afghan and Tajik pamirs. Hum Ecol 37(6):677–690

Kerven C, Alimaev I, Behnke R, Davidson G, Franchois L, Malmakov N, Mathijs E, Smailov A, Temirbekov S, Wright I (2004) Retraction and expansion of flock mobility in Central Asia: costs and consequences. Afr J Range Forage Sci 21(3):91–102

Kraudzun T (2011) From the Pamirs frontier to international borders: exchange relations of the borderland population. In: Bruns B, Miggelbrink J (eds) Subverting borders. Doing research on smuggling and small-scale trade. VS Verlag für Sozialwissenschaften, Wiesbaden, pp 171–191

Kraudzun T (2012) Livelihoods of the 'New Livestock Breeders' in the Eastern Pamirs of Tajikistan. In: Kreutzmann H (ed) Pastoral practices in High Asia – agency of 'development' effected by modernisation, resettlement and transformation. Springer, Dordrecht, pp 89–107

Kraudzun T, Vanselow KA, Samimi C (2012) Realities and Myths of the Teresken Syndrome – an evaluation of the exploitation of dwarf shrub resources in the Eastern Pamirs of Tajikistan (submitted)

Kreutzmann H (1996) Ethnizität im Entwicklungsprozeß: die Wakhi in Hochasien. Reimer, Berlin

Kreutzmann H (2003) Ethnic minorities and marginality in the Pamirian Knot. Survival of Wakhi and Kirghiz in a harsh environment and global contexts. Geogr J 169(3):215–235

Kreutzmann H (2004) Mittelasien. Politische Entwicklung, Grenzkonflikte und Ausbau der Verkehrsinfrastruktur. Geographische Rundschau 56(10):4–9

Kreutzmann H (2006) The significance of geopolitical issues for development of mountainous areas of Central Asia. Strategies for development and food security in mountainous areas of Central Asia. AKDN, Dushanbe

Kreutzmann H (2009) Transformations of High Mountain pastoral strategies in the Pamirian Knot. Nomadic Peoples 13(2):102–123

Kreutzmann H (2011a) Pastoralism in Central Asian mountain regions. In: Kreutzmann H, Abdulalishoev K, Zhaohui L, Richter J (eds) Pastoralism and rangeland management in mountain areas in the context of climate and global change., pp 38–63

Kreutzmann H (2011b) Pastoralism in the Pamirs. Regional contexts, political boundaries and market integration in Central Asia. In: Gertel J, Le Heron R (eds) Economic spaces of pastoral production and commodity systems: markets and livelihoods. Ashgate, Aldershot, pp 175–194

Kreutzmann H (2012) Pastoralism: A Way Forward or Back? In Kreutzmann H (ed.) Pastoral practices in High Asia, Springer Netherlands, pp. 323–336

Ludi E (2003) Sustainable pasture management in Kyrgyzstan and Tajikistan: development needs and recommendations. Mt Res Dev 23(2):119–123

Maanaev E, Ploskikh V (1983) Na 'kryshe mira' (Istoricheskie ocherki o pamiro-alayskikh kirgizakh). Mektep, Frunze

Maselli D (2012) Promoting sustainable mountain development at the global level. Int Mt Soc 32(S1):64–70

Mearns R (1996) Commons and collectives: the lack of social capital in Central Asia's land reforms. Institute of Development Studies, Brighton

Meinzen-Dick RS, Pradhan R (2001) Implications of legal pluralism for natural resource management. IDS Bull 32(4):10–17

Miehe G, Winiger M, Böhner J, Yili Z (2001) The climatic diagram map of High Asia. Purpose and concepts. Erdkunde 55(1):94–97

Ostrom E, Burger J, Field CB, Norgaard RB, Policansky D (1999) Revisiting the commons: local lessons, global challenges. Science 284(5412):278–282

Peyrouse S (2009) The multiple paradoxes of the agriculture issue in Central Asia. EU-Central Asia Monitoring

Rahimon RM (2012) Evolution of land use in nomadic culture in Central Asia with special reference to Kyrgyzstan and Kazakhstan. In: Squires VR (ed) Rangeland stewardship in Central Asia. Springer, Dordrecht, pp 51–70 (Chapter 3, this volume)

Robinson S, Whitton M (2010) Pasture in Gorno-Badakhshan, Tajikistan: common resource or private property? Pastoralism 1(2)

Robinson S, Higginbotham I, Guenther T, Germain A (2008) Land reform in Tajikistan: consequences for tenure security, agricultural productivity and land management practices. In: Behnke R (ed) The socio-economic causes and consequences of desertification in Central Asia. Springer, Dordrecht, pp 171–203

Robinson S, Whitton M, Biber-Klemm S, Muzofirshoev N (2010) The impact of land-reform legislation on pasture tenure in Gorno-Badakhshan: from common resource to private property? Mt Res Dev 30(1):4–13

Schlager E, Ostrom E (1992) Property-rights regimes and natural resources: a conceptual analysis. Land Econ 68(3):249–262

Schoch N, Steimann B, Thieme S (2010) Migration and animal husbandry: competing or complementary livelihood strategies. Evidence from Kyrgyzstan. Nat Resour Forum 34(3):211–221

Steimann B (2011) Contested wealth. Uncertainty and inequality in Kyrgyz pastoralism. Geographische Rundschau 63(7–8):54–59

Undeland A (2005) Kyrgyz livestock study: pasture management and use. World Bank, Bishkek

UNEP (2002) Global environment outlook 3 – past, present and future perspectives. Earthscan Publications, London/Sterling

United Nations (1992) Managing fragile ecosystems: mountain sustainable development by the United Nations. Chapter 13 of Agenda 21

Vanselow KA (2011) The high-mountain pastures of the Eastern Pamirs (Tajikistan) – an evaluation of the ecological basis and the pasture potential. PhD dissertation, University of Erlangen-Nuremberg, Erlangen

Vanselow KA, Kraudzun T, Samimi C (2012) Grazing practices and pasture tenure in the Eastern Pamirs: the nexus of pasture use, pasture potential and property rights. Mt Res Dev 32(3):324–336

von Benda-Beckmann F (2002) Who's afraid of legal pluralism? J Leg Pluralism Unoff Law 47:37–83

Walter H, Breckle S (1991) Ökologie der Erde Bd3. Spezielle Ökologie der gemäßigten und arktischen Zonen Euro-Nordasiens. Zonobiom VI – IX. Fischer, Stuttgart

Wilson T (1997) Livestock, pastures, and the environment in the Kyrgyz Republic, Central Asia. Mt Res Dev 17(1):57–68

Wolfgramm B, Shigaeva J, Nekushoeva G, Bonfoh B, Breu T, Liniger H, Maselli D (2010) Kyrgyz and Tajik land use in transition: challenges, responses and opportunities. In: Hurni H, Wiesmann U (eds) Global change and sustainable development – a synthesis of regional experiences from research partnerships, vol 5, Perspectives of the Swiss National Centre of Competence in Research (NCCR) North-South. University of Bern/Geographica Bernensia, Bern, pp 241–254

Chapter 5
Better Land Stewardship to Avert Poverty and Land Degradation: A Viewpoint from Afghanistan

Mohammad Hossein Emadi

Abstract This chapter attempts to demonstrate the nature of land degradation and represent its mutual relation with poverty in rural Afghanistan and tries to show how the notion of land stewardship could introduce new solutions to the issue. It explores the complexities and diversity which exist in natural resource utilization with more focus on rangeland-based agriculture and livestock raising. It shows how 'Post-Taliban Afghanistan' faces a wide range of complex issues including deep rural poverty and overexploited natural resources. Rural livelihoods and food security have been threatened by the degradation of the natural resource base. The land dispute is one of the most important and highly growing politicized issues in Afghanistan, it being across tribal lines and a major causative factor of local insecurity. The root causes of the problematical situation facing Afghanistan are outlined.

Keywords Poverty • War • Tribalism • Kuchi • Nomads • Rangelands • Fuelwood • Land degradation • Livestock governance • Legal framework • Deforestation • Pro-poor • Drug trafficking • Terrorism

Key Points

- War and drought has contributed to deforestation, destruction of traditional irrigation systems, soil erosion, desertification, and loss of plant and animal biodiversity. Population growth, changes in migratory patterns, and poverty are leading to overexploitation of natural resources. These factors undermine land fertility, increase the risks of landslides, floods, recurrent droughts, and threaten the stability of already fragile ecosystems. A combination of all these factors reduces the ability of the agricultural sector to produce adequate food to feed the entire population.

M.H. Emadi (✉)
International Consultant and Senior Advisor to the Minister of Agriculture, Irrigation and Livestock, Kabul, Afghanistan, No 206, Golestan 5th Street, Pasdaran Ave, Tehran, Iran
e-mail: mh.emadi@mail.gov.af

V. Squires (ed.), *Rangeland Stewardship in Central Asia: Balancing Improved Livelihoods, Biodiversity Conservation and Land Protection*,
DOI 10.1007/978-94-007-5367-9_5, © Springer Science+Business Media Dordrecht 2012

- Despite some recent improvements on strategy and policy formulation, the current approach to land management is not guaranteed to improve the situation. Whilst more reliable data and information is needed to define the cause of the problematic situation, the mainstream view believes that there is indisputable downward trend in the state of Afghanistan's natural resources. There are three main influencing factors: harvesting of timber and shrubs by local people, conversion of forest and rangelands to farming land, and overgrazing by livestock.
- Other underlying causes are socio-economic factors, poverty, insecurity, population growth, unclear land tenure, lack of effective governance, and no effective policy or legal framework. One additional biophysical causative factor is also acknowledged—climate change. The growth of rural settlements, unclear land tenure, and the over-utilization of the resources has led to the customary transhumance of pastoralist nomads (kuchi) being put under huge pressure, resulting in the much publicised, highly political, and worsening security situation in many parts of the fragile country as an international issue.
- The international community and the Government of Afghanistan, who mobilised resources to tackle the poverty and establish a viable, defined legal status of land ownership and natural resources management, need to foster biodiversity conservation and land stewardship for sustainable change in holistic manner, on the other hand.

1 Introduction

Lack of available resources and infrastructure facilities and weak security and governance are the major issues that the government faces. In a country where over 80% of the population relies directly on the natural resource base to meet its daily needs, widespread environmental degradation poses an immense threat to livelihoods. It is generally recognised that countries like Afghanistan, with high rates of poverty, are often faced with high levels of deforestation and natural resource degradation, which confronts policy formulation with a paradox. The preservation of degrading land and its resources for sustainable use are the most important environmental problems which directly influence the rural population in Afghanistan, although we should consider that Afghans are facing a wide range of issues. On the one hand, policymakers insist on pro-poor policies and exploring ways of access to natural resources to benefit the poor. On the other hand, sustaining the degraded land and resource is essential, and it needs to be protected from overexploitation by the poor, who in most cases have desperate need for and dependency on natural resources. The process of policy development and change in such a situation is very difficult, particularly in Afghanistan which is in a post-conflict situation and is subject to a very active conflict in the southern part of country.

Poverty and land degradation coexist in most of central Asian countries including Afghanistan. Over 80% of the Afghan population lives in rural areas practising agricultural and related rural activities that rely heavily on the use of land and natural resources. The decades of conflict, ongoing instability, lack of effective governance and service delivery, socio-economic insecurity and overall poverty, susceptibility to droughts and other natural hazards, population increase, and influx of displaced and returning population, have all exacted a heavy toll on the environment and land degradation of the country. As a result, the country's vulnerability to natural disasters and food shortages has increased. Overall, the natural resource base continues to suffer due to:

- Competing land use (agriculture, human settlements, forests, rangeland, wetlands, and protected areas) amongst local people, warlords, and new role players of political power structure
- The negative impact of war, increasing population (with very high rate of growth), human settlements, periodical drought, overexploitation, and landmines on natural resources
- Ambiguous legal status of land ownership and access to natural resources (water, forests, rangeland, biodiversity, wetlands, and protected areas)
- Lack of an enabling policy, legislative, and regulatory framework for land management, along with their weak governance and management with emphasis on long-term outlook and next generation

Afghanistan's population growth rate is currently one of the highest in the world; recent UNFPA figures would suggest a current rate of 3.6%—this extrapolates to a projected doubling of Afghanistan's population by 2030. This population pressure is and will manifest itself in several key respects concerning rangelands: (a) growth of rural settlements and emergence of new rural settlements, (b) expansion of larger urban settlements into rural areas, and (c) increased demand on renewable resources. Population pressure underpins almost all the socio-economic threats to the sustainability of Afghanistan's renewable resources.

Agriculture, not including poppy cultivation, generates about 40% of the GDP and employs about 70% of the labour force and is the a major source of livelihood in the country (Emadi 2011). The degradation of the natural resource base, therefore, directly and severely impacts the livelihoods of the majority of the Afghan population as well as the country's economic development as a whole. Particularly affected are the poor and most vulnerable, such as households headed by women or with physically disabled members, landless households, or those farming on only small rain-fed plots, many of whose rights are thus unfulfilled.

To be sustainable, economic growth cannot be achieved at the cost of environmental and natural resource degradation. Establishing clear environment–poverty linkage in the context of Afghan development, improving natural resource management with due consideration for the vulnerable population, preventing further degradation of the environment, and improving and maintaining the integrity of ecosystem services are therefore the key to lasting recovery, human security, and the sustainable development of Afghanistan.

2 Land Status and Natural Resources in Afghanistan

Of Afghanistan's 655,000 km^2 of total land area, only 12% (7.9 million hectares) is arable and 4% irrigated. An additional 45% is rangeland under permanent pastures (FAO 1993), less than 1.5% is under forest cover (FAO 2005) with the remaining 39% being mountainous so natural resources and forest area are limited (Figs. 5.1 and 5.2). Forest occupies only 2% of the total area. Most of the country's valuable forests have been degraded during the last 25 years of war and social unrest. It is estimated that between 1978 and 2002, the area under conifer forests in the eastern part of the country was reduced by 50% (UNEP 2003).

Today, most of the country appears to be subject to some degree of land degradation. As it presented in the Fig. 5.1, much of the land surface in Afghanistan is used as rangeland for grazing livestock. The potential for regenerating vegetation is likely to be seriously affected by heavy fuel wood collection or timber harvesting that far outstrips woodland regeneration and by browsing and grazing domestic livestock. Soil erosion is also a serious problem due to the loss of protective vegetation cover. Assessing the incomplete evidence available in the late 1970s, the Food and Agriculture

Fig. 5.1 Rangelands occupy a large part of Afghanistan. Much of the land is mountainous

Fig. 5.2 Mountainous areas generally lack forest cover but provide summer grazing and are important watersheds

Organization of the United Nations (FAO) concluded that most of north, central, and eastern Afghanistan was wooded until early in the nineteenth century (Emadi 2008a, b). Large areas of rangelands (approximately 45% of the total territory) are considered barren or waste land and are also used for grazing, particularly in winter. No comprehensive assessment of the current status of the rangelands and the (reversible or irreversible) effects of drought on rangeland has been carried out; even though a number of documents claim that there is extensive overgrazing in Afghanistan, there is no systematic evidence to corroborate this. Increasingly, pastures have been converted to rain-fed crop production, with devastating consequences.

Sustainability of Afghanistan's pasture can also be viewed from two distinct but closely interrelated dimensions: socio-economic sustainability and biophysical sustainability. Clearly, biophysical sustainability underpins socio-economic sustainability in that stocking rates are dependent on levels of available edible biomass. From such a perspective, large parts of Afghanistan could be classified as a 'non-equilibrium' ecosystem citing intense fluctuations in livestock numbers as a result of variable precipitation levels (Figs. 5.3 and 5.4).

The most important and growing concern confronting natural resources in Afghanistan recognised as quality degradation[1] which is reported to be much worse than it was before. According to Emadi (2008a), the main causes for shrinkage of forest area could be:

[1] Notably manifests itself as loss of productivity.

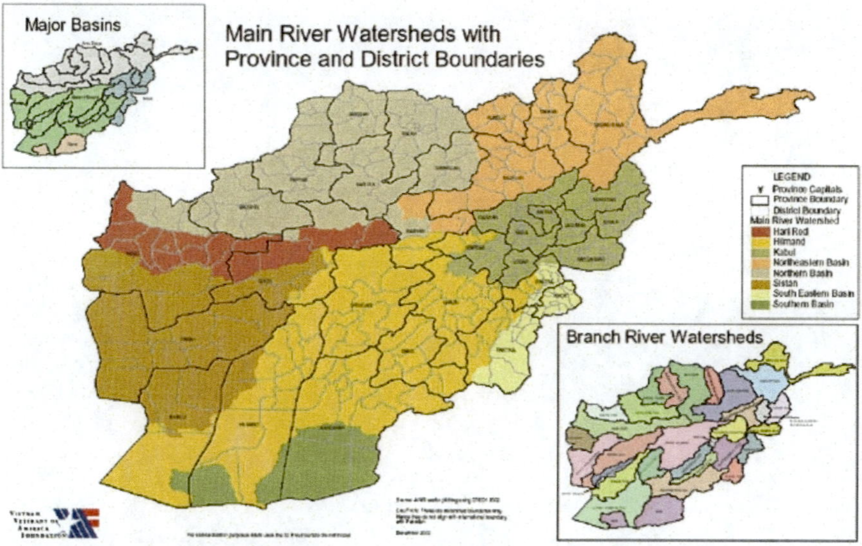

Fig. 5.3 Water is important in this high, arid country and hydroelectric power generation is a vital sector in the country's economy

Fig. 5.4 Snow-capped mountains and freshwater lakes are part of the important resource base to be protected for future generations

- Severe drought from 1997 to 2001 and again in 2005–2006
- Overharvesting of fuelwood, timber, and non-wood forest products (NWFPs) by local people as their main source of energy and income

Fig. 5.5 Large flocks of sheep and goats graze sparse forage on badly degraded hillsides

- Disintegration of the indigenous system of natural resource management and its institutions during the last five decades
- Collapse of the formal and governmental regulatory institutions for protecting forests over the last two decades
- Bombardment of the forest during the Russian occupation
- Recent access to forests for road and regional markets especially for massive export of timber to Pakistan as the source of cash for warlords during the civil war and the Taliban regime (Fig. 5.5)

To date, within the field of natural resource management (NRM), the initial focus of intervention has tended to be based on (a) an uncoordinated approach of looking at land and (b) more focus on forestry rehabilitation and management, whilst there has been little progress in rangeland management, despite its well-recognised significance in preserving the country's natural resource base (Fig. 5.6).

Collection of shrubs and other woody matter mainly for fire fuel is a major issue (Fig. 5.4). This can be broken down to external and internal perpetrators: internal, the local communities primarily on a subsistence level, and external, commercial operators exploiting the rangeland resource for financial gains outside of the local community.

Increasingly, the ploughing up and conversion of rangelands to rain-fed farming is becoming a significant issue 'with devastating consequences' (Emadi 2011) and is a major causative factor of rangeland degradation in Afghanistan especially across the northern rangelands. There is some evidence that ploughing rangeland is often a means of securing land tenure. In areas that receive less than about 300 mm of rainfall per year, ploughing up of rangelands for conversion to rain-fed farming is undisputedly a poor strategy and has already led to desertification of rangelands in northern and western provinces (Fig. 5.7).

Population growth resulting in expanding rural communities and increased competition for use of the rangeland resource has led to increasing and much publicised conflict

Fig. 5.6 Collection of woody plants is denuding the hillsides and contributing to accelerated land degradation but people lack alternative fuels

Fig. 5.7 Conversion of rangeland to cropland is devastating large areas of the country (See also Jackson, Chap. 6, this volume)

between local (sedentary) communities and nomadic pastoralists whose customary transhumance patterns (of migration between summer and winter grazing areas) see them competing for use of the same rangeland resources (Jacobs, Chap. 6).

Environmental degradation in general and quality degradation in particular are a manifestation of deeper causes relative to weak institutions, conflicting and unequal access to ownership, poverty, population pressures, urbanisation, and insufficient training in good environmental and natural resource management. The UN Common Country Assessment (CCA) for the Islamic Republic of Afghanistan (2004) identified the following root causes for development challenges in Afghanistan, including environmental challenges:

- A young governance structure still in the process of establishment and/or early stages of development, and unable to curb political insecurity, obsolete industrial and agricultural technology, and practices that harm the environment
- A culture of unsustainable resource use that must be reversed
- Wide variations in socio-economic indicators, by gender, region, and the rural–urban divide
- Sociocultural traditions that result in widespread marginalisation based on gender, social status, or ethnicity

The report points out that for communities there is lack of awareness of sound environmental practices, which needs to be addressed at the same time as improving their livelihoods and economic conditions. Communities must become owners of their natural resource base and feel that they actively participate in national economic growth. Otherwise, poverty and environmental damage can be caught in a downward spiral. Also the lack of reliable statistics for several critical sectors, including the environment, is considered a major impediment in tackling the root causes for improving natural resource management.

3 State and Land Strategy in Post-Taliban Era

The government of Afghanistan fully recognises that failure to address land degradation will negatively affect the long-term growth of the country as well as meeting the Millennium Development Goals (MDGs). However, the government's institutional and policy framework on the environment is fairly young and is still very much in the process of formulation and development. Since the Afghanistan Compact 2006—a political commitment of both Afghanistan and the international community to work towards 5-year high-level benchmarks—was adopted, the government has embarked on a comprehensive program of national planning and development through the implementation of the interim Afghanistan National Development Strategy (i-ANDS) (2006–2007) and full ANDS (formulated March 2008), which are fully aligned with Afghanistan's MDGs. The environment is a cross-cutting component in the i-ANDS—its 5-year strategic benchmark on the environment—and was adopted as follows:

Afghanistan endorsed the MDGs in 2004, much later than other countries due to the ongoing conflict at the time of the Millennium Summit in 2000. As a result, Afghanistan has an extended deadline (2020 as opposed to 2015) and due to the particular post-conflict challenge of the country, it has one added goal—*Goal 9:* Enhancing Security. In line with Afghanistan's MDGs, environmental regulatory frameworks and management services will be established for the protection of air and water quality, waste management and pollution control, and natural resource policies will be developed and implementation started at all levels of government, as well as at the community level, by 2007[12]. i-ANDS recognises the following constraints to achieving this benchmark:

- Lack of government ability to integrate sustainable development approaches into the national development framework
- Underdeveloped and unenforceable environmental legislation and regulatory framework
- Limited public awareness of environmental and natural resource issues
- Unsustainable exploitation of and inequitable access to natural resources exacerbated by insecurity

Furthermore, i-ANDS lists the following areas as part of the government's sectoral strategic focus on the environment that requires support from the international community:

- Clarification of responsibilities in the environmental arena and capacity enhancement of the National Environmental Protection Agency (NEPA)
- Development and implementation of a legal and regulatory framework that ensures sustainable use of natural resources
- Establishment of environmental and natural resource management capacity within line ministries
- Initiation of community-based management of natural resources
- Introduction of environmental education and vocational training
- Promotion of regional cooperation on environmental and natural resource management

4 Natural Resource Management Strategies in the Post-Taliban Period

Since the Bonn Agreement in December 2001, MAIL, supported by donors, has formulated a number of policy and strategy documents which have built cumulatively upon one another to create a comprehensive and coherent framework for the development of the agriculture sector and natural resource management (NRM) subsector. Key documents amongst these are the Agricultural Sector and Natural Resources Policy Paper (2004) and the 2005 Agriculture Sector Development Strategy (the Master Plan), the MAIL/ANDS Agriculture Strategy Paper (2007) and, in partnership with the Ministry of Rural Rehabilitation and Development (MRRD), the 2008 Agriculture and Rural Development Sector Strategy. Most recently, the National Agriculture Development Framework (NADF) (2009) superseded and combined all the previous strategic frameworks into a comprehensive inventory of key issues for agricultural development, divided into four main pillars/programs (natural resource management, agriculture production and productivity, economic regeneration, and program support and change management), each with a related program document.

In addition and separately to the above narrative, a policy and strategy for forest and range management was prepared in 2005 by consultants from UNEP, approved in 2006 by the cabinet and a National Land Policy (2007) developed by an inter-ministerial committee chaired by the Ministry of Justice, which tackled the central issue of land tenure.

Consistent with the tenets of Afghanistan's environment law and upcoming natural resource laws and regulations, the Ministry of Agriculture, Irrigation and Livestock (MAIL) has designed a natural resource management program with a community-based approach that positions communities throughout Afghanistan to assume responsibility for practising productive and environment-enhancing management of natural resources for their own benefit. The value of this approach hinges crucially on the involvement of communities, not just in the implementation of natural resource management, but in its planning, with assistance from the government and other partners (Emadi 2008a, b). Thus, the natural resource management program relies on technical, legal, policy, and institutional changes and the creation of functional relationships amongst the government, communities, and other partners.

4.1 Objectives

Communities and institutions throughout Afghanistan will establish and maintain natural resources and define regimes of utilization that achieve a balance between maximisation of production and productivity in all agricultural land uses and effective maintenance and enhancement of natural resource bases.

This objective leads to a set of activities and results that will enable communities and government institutions at provincial, district, and central levels to establish systems of governance and management, to build their capacity to resolve conflicts over natural resources, and to develop and maintain natural resources. The following components, listed in order of priority, comprise the natural resource management program activities (Emadi 2008a, b).

4.2 Proposed Components and Output

Legal and institutional framework: Government institutions and communities throughout Afghanistan clarify rights to natural resources, in accordance with laws, with the support of national and provincial authorities.

Infrastructure: Inputs necessary for natural resource management activities such as materials and equipment for construction.

Baseline assessment: Community, district, provincial, and national authorities complete a baseline assessment on the diversity, availability, and distribution of natural resources at community, provincial, and national levels.

Implementation of natural resource conservation systems: Community-, district-, provincial-, and national-level authorities put in place and implement the systems to protect natural resources from natural threats and restore and enhance the natural resource base.

Sustainable improvement of livelihoods: Communities and private sector entities generate income from the harvest and commercialisation of natural resource products while ensuring the natural resource base is protected and maintained.

Facilitate communities to manage natural resources sustainably: Communities bring natural resources under working, sustainable management regimes, with the guidance and support of government and support agencies. Communities are aware of the importance of the sustainable management of natural resources and are motivated to establish regimes of utilization that achieve a balance between maximisation of production and productivity in all agricultural land uses and effective maintenance and enhancement of natural resource bases.

In MAIL strategy documents, it is mentioned that the NRM development projects are aimed at improving the livelihood and economic conditions of rural families and enhancing household food security with their active participation in national economic growth. The aforesaid goals are intended to be guaranteed through the development of improved, efficient, and sustainable use of water resources, agriculture, food processing, forestry, rangeland, and other subsector development (Emadi 2008b). The NRM sector sets some indicators to evaluate its activities in future.

These indicators are the following:

- Enactment, realisation, and enforcement of natural resource laws and regulations by MAIL and relevant sectors of the government. These were partially achieved by 2010 and still continue.
- A decrease in conflicts over natural resources, including land, and successful establishment and maintenance of regimes of utilization for different types of natural resources.
- Central-, provincial-, and district-level agricultural departments have the ability to co-ordinate amongst partners involved in natural resource management.
- To provide appropriate and effective participatory and technical outreach to communities in support of the implementation of natural resource management projects.

5 Latest National Plan for Sustainable and Integrated Rangelands Management

Preparation of the National Plan for Sustainable Rangelands Management was initiated by the MAIL, in late 2010, as a response to the indisputable degradation and eating away of Afghanistan's rangelands and the corollary negative socio-economic impacts this was having on the country. It is one of four main components of the *Environmental Conservation National Priority Program*, with an overall goal of '*Sustainable and productive use of lands adopted throughout Afghanistan*'. The author has initiated the national plan to provide both a framework and a road map for MAIL and its development partners to facilitate a comprehensive integrated approach to land management. The process was explored and discussed with most of international partners and the representative of the local communities in Afghanistan throughout series of meeting and workshops. Based on a 1-year debate and negotiation amongst all stakeholders, three complementary subcomponents were explored and mentioned as follows:

***Subcomponent 1**: Rangeland Assessment and Monitoring Systems*
It is apparent that there is a dearth of sound information on Afghanistan's rangelands; gauging their current status is therefore a critical first step in the process.

1.1: Status assessment system – to gauge the current status of Afghanistan's rangelands in terms of natural resource availability, land use patterns, local knowledge and institutions, etc.

1.2: Monitoring system – local communities trained in the use of simple collection techniques through a user-friendly guide/manual

1.3: Geographical information system – for the compilation, analysis, and mapping of the community-generated data to support the entire program

***Subcomponent 2**: Practical 'On-the-Ground' Solutions for Decreasing Rangeland Degradation*
It is widely accepted that the only *sustainable way* for decreasing rangeland degradation is through participatory approaches. Instruments for decreasing rangeland degradation sit along a sliding scale from the radical to the conservative—the National Plan advocates an integrated approach using an array of these instruments.

2.1: Consolidation of existing approaches for decreasing rangeland degradation, building on lessons learned from MAIL's development partners on the ground in Afghanistan, e.g. UNEP, AKF, and regionally, e.g. ICIMOD's work in the Hindu Kush-Himalayas

2.2: Rolling out of consolidated approaches resulting in a unified and coherent effort to decreasing rangeland degradation and loss

***Subcomponent 3**: A Workable Legal and Regulatory Framework for Rangelands Management*
Policy, legal, and regulatory frameworks are seen as the key instruments for implementing the National Plan. The key underlying principle to the framework is that the government of Afghanistan is the indisputable legal owner of all rangelands.

- MAIL and relevant sectors of the government. These were partially achieved by 2010 and still continue.
- A decrease in conflicts over natural resources, including land, and successful establishment and maintenance of regimes of utilization for different types of natural resources.

- Central-, provincial-, and district-level agricultural departments have the ability to co-ordinate amongst partners involved in natural resource management and to provide appropriate and effective participatory and technical outreach to communities in support of the implementation of natural resource management projects.

6 Overall Picture

After the fall of the Taliban, the international community and government of Afghanistan have a significant challenge to achieve the MDG that needs equitable and sustained pro-poor growth over a long period of time. To tackle the poverty which is deeply embedded in rural areas, rural development as a strategy is placed at the centre of pro-poor growth as national policy. Afghanistan's rural development needs investment and considerable financial and human resources in order to translate these policies into concrete actions. Therefore, the government with the support of the international community has designed and begun to implement a number of national priority programs. Besides the international aid and support to build sustainable development in Afghanistan, there is an absolute need for economic regeneration and employment based on resource mobilization—resources which have been badly exploited during the last three decades. To tackle this issue and mobilise natural resources for more employment opportunities, domestic products, and services, policymakers are confronting serious challenges (Emadi 2008a, b):

- Poverty that is deeply interconnected and entwined with lack of security, lack of infrastructure, and resource distribution.
- Poverty embedded in rural areas; rural life is deeply dependent on natural resources.
- There is a need for economic/income regeneration and employment based on resource mobilization in rural Afghanistan to eradicate abject poverty.
- The government (with the weak support of the international community) has designed and begun to implement a number of separate national programs on natural resource management.
- These programs, which are based on conventional technology transfer and government control over resources, have not been significantly successful nor have they achieved their targets.
- The land/pasture dispute is one of the most important and growing issues in Afghanistan; It has caused many local security issues in recent years and is linked with tribal identity and international drug trafficking and terrorism.
- Policy adjustment is required towards a more pro-poor and participatory approach in the context of a society mired in severe conflict with a shortage of functioning governance at a variety of levels.

7 Further Challenges for Land Stewardship in Afghanistan

Effective and holistic approach of land stewardship helps maintain and restore the function of the natural resources as a whole. Based on the practical results and outcomes of the last several years of natural resource management activities in the field, this chapter argues that the challenges in introducing and implementing a pro-poor land stewardship point of view and policy in such situations are more demanding in the field than expected. The paradigm of land stewardship, where local communities are granted varying degrees of management power in managing land and resource, advocates that the government initiates a process of transferring user rights and responsibilities to communities but also that they transfer those rights in a way which ensures that the communities will use that land in a productive and sustainable manner.

The main counterargument is that only if there is a *long-term* interest in the productivity of the land, such that a landowner of the land might have, then management can be based on stewardship and sustainable practices. Government ownership can in the mind of local people mean that it belongs to no one at all and therefore might result in a 'free-for-all' to exploit the resource which refers to what is known as the 'tragedy of common'. For local communities, ownership of rangelands is less consequential than deciding who legitimately holds the user rights to particular parcels of rangeland during specific period of the year. In summary, whatever the management relationship between community and government, there seems little to be gained from wrestling with the land tenure issue at the same time or at this particular point in time; it merely creates more legal complications and removes safeguards the government automatically has through legal ownership.

According to the above mentioned, there are some fundamental barriers to the adoption of a land stewardship paradigm in the context of Afghanistan including:

- Lack of a clear picture and certainty over the future and control over the resources amongst the local people, that makes the situation fuzzy and competitive in a negative manner
- Low level of trust and confidence amongst the local people towards the government and its agencies as a neutral regulator
- Embedded poverty amongst local families which enforces them to compete intensively for access to the degraded land and resources without long-term thinking of overall results
- Weakness of skills and attitudes amongst change agents who are going to introduce the notion of land stewardship amongst local communities

Being successful in implementing such human-focused development and pro-poor approaches, there is a need for considering the above-mentioned contextual issues and introducing radical changes on conceptualising and formulating comprehensive approaches to tackle the interrelated complex issues and address such challenges.

In such a complex context, the government and international forces are facing three major issues including security, governance, and development all at once and the problems of governance associated with the dissatisfaction of citizens and increasing insecurity. The thrust of this chapter, which is based on concrete examples on the ground, suggests the generalisation of natural resource management in such a socio-economic context through (a) simplified, (b) standardised, and (c) top–down, natural resource management approaches which could not be applicable and effective any more. There is a need for reframing the conceptual and administrative framework, which could improve the situation and resolve the sectoral policy-making process to be more contextual, adjustable, and participative. Much has been written about 'community-based' management of rangelands—but what does this approach entail exactly? There are numerous past and ongoing projects in Afghanistan which pertain to community-based management approaches, but a general lack of synthesis across the board, with government body (MAIL) itself often a part player as its better organised and resourced development partners run the agenda. The legal and regulatory framework behind the community management of rangelands needs straightening out and taking forwards.

One of the examples is inclusion of indigenous knowledge of local people which could provide the starting point for co-management for conservation, protection, and rehabilitation of land; all strengthened and complemented by externally provided technical scientific skills as and when necessary.

Since the fall of the Taliban, there has been a disproportionate amount of focus/donor interest in forests and protected areas and not enough attention paid to rangelands, despite the 'well-recognised significance' of rangelands in preserving the bulk of the country's natural resource base, the fact that rangelands account for more than 45% of Afghanistan's total land area and are the major source of income (directly and indirectly) for nomadic pastoralists and rural people (Jacobs, Chap. 6). The rehabilitation and sustainable management of rangelands is a national need and should be a priority focus of International intervention. However, where development partners have turned their attention to rangelands, there has so far been a 'disheartening' lack of coordination amongst them or with the Afghan Government.

To tackle the issue, there are further challenges and requirements that need to be overcome:

- More coordination and harmony amongst all 14 technical and international agencies that are collaborating in land and environmental issues in Afghanistan with MAIL is needed.
- A more comprehensive approach to planning is needed to integrate rural development and natural resource management to be able to tackle the complex, diverse, and deeply rooted issues.
- Community forestry in contemporary Afghanistan is promising and could play a major role in targeting natural resource management, but it needs radical changes in approaching people and ways of integrating people in the process of co-management of the land and resources.
- Rehabilitation and empowerment of government staff as a basis for the community forestry approach.

- Recognition of the indigenous knowledge of local people and also searching for existing local institutions for natural resource management is essential for people's active participation
- Proper, clear, and urgent legislation for a land tenure system is needed so that recognition of community property rights based on customary rights (*Urf*) and *Sharia* (Islamic religious law) is considered (based on the constitution).
- Development of associated infrastructure and small–medium enterprises is fundamental and a commodity chain analysis on timber profits and market development reform for NWFPs is needed.
- Technological assistance and training for value-added processing of forest products should be simple, affordable, and sustainable.

7.1 Summary

The main outstanding issue for rural land policy (especially rangelands) to address in Afghanistan is to sort out how pastures are owned, accessed, and managed. Current policy and law relating to pastures in Afghanistan and many other countries in the Central Asian region is outdated and unworkable, and its paradigms are hotly contested on the ground (Robinson, Chap. 11). A new policy *process* is needed to support the *framework* already devised (see above) for rural land relations in general. This is in order to arrive at fair, peaceful, and workable arrangements. Modern national land policy is not easily arrived at through centralised paper exercises. These tend to reproduce the same systems, under different names. A fresh approach to land policy requires practical investigation of new systems—and with land users themselves. This means adopting a bottom–up, participatory, and learning by doing approach to those issues which are proving most difficult to resolve. Rights over pasture use are one such area—and in fact the most important rural tenure issue in Afghanistan. Competition for land is intensifying between agriculturalists and pastoralists, within a national political context. Former customary agreements for seasonal grazing rights are now being challenged. The logical framework for carrying out a learning-by-doing approach is through discrete community-based pilot initiatives; research and consultation is not enough, and policy and law should not be finalised until lessons have been drawn from these pilots. Community based means including the resident and the nomadic communities when the subject is pasture access. A clear institutional focus and political commitment is required to support piloting of new approaches and to bring the lessons into policymaking and new legal drafting to entrench the policies and new practices.

The Central Asian context of pastoralism, of which Afghanistan is a part, has some of the core elements of sustainable pastoralism, such as (i) pastoralism as an evolutionary adaptation of mankind to the use of rangelands, (ii) rangelands as resilient and varied ecosystems, (iii) the rationality of pastoralism, (iv) the importance of mobility, (v) the importance of adapted species, (vi) the important contribution of women to the pastoral economic system, (vii) the wealth of indigenous knowledge, and (viii) the lack of sufficient realization amongst governments of the rationality and value of pastoral systems to the respective countries. This is why Afghanistan is such a useful case study.

The author and the book's editor hope that the analyses in this and in other chapters can help to advance the cause of better land stewardship in the Central Asian region.

References and Further Reading

de Weijer F (2007) Afghanistan's kuchi pastoralists: change and adaptation. Nomad Peoples 11(1):9–37

Emadi MH (2008a) Influencing factors on rural household income in Afghanistan. Research report, published by AIRD Afghanistan

Emadi MH (2008b) Community based natural resource management. In: Post conflict Afghanistan; changes and challenges. Paper presented on International Grass land conference and International Rangeland conference (IGC_IRC) Hohhot, China, July 2008

Emadi MH (2011) Natural resource management and poverty in post-Taliban Afghanistan. Int J Environ Stud (IJES) 68(3):267–279, Rutledge

FAO (1993) FAO land cover map. FAO, Rome

FAO (2005) Global forest resource assessment 2005. FAO, Rome. Available online at http://www.fao.org/forestry/site/32179/en

Human Development Report (2009) Afghanistan, the human development index. www.undp.org/en/countries/country…/cty_fs_AFG.html

Interim Afghan Development Strategy (i-ANDS) (2006–2007) Afghanistan national development strategy summary report – an interim strategy for security, governance, economic growth and poverty reduction (i-ANDS). Office of President, Afghanistan, p 27

Jacobs M, Schloeder CA (2012) Extensive livestock production: Afghanistan's kuchi herders, risks to and strategies for their survival. In: Squires VR (ed) Rangeland stewardship in Central Asia. Springer, Dordrecht, pp 109–128 (Chapter 6, this volume)

Robinson S, Wiedemann C, Michel S, Zhumabayev Y, Singh N (2012) Pastoral tenure in Central Asia: theme and variation in the five former Soviet Republics. In: Squires VR (ed) Rangeland stewardship in Central Asia. Springer, Dordrecht, pp 239–274 (Chapter 11, this volume)

Roe A (2011) Kuchis and water use in Afghanistan Afghanistan: human development report 2011. Center for Policy and Human Development, Kabul University 16p

Schütte S (2012) Pastoralism, power and politics: access to pastures in Northern Afghanistan. In: Kreutzmann H (ed) Pastoral practices in High Asia, Advances in Asian human-environmental research. Springer Science + Business Media B.V, Dordrecht, pp 53–70

UN (2004) Common Country Assessment (CCA) for the Islamic Republic of Afghanistan, October 2004, Ministry of Agriculture

UNEP United Nations Environment Program (UNEP) (2003) Afghanistan post-conflict environmental assessment. 176. United Nations Environment Programme, Geneva, p 11

Chapter 6
Extensive Livestock Production: Afghanistan's Kuchi Herders, Risks to and Strategies for Their Survival

Michael J. Jacobs and Catherine A. Schloeder

Abstract The focus in this chapter is on those Kuchi (Afghan nomads and transhumant herders) who are still engaged in extensive livestock production, and it explores the challenges that they face today in maintaining their livelihood with emphasis on their greatest livelihood risks: insecurity and conflict, land access and land appropriations, and diminishing water and winter fodder sources. A plan is presented for dealing with these risks that can be executed without delay.

Keywords Land user rights • Clans • De-mining • Water supply • Forage • Rents • Conflicts • Droughts • Land degradation • Fuelwood • Biodiversity • Poverty • Sedentarization • Policy • Risk • Vulnerability • Sustainable development

Key Points

- Afghanistan's rangelands, comprising 75–80% of the landscape, are quite diverse as a consequence of much variability in climate, soils, latitude, and elevation across the country. About 80% of the country's population still resides in rural areas, and of these, 5% is considered to be actively engaged in extensive livestock production. Afghanistan's extensive livestock producers, or herders, can be characterized as either nomadic or transhumant.
- Extensive livestock production in Afghanistan has historically been fraught with risks. Beginning in the 1980s with the civil war with the Soviets, however, the

M.J. Jacobs (✉)
Department of Ecosystem Sciences and Management,
Texas A and M University, Texas, USA
e-mail: mjjacobs@tamu.edu

C.A. Schloeder
Department of Ecosystem Sciences and Management, Texas A and M University,
PO. Box 445, Fortine, MT 59918-0445, USA
e-mail: c.schloeder@tamu.edu

V. Squires (ed.), *Rangeland Stewardship in Central Asia: Balancing Improved Livelihoods, Biodiversity Conservation and Land Protection*,
DOI 10.1007/978-94-007-5367-9_6, © Springer Science+Business Media Dordrecht 2012

Kuchi experienced new risks resulting from a breakdown in governance and civil order throughout the country and the widespread use of landmines. Today, efforts to rebuild the country's economy and judicial system are such that the Kuchi face new risks as well as many of the old ones. Of all the risks they experience, insecurity and conflict, and land conversion and land appropriations, ranked the highest in importance.

- The continued lack of governance and judicial mechanisms has made it impossible for Kuchi and villagers alike to file any grievances, seek compensation for their losses, or obtain justice. Furthermore, years of continual conflict have eroded many of the positive relationships that once existed between certain clans and village communities leading to further insecurity and conflicts, fewer opportunities for justice and an increase in human casualties. The Kuchi reported in 2009 that they can lose much or all of of their herd each year to criminals and commanders charging illegal fees and that during times of drought, illegal fee assessments can increase by up to 50% when no other options available. They also face the possibility of new conflicts if they have to find an alternative migration route.
- Despite the Kuchi's invaluable contribution to Afghanistan's national economy, they continue to be one of the most marginalized cultures in the country today. Over half of the total population is now settled after massive livestock losses. For those families still engaged in herding, the risks they face have led to poverty and food insecurity being the norm for most. While periodic droughts will almost certainly continue to create risk for the Kuchi, preventing them from making their annual movements will lead to greater negative consequences for humans and animals alike.

1 Introduction

Afghanistan's rangelands, comprising 75–80% of the landscape, are quite diverse as a consequence of much variability in climate, soils, latitude, and elevations across the country (Breckle 2007; Thieme 2006). To date, over 4,000 plant species have been recorded for the country with these representing flora from a range of vegetation types including alpine, *Pistachia* woodlands, steppe, shrublands, and desert (Flora Iranica 2009; Schloeder and Jacobs 2010). A long history of extensive livestock production has no doubt played a role in shaping Afghanistan's vegetation types, but it is not clear when exactly this livelihood became common. It has been suggested that in Asia, livestock production may have begun as far back as 1000 BCE following the domestication of sheep and the collapse of irrigation systems (Khazanov 1984). Other anthropogenic influences of consequence are species eradications and habitat degradation resulting from the harvest of a variety of plant matter used for fuel, building materials, and traditional and medicinal uses, in part or in their entirety, and the settlement and cultivation of rangeland areas[1] (Bedunah 2006).

[1] Schloeder and Jacobs (pers. obs).

Fig. 6.1 Kuchi graze their flocks in the high altitude Hindu Kush and then return to lowlands for the winter

Due to the continued lack of development in Afghanistan, 80% of the country's population still resides in rural areas, and of these, 5% is considered to be actively engaged in extensive livestock production (de Weijer 2005; MRRD 2007). Afghanistan's extensive livestock producers, or herders, can be characterized as either nomadic or transhumant. The transhumant herders of Afghanistan are those that have permanent dwellings to return to with their livestock after spending the summer in the rangeland areas of the Hindu Kush Mountains (Fig. 6.1).

These Kuchi include clans from the Pashtun, Tajik, Uzbek, Arab, Aimaz, Hazara, Khirghiz, Turkmen, and Baluchi ethnic groups. In contrast, Afghanistan's nomadic herders include clans without permanent dwellings to return to and who often compete for prime wintering sites each year after returning from summer grazing areas in the mountains (Schütte 2012). These include clans of the Pashtun and Baluchi ethnic groups. With the exception of the Khirgiz, Afghanistan's extensive livestock producers are commonly referred to as Kuchi. Kuchi is a Farsi verb meaning *to move*. Not all Kuchi move, however, as a consequence of having lost their herds to drought, insecurity, and land appropriations (UNHCR 2006) (Fig. 6.2).

In this chapter, we will focus on those Kuchi who are still engaged in extensive livestock production and explore the challenges that they face today in maintaining their livelihood with emphasis on their greatest livelihood risks: insecurity and conflict, land access and land appropriations, and diminishing water and winter fodder sources. We will then present a plan for dealing with these risks that can be executed without delay. Much of the un-referenced data presented in this chapter are from our work with the Afghanistan PEACE project[2] (PEACE hereon). PEACE is a USAID-funded project initiated in 2006, with the mandate to improve extensive livestock production and range management in Afghanistan. It is a multiprogram project that includes the institutionalization of rangeland monitoring and livestock market price technologies; rangeland and flora assessments; community and risk management assessments; professional support for drafting legal frameworks, policies, and legislation; and a conflict

[2] PEACE: Pastoral engagement, adaptation and capacity enhancement see http://www.afghanpeace.org

Fig. 6.2 Kuchi on the move

resolution program. The programs were undertaken in collaboration with various Afghanistan government institutions, nongovernment organizations (NGOs), Kuchi leadership, and community leaders. The project is fully implemented by Texas A&M University under a subcontract agreement with the University of California-Davis.

2 Livestock Production and Its Challenges

In 2009, the Ministry of Rehabilitation and Rural Development reported that the Kuchi owned 75% of the livestock in Afghanistan (MRRD 2009). Also in 2009, we reported that the Kuchi were responsible for supplying over 72% of the shoats (e.g., sheep and goats) sold in Afghanistan's major livestock markets in 2008 (Schloeder and Jacobs 2009). Estimated income generated from the sale of these shoats was $443 million suggesting that the economic contribution of the Kuchi to Afghanistan is not insignificant. The Kuchi are the primary suppliers of *qrut*, a dried yogurt product (Barfield 2004); skins and wool (de Weijer 2005); and young stock sold to farmers as they move through areas (Thompson 2007). Additionally, Tajik and Uzbek Kuchi are the main suppliers of the *qarakul* pelt industry which experienced a resurgence across northern Afghanistan in 2005 with an estimated $20 million in generated income (MTND 2009). The Kuchi have been also known to play a major role in trade by bringing goods into remote mountainous areas in exchange for agricultural products (Barfield 2004).

Fig. 6.3 Conversion of the more productive rangeland to cropland is a big concern. It reduces forage supply and cuts off access to other land

Extensive livestock production in Afghanistan has historically been fraught with risks. Beginning in the 1980s with the civil war with the Soviets, however, the Kuchi experienced new risks resulting from a breakdown in governance and civil order throughout the country and the widespread use of landmines. Today, efforts to rebuild the country's economy and judicial system are such that the Kuchi face new risks as well as many of the old ones. Some of the more significant risks the Kuchi are encountering today include the following:

- *Insecurity and conflict* – thefts, abductions, attacks, assessment of fees, and earlier than normal departures from grazing lands
- *Land conversions/appropriations from cultivation, village expansion, (re)settlement programs (for IDPs and refugees), and greed on the part of commanders and warlords* – assessment of fees, loss of critical grazing lands, competition with villagers for remaining grazing lands, confinement of herds on grazing lands insufficient in size, earlier than normal departures from grazing lands, and the need to find alternative migration routes (prior to and during migration)
- *Diminishing access to water* – earlier than normal departures from grazing lands, and the need to find alternative migration routes (prior to and during migration)
- Competition for winter fodder – loss of fodder resource, armed conflicts, and assessment of fees

Of all the risks they experience, insecurity and conflict, and land conversion and land appropriations, ranked the highest in importance during a risk assessment exercise with different Kuchi groups (Desta 2009) (Fig. 6.3).

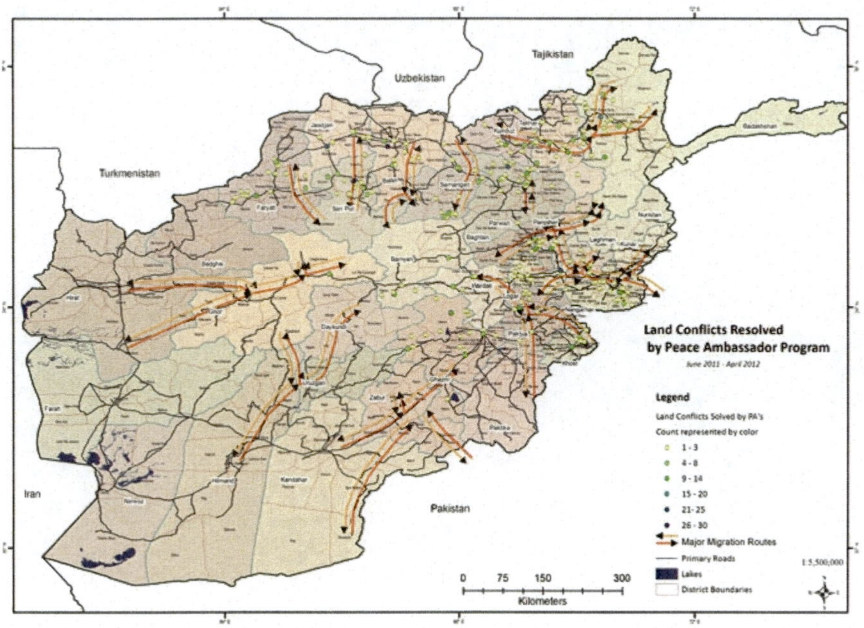

Fig. 6.4 Migration routes have changed and much negotiation has been involved in getting agreement about new routes. PEACE has negotiated these routes

The continued lack of governance and judicial mechanisms has made it impossible for Kuchi and villagers alike to file any grievances, seek compensation for their losses, or obtain justice. Furthermore, years of continual conflict have eroded many of the positive relationships that once existed between certain clans and village communities leading to further insecurity and conflicts, fewer opportunities for justice, and an increase in human casualties. The Kuchi have also had to deal with general insecurity as insurgent activity continues to plague the country. They are known to get caught up in night raids simply because of the Taliban having moved into areas near their residences or camps. They have also had to abandon areas prematurely and limit their movements during or in anticipation of a conflict. Conflicts between commanders are issues of concern as well because the victor may not always end up being the one in control previously. The risks in this instance range from the theft of one's entire herd, the assessment of an access fee, or the assessment of a higher than "normal" access fee. The Kuchi reported in 2009 that they can lose of up to 20% of their herd each year to criminals and commanders charging illegal fees and that during times of drought, illegal fee assessments can increase by up to 50% and that one can lose their entire herd if there are no other available options (Desta 2009). They also face the possibility of new conflicts if they have to find an alternative migration route (Fig. 6.4).

The lowland areas of Afghanistan are the main wintering areas for the Kuchi. Faryab, Badghis, Sar-e Pul, Kunduz, and Badakhshan provinces are the primary source of shoats for the larger rural and urban markets in Balkh, Nangarhar, and Kabul.

In the last 20 years, these provinces have been repeatedly hit by drought resulting in massive livestock losses, and in the last 15 years, large swaths of land have been appropriated at an unprecedented rate by farmers and commanders attempting rain-fed cultivation. The consequence of these events has been the loss of important winter grazing areas and the degradation of rangeland areas where cultivation is being attempted in areas with unsuitable soils and rainfall conditions. Degradation in the lowland areas takes various forms including a complete loss of plant cover to conversion to annual plant cover comprised of poor quality, early seral stage plant species, some of which are toxic.

In 2008, herders in Baghlan and Samangan provinces reported, during interviews in the field, that 90% of their does had spontaneously aborted during the previous winter after being fed dried native forages that had been collected for winter fodder. An examination of the remaining winter stockpile revealed an abundance of *Leptalium filifolium*, a species in the Cruciferae family known to contain compounds leading to spontaneous abortions. A study conducted by PEACE on rangeland diet quality in Baghlan, Samangan, Faryab, Kunduz, and Balkh indicated that the percentage of crude protein and total digestible nitrogen were less than the minimum required for lactating ewes and does (Schloeder et al. in prep). Low milk yields affect offspring survival and future body condition as well as the families that own the animals because of their dependency on dairy products for food and revenue. Low milk yields also place an additional economic burden on the family to provide an alternative food source. Rangeland surveys in these same areas also indicated a preponderance of low quality and toxic species of the Brassicaceae, Euphorbiaceae, Papaveraceae, and Leguminosae families.[3] Low quality diets and exposure to toxic plants can also lead to declines in body condition, increases in livestock mortality rates, and an increased susceptibility to parasites and diseases.

Currently, the bulk of drinking water for the Kuchi and their animals comes from rain and river sources.[4] The choice of historical migration routes for any herder in the world is always strongly linked to reliable water sources in addition to reliable forage sources. The Kuchi's ability to access reliable water while en route to seasonal grazing areas is increasingly becoming more difficult as routes become blocked or more complicated to navigate with the consequence that some have seen their entire livelihood vanish in just a matter of days due to massive die-offs from dehydration. Others have resorted to moving their animals by truck around some of the more challenging conflict areas, including from the Jalalabad area to Dasht-e Lalai north of Kabul and the Kunduz area to Shewa Kalan in Badakhshan. This is an expensive endeavor, however, and one that often leads to stress in response to long transport times, high temperatures, and the lack of water and feed while en route. Elevated stress levels often present as health issues (e.g., pneumonia), declines in body weight, and, in the worse case, high mortality. Given the cost of transporting animals, this is an option that only a few herders have chosen to date. A related issue is that the lack of potable water can lead to chronic health issues for humans and

[3] Afghanistan PEACE Project: http://cnrit.tamu.edu/peace/surveys.html

[4] Afghanistan PEACE Project: http://cnrit.tamu.edu/peace/kuchimaps.html

animals alike. As a mobile society, this then becomes an issue of treating one's own health or the health of their animals. This is a challenging if not impossible undertaking given that government services are frequently unavailable along migration routes and in summering areas and more specifically, they are not set up to service the needs of a mobile society.

Afghanistan was once one of the most heavily mined countries in the world, with grazing land accounting for 75.6% of all mined areas (UNMAP 2009). De-mining efforts are prioritized according to the number of human and livestock casualties. As a consequence, the majority of de-mining efforts to date have focused on settlement areas, roads, and infrastructure where there is regular, frequent, and abundant human activity. The Kuchi remain vulnerable, as a consequence, because of their reliance on rangelands and because it can take time to reach the critical threshold of casualties triggering a de-mining response given that they tend to use most areas on a seasonal basis only. On well-known migration routes, the Kuchi know which areas to avoid. When forced to change routes, their vulnerability increases substantially with dire consequences for humans and livestock alike. Changing migration routes while in transit can also lead to new conflicts, an assessment of fees or higher than "normal" fees, and increase the herders' chances of encountering criminals or insurgent activity as a result of not knowing the new landscape they are forced to negotiate through. De-mining activities also take a toll in that many types of activities can result in rangeland degradation particularly if they involve large machinery turning and scraping the upper soil layers.[5]

3 Organization and Status of the Kuchi

The Kuchi are organized by clans with a clear leadership structure beginning at the clan level and extending first to the district and then the provincial level. It should be mentioned that often regional leadership does not fall neatly into districts or provinces (Barfield 2004). The extent of clan leadership is independent of geographic area and related instead to close family and political ties. In 2006, a more formal structure of organization was instituted by the president of Afghanistan, for the Kuchi. This involved the establishment of the Independent Department of Kuchi, since renamed to Independent General Directorate of Kuchi (IGDK), as a mechanism for facilitating government representation and for solving problems that extended beyond a clan's control. The IGDK operates as a quasi-ministry but one with the mandate of reporting to the president directly and a single agenda: to address all social, economic, and political issues affecting the Kuchi across Afghanistan, regardless of ethnicity. When IGDK was a newly formed institution, it had little capacity or capital to really help the Kuchi people; however, as of January 2010, it has 31 local Kuchi directors in place to serve as representatives for

[5] Schloeder and Jacobs (pers. obs).

31 of the 34 provinces in Afghanistan. The Kuchi provincial directors (KPDs) manage business by shura, a meeting of clan and district leaders who meet on a monthly basis to discuss and manage issues related to the Kuchi within their jurisdiction. They also interact regularly with the IGDK through the transfer of information and planning and management purposes.

A risk assessment of herders still engaged in extensive livestock production revealed that working as field laborers was one of the primary main ways of generating additional income or food throughout the year, after the sale of animals and animal by-products, with total earnings from labor wages currently contributing only a little to annual income or daily diet (Desta 2009). Total earnings from the sale of animals and animal by-products were reported to be lower now than in the past because of a reluctance to sell as livestock numbers continue to dwindle, low market prices (for wool, cashmere, and hides), and lack of access to the markets. Some reported owning equipment which they rented out as they passed by communities, but this was considered the exception rather than the norm. The participants in the assessments also reported that when migrating, it was difficult to negotiate a fair price for live animals or hides because the lack of communications made it difficult to keep up with current market prices. Other issues the Kuchi encounter are lack of access to basic services including veterinary, education, clinics, emergency food aid, and limited winter forage stocks.

Despite the Kuchi's invaluable contribution to Afghanistan's national economy, they continue to be one of the most marginalized cultures in the country today (MRRD 2009; Bedunah 2006). Over half of the total population is now settled after massive livestock losses, and in 2006, the Kuchi represented 70% of the internally displaced populations (IDPs) in Afghanistan (MRRD 2007; UNHCR 2006). Without a livelihood, these IDPs are destined to become permanently dependent on donations, begging, and daily wages to survive. For those families still engaged in herding, the risks they face have led to poverty and food insecurity being the norm for most. While periodic droughts will almost certainly continue to create risk for the Kuchi, preventing them from making their annual movements will lead to greater negative consequences for humans and animals alike. Additionally, opportunities for villagers to economically diversify will decline or disappear entirely. The greatest threat, however, is the end of the Kuchi's livelihood entirely and consequently, the decimation of Afghanistan's small livestock industry.

4 Strategy to Ensure Endurance of Afghanistan's Extensive Livestock Production Sector

Livestock production plays a major role in the stability of Afghanistan's economy, and in terms of food security, meat and dairy represent an important source of nutrition. Governance and policy issues, access to basic public services and emergency aid, portfolio diversification, conflict management, and social and human capital development are crucial to ensuring the survival of the Kuchi's pastoral livelihood and guaranteeing food security for them and the nation. Addressing these needs is also essential to the success of the government programs described by Emadi (Chap. 5) and here.

This plan is divided into two components: one for dealing with emergency situations and the other enabling the transition to a more sustainable livelihood. The foundation of this plan is that livestock production is the primary and most important livelihood of the Kuchi people and that the Kuchi people are the primary livestock producers in Afghanistan, and as such, maintaining this livelihood is critical to the economic security of Afghanistan as well as being critical to the cultural and social identity of the Kuchi people.

5 Emergency Plan: Climate–Related Coping Strategies

There are three activities designed to provide emergency assistance to the Kuchi people in the event that they experience an unexpected, climate-related risk (e.g., cold weather, drought, heat wave).

Emergency food distribution should occur in areas where nomadic herders are concentrated. Nongovernmental organizations (NGOs) and the World Food Programme (WFP) have been primarily responsible for emergency assistance, but reports suggest that the Kuchi are not being served in the same way that villagers have been served. Most likely, the reason for underserving the Kuchi community is that the NGOs and WFP tend to work with village shuras, most of which have poor if any Kuchi representation. The lack of Kuchi representation on shuras leads to an underestimate of the number of families and family size and their approximate location. Collaborating with the IGDK and KPDs will help alleviate this problem. In support, PEACE has compiled and mapped information from surveys that will prove useful.[6] With this information, emergency responders should be able to improve the food distribution situation tremendously during climate-related emergency situations.

Winter livestock feeding should be conducted in areas that have a history of requiring emergency winter assistance and where Kuchi are known to concentrate when anticipating a bad winter. Advance knowledge of the numbers of livestock that will require feeding is also of utmost importance. PEACE has compiled and mapped information from surveys with the Kuchi on the location and approximate numbers of Kuchi livestock throughout the country, seasonally, that will be of help in this effort. Collaborating with the IGDK and KPDs will also be of help. Winter livestock feeding will require significant planning in advance in order to be effective. Critical components to consider when planning include (1) locating and purchasing feed/forage well in advance, generally no later than the end of September; (2) advance

[6] Afghanistan PEACE Project: http://cnrit.tamu.edu/peace/kuchimaps.html

identification of distribution centers to maximize the effectiveness and ease of distribution; and (3) transportation of feed materials to the distribution centers in a timely manner to avoid weather-related transportation issues. Airdrops are an option for more isolated situations but would be more costly. This method should be considered when all other options are exhausted.

Commercial destocking of livestock is a practice that the Kuchi have mentioned that they are currently engaged in during adverse climatic conditions when they see no other alternative available left to them. Normally, the Kuchi do not sell animals during the winter or spring unless they find they are unable to feed them and know that they will die. When traders know that the Kuchi need to sell their livestock, they will naturally take advantage of this situation and offer very low prices for the animals. In general, the price received by traders is insufficient to buy enough forage for the remaining herd. One way to help Kuchi livestock producers in this instance would be to subsidize the purchase price of the animals so that they get a fair price. Subsidies have the additional advantage of giving an individual another means of income generation, whether they are Kuchi or not. Nomadic herder-assisted programs in Kenya and Ethiopia have adopted such a technique during emergency situations (Coppock 2010; Coppock et al. 2009). Subsidies can come in the form of no-interest loans or by covering transportation costs of the trader. To succeed, traders must be required to sign a binding agreement that guarantees a fair purchase price and that the animals will be purchased directly from a Kuchi herder, with the purchase location and herders specified in advance.

Noncommercial destocking is another viable option during an emergency situation. This method is a bit less challenging to accomplish in that it requires that animals are slaughtered and the meat dried for later consumption or sale. In southern Ethiopia and northern Kenya, partnerships have been developed between nomadic herders and government and nongovernment institutions that enable the herders to slaughter, dry, and sell the meat to local schools, charities, and other institutions at a fair price (Coppock et al. 2009).

Destocking and restocking elsewhere in Afghanistan is another method that could be adopted in an emergency situation. With this method, animals would be purchased at a fair price in areas where they are in need of emergency destocking and then transported to areas (where the risk is minimal or significantly less) and given to those Kuchi who are in need of restocking after losing their animals for various reasons (including climate-related and security and risks, to name a few) at a minimal cost. To undertake this effort, there are two issues that would have to be addressed prior to their transport: (1) the Kuchi receiving the animals would have to demonstrate that they have sufficient feed/forage to maintain the animals until the spring (the minimal cost), and (2) the transportation method would have to be agreed upon well in advance and ready to mobilize at a moment's notice. Given the lack of private commercial flights in the south, a military-assisted airlift would be the most efficient method available for this region. The IGDK and KPDs can assist in identifying the Kuchi that are being most affected and who would most benefit from restocking.

Veterinary services would need to be accelerated and have a focal point during emergency situations to avoid the spread of disease within a malnourished livestock population, and to avoid its spread to those areas where they might be relocated. Treatment would also be required prior to noncommercial destocking to ensure that the meat is fit for consumption.

6 Transition Plan: Risk Management - Prevention and Mitigation Strategies

There are four primary activities that will help the Kuchi people eliminate, limit, or manage the risks they face on a day-by-day basis with the result that they will become less likely to need to cope with emergencies when they arise. If successful, the risk management strategies detailed here will also serve to improve Afghanistan's livestock economy.

Economic Diversification in the form of outreach programs is one strategy for resolving the risks that the Kuchi face on a daily basis. Outreach programs are currently in effect in several other countries, with much success, with pastoralists facing similar risks (Coppock et al. 2009). Outreach programs involve the formation of collective action groups (CAG) whose mission is to identify and prioritize which functions are of import to the community that will lead to diversification. Typically, a savings-led credit function is the first to be adopted because this function serves as the primary means of capital for diversification. Literacy and numeration programs are typically another first function to adopt because of the low literacy rate among the Kuchi.

It is essential, however, that CAGs take on only a few functions at first in order to succeed and to have the opportunity to learn from their experiences without feeling overwhelmed by the process. In addition, those interested in forming CAGs must work to gain the trust of their own community. Once a CAG determines it has the necessary human and social capacity, in addition to demonstrating it has the start-up financial capital, it can expand its group to include government and nongovernment members with clear guidelines outlined and agreed upon in advance on the responsibilities and roles of each partner. Typical functions adopted at this stage include working collaboratively to devise strategies to improve the marketing of livestock and livestock-related products, the establishment of joint shuras for resolving conflicts, developing programs to add value to existing products, designing and adopting community-wide human capital programs, and the development of programs for the dissemination of information to the community, to name a few. To do these, CAGs will require the participation of donors and government stakeholders, in the form of materials, education, training, and mentoring, with guidance by the coordinating body.

Another responsibility of the CAG is to agree to a participatory action research (PAR) component. PAR is essential to the success of any action group because it provides valuable feedback to the group and to other groups as well. It is also a way of imparting information to others who are attempting to form their own action groups and to the world at large on what ideas were chosen or rejected over others and why. By definition, it means the observation and documentation of the action-group decision-making process, problems encountered, and how they were resolved. It also involves the collection of baseline information followed by the monitoring of indicators to assess progress, failures, and successes. In essence, this is no different than any monitoring and evaluation process, with the exception that the information eventually gets published in refereed journals for the benefit of a much wider audience than just Afghanistan alone.

Improved marketing and value addition within the framework of economic diversification will improve the situation for the Kuchi by adding value to existing products through fattening programs, improvements in methods used to process and collect products (e.g., wool, cashmere), and improved marketing techniques to attract or engage traders who will engage in business partnerships that will ensure a higher value product.

Human Capital: Education and Health involves education and health programs to the general population and CAGs to not only improve the existing situation but to build the capacity of all Kuchi people over time. Programs that are needed at this time include general health programs, maternal and prenatal programs, and education programs for children, women, and the illiterate. To implement programs of this nature, implementing partners will need to work closely with the IGDK, its provincial offices, and the CAGs to identify the best methods for delivery of services and materials and training.

Adoption of Pro-extensive Livestock Development Policies involves a suite of activities aimed at improving management Afghanistan's rangeland areas, improvements in the delivery of standard veterinary services, strengthening of existing laws and adoption of laws which improve the situation of Afghanistan's rangelands.

(A) Research and Monitoring – Research to collect information essential to the development of effective management plans, as are design, monitoring, and evaluation protocols; rehabilitation of areas back to productive rangeland; and improved fodder production. These activities will all involve both government and nongovernment institutions working in close collaboration. It will also involve active participation by the Kuchi, in the selection, design, and all implementation phases.

(B) Rangeland Law – Afghanistan's rangelands are still being plowed under, at an alarming rate, despite evidence that they will not support any type of rain-fed agriculture. Afghanistan is in need of laws designed to protect its remaining rangelands, as described by Emadi (Chap. 5). To ensure the adoption of an effective law, there must be a coordinated effort to involve the Kuchi and other stakeholders during the drafting process, rather than only through the public comment process. This will require engagement and coordination with the IGDK and KPDs as soon as possible.

(C) Veterinary and Extension – Access to veterinary services throughout Afghanistan regardless of ownership would ensure healthier livestock, delivery of a better product to the market, and more income to the seller. There are several organizations currently involved in developing the capacity of Afghanistan to deliver veterinary services to livestock producers. Unfortunately, the Kuchi have been largely overlooked to date, but there are some efforts under way to improve this situation. With respect to extension services, these are mostly lacking for the entire country. In order to reach the primary livestock stakeholder in Afghanistan, veterinary and extension programs need to adopt a policy of training Kuchi as veterinary technicians and extension workers, in addition to designing a program that ensures the delivery of quality pharmaceutical products to a mobile community. This will require engagement by the Kuchi in the selection and coordination process.

Conflict Management involves programs aimed at building the capacity of Kuchi and non-Kuchi leaders to solve some of their more immediate and challenging land access and appropriation issues. The idea is that if the leaders can solve most conflicts themselves, in collaboration with communities, then at least one barrier to improving livestock production and ensuring food security can be removed. PEACE initiated a conflict management program in 2007, and it continues today after evolving in stages (Jacobs et al. 2009a, b). It started with capacity building of KPDs, then expanded to include district and clan-level Kuchi leaders, and from there, it grew to include Kuchi and non-Kuchi leaders from specific conflict regions. Institutions involved in this effort included the IGDK, the Peace Commission – Office of Tribal Affairs (OTA), and the Sanayee Development Organization (SDO). The foundations of the program were simple – involve the government, involve respected local leaders, provide the necessary trainings and skills, keep the demands simple, and encourage equal representation during meetings. The goals were simple as well – build capacity of the participants to communicate more effectively, serve as a conflict resolution mediator, and encourage equal representation during meetings. The outcomes from the program to date are many (Fig. 6.5 and Table 6.1), and these will continue to grow with time and extension of the effort:

- Peaceful resolution of more than 1,000 social and land/resource-based conflicts
- Improved relations with local communities
- Improved relations with government and local officials
- Consensus that the government has created many of the problems they were experiencing over land tenure and the use of public rangelands
- Consensus that political leaders do not want them to settle land access issues at the community level, that high-powered Kuchi and Hazara government leaders were responsible for perpetuating the conflicts to further their own political agendas
- Acknowledgement that history and complexity of issues mean that a quick resolution is not possible
- Consensus that fighting has not helped them to solve any of their conflicts and should therefore be discouraged
- Improved relations with district and clan leaders

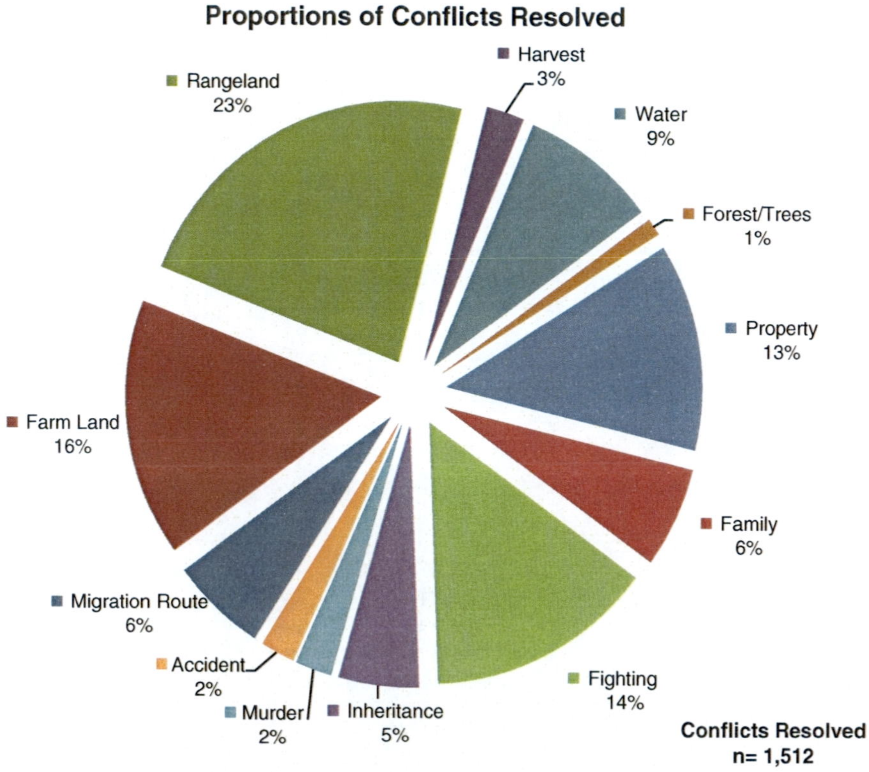

Fig. 6.5 Types and proportions of conflicts resolved by Peace Ambassador Program between June 2011 and April 2012. Types of conflicts: *land* (rangeland, farmland, migration route, harvest, water, forest/trees) and *social* (property, accident, murder, inheritance, fighting, family)

- Increased support for the role of the IGDK and its provincial offices
- Transfer of knowledge of Kuchi-related issues from the level of constituents to the IGDK
- Demonstration of knowledge of Kuchi-related issues and solutions to Afghanistan parliament and president

Success of the conflict management program is determined to be a function of the following:

- Training was adaptive, incorporating the needs of illiterate participants.
- Quarterly presentations and skits enabled participants to learn from each other.
- The mediators included respected leaders from both sides of the issue who were committed to a peaceful solution.
- The communities were the ones who chose which conflicts they wanted solved.
- Participants from both sides of the issue engaged equally in the process.
- Participants had a voice in finding and choosing the solution(s).
- Agreements were binding.

Table 6.1 Land-related conflicts, between different user groups, which have been resolved by Peace Ambassador Program between June 2011 and April 2012

Province	Between villagers	Between Kuchi	Between villagers and Kuchi
Badakhshan	49	3	23
Baghlan	22	24	47
Balkh	40	10	15
Bamyan	16	0	3
Faryab	24	23	15
Ghazni	29	4	14
Jalalabad	14	3	23
Jawzjan	34	6	8
Kabul	20	6	10
Kapisa	19	3	8
Khost	1	4	10
Kunduz	13	15	6
Laghman	7	6	3
Parwan	5	3	5
Samangan	27	12	41
Sar-e Pul	10	11	38
Takhar	37	14	19
Wardak	14	17	30
Total	**381**	**164**	**318**

This is a different approach to the traditional problem-solving method used in Afghanistan that of presenting your problem to the local shura, or assembly of leaders, and having the shura decide the outcome. While this system has proven to be of much use to Muslims around the world (UNDP 2007), it fails in Afghanistan when the issue involves multiple ethnic groups or identities. The reason is that most shuras are comprised of local representatives only. Additionally, Afghanistan's shura leaders tend to be political appointments with no formal training in conflict resolution. As a consequence, their judgments tend to be biased when the issue involves noncommunity members.

6.1 Strategy Issues

The Kuchi represent a largely mobile community 6–9 months out of the year. Currently, most programs in Afghanistan are designed for delivery to sedentary communities. In addition, most organizations work directly with local governments and shuras in each province, when designing and implementing a program. The Kuchi are now organized under the office of the IGDK; however, they are still unskilled in the design and implementation of programs specific to their needs.

There is also a lack of coordination and poor communication between the different government and nongovernment organizations for some of the interventions proposed here. All of these issues make delivery to the Kuchi generally challenging.

The programs described here are either designed to address these issues or will require mechanisms built into them to overcome them. One required mechanism is that this plan be implemented as a coordinated effort between all relevant government institutions, NGOs, the IGDK, and all Kuchi provincial offices. To participate, these organizations must agree to using a participatory process to identify, design, and implement any program. Another requirement is that one institution only, with experience and familiarity with the Kuchi situation such as the IGDK, function as the coordinating body under which the various stakeholders operate. Monitoring and evaluation of emergency and nonemergency efforts are also critical to assess success and to improve upon each intervention measure over time. In all of the situations described above, a monitoring and evaluation component should be built in to ensure the intervention measure is being implemented and adapted where needed.

Mentoring is another critical component. Until capacity building is achieved, support is needed to encourage participation and deal with frustrations and feelings of being overwhelmed.

7 Summary

Extensive livestock production in Afghanistan is currently at great risk. Access issues, appropriations of land, insecurity, conflict, and lack of winter fodder are at the heart of many of the Kuchi's problems. These same issues are responsible for many of the current environmental problems and threaten many ecosystem services in addition to their threat to food security for the country. There are numerous interventions that, if approached correctly, can improve the situation while providing opportunities for better land stewardship. Immediate interventions, such as emergency food distribution, winter livestock feeding, and destocking/restocking programs, are some of the more obvious solutions to help the Kuchi cope with their most urgent issues. Additional benefits from these include economic opportunities for Kuchi and non-Kuchi alike. Some of the longer-term solutions to eliminating, limiting, or managing risks include programs aimed at economic diversification, investment in human capital, development of pro-extensive livestock production policies, and conflict management. Collectively, these interventions will promote an industry which is vital to Afghanistan and its people; they will also promote efforts to improve land stewardship.

All interventions described here will require an investment in time, however, because of the need to transfer skills to Afghan communities and the local leadership. Likewise, earning trust takes time and much commitment. Mentoring is also a vital component to any program, as well as ensuring a participatory process. These constraints all highlight the need for donors to understand and commit to longer funding and program cycles. Sustainable development is rarely a short-term, high-impact endeavor. The benefit will be that we will have provided a large number of Afghans with the skills needed to solve their own issues, at all levels of society and government. This will eventually translate to better human and environmental conditions in Afghanistan.

References and Further Reading

Barfield TJ (2004) Nomadic pastoralists in Afghanistan: reconstruction of the pastoral economy. Bank Information Center, Washington, DC, 13 pp. Online: http://www.bicusa.org/en/Region.Resources.5.aspx

Bedunah DJ (2006) An analysis of Afghanistan's rangelands and management issues for the development of policy and strategies for sustainable management. 5 January 2006, Chemonics International, Inc. Online: http://pdf.usaid.gov/pdf_docs/PNADG285.pdf

Breckle SW (2007) Flora and vegetation of Afghanistan. Basic Appl Dryland Res 1(2):155–194

Coppock DL (2010) Action research, knowledge and impact: experiences of the global livestock CRSP pastoral risk management project in the Southern Ethiopian rangelands. Global Livestock Collaborative Research Support Program, University of California, Davis

Coppock DL, Desta S, Tezera S, Gebru G (2009) An innovation system in the rangelands: using collective action to diversify livelihoods among settled pastoralists in Ethiopia. In: Waters-Bayer A, Wettasinha C, Njuki J, Sanginga P, Kaaria S (eds) Innovation Africa: enriching farmer's livelihoods. EarthScan Publications, London, pp 104–119

de Weijer F (2005) Conference on Afghanistan pastoralists (Kuchi). Report of proceedings 15–17 November 2005. United States Agency for International Development. Online: http://cnrit.tamu.edu/peace/pdfs/AfghanPastorlistConference01.06.pdf

Desta S (2009) Risk management strategy for Kuchi Herders in Afghanistan. Afghanistan PEACE project. Online: http://cnrit.tamu.edu/peace/pdfs/PEACE%20Risk%20Assessment%20Report%20Jan%202009.pd

Desta S, Schloeder C (2010) Report on initiating a pilot community-based risk management project with the Kuchi in Northern Afghanistan. Available at: http://cnrit.tamu.edu/peace/riskmgmt.html

Emadi MH (2012) Better land stewardship to avert poverty and land degradation in Afghanistan. In: Squires VR (ed) Rangeland stewardship in Central Asia. Springer, Dordrecht, pp 91–108 (Chapter 5, this volume)

Flora Iranica (1963–2009) In: Rechinger KH (ed) Flora Iranica volumes 1–179. Graz, Akademische Druck- u. Verlagsanstalt

Jacobs MJ, Naumovski I, Schloeder C, Dalili RM (2009a) Empowering Afghan Herders to build peace. Research brief 09-01-PEACE. Global Livestock Collaborative Research Support Program (GL-CRSP), University of California, Davis, California

Jacobs MJ, Naumovski I, Schloeder C, Dalili RM (2009b) Peace ambassadors: an innovative approach to community-level conflict resolution in Afghanistan. Research brief 09-02-PEACE. Global Livestock Collaborative Research Support Program (GL-CRSP), University of California, Davis, California

Khazanov AM (1984) Nomads and the outside world. Cambridge University Press, Cambridge

MRRD: Ministry of Rehabilitation and Rural Development and the Central Statistics Office (2007 June) National Risk and Vulnerability Assessment 2005. Online: http://home.wfp.org/stellent/groups/public/documents/ena/wfp193560.pdf

MRRD: Ministry of Rural Rehabilitation and Development and Central Statistics Office (2009) National risk and vulnerability assessment 2007/8: a profile of Afghanistan. ICON-Institute GmbH & Co KG Consulting Gruppe, Cologne

MTND: Meat Trade News Daily (2009) http://www.meattradenewsdaily.co.uk/news/110609/afghanistanproblems_in_the_wool_industry.aspx

Schloeder CA, Jacobs MJ (pers. obs) Afghanistan PEACE project. Online: http://cnrit.tamu.edu/peace/

Schloeder C, Jacobs MJ (2009) Afghanistan livestock market assessment: report on Afghanistan livestock market dynamics October 2008–2009. Afghanistan PEACE project. Online: http://cnrit.tamu.edu/peace/pdfs/PEACE%20Livestock%20Market%20Synthesis%20Report%20June%202010%20web%20version.pdf

Schloeder C, Jacobs M (2010) Complete list of flora of Afghanistan: compilation of records from various sources. Afghanistan PEACE project. Online: http://cnrit.tamu.edu/peace/plantspecies.html

Schloeder CA, Jacobs MJ, Sherzad MD (In preparation) Analysis of rangeland diet quality in Afghanistan

Schütte S (2012) Pastoralism, power and politics: access to pastures in northern Afghanistan. In: Kreutzmann H (ed) Pastoral practices in High Asia: agency of 'Development' effected by modernization, resettlement and transformation. Springer Science and Media, Dordrecht, pp 53–70

Thieme O (2006) Afghanistan, country pasture/forage resource profiles. Food and Agriculture Organization, Rome. 19 pp. Online: http://www.fao.org/ag/AGP/AGPC/doc/Counprof/PDF%20files/Afghanistan.pdf

Thompson E (2007) Water management, livestock and the opium economy: marketing of livestock. Applied Thematic Research into Water Management, Livestock and the Opium Economy October 2007

UNDP (2007) Afghanistan human development report 2007: bridging modernity and tradition – the rule of law and the search for justice. Human Development Index, Kabul

UNHCR: Office of the United Nations High Commissioner for Refugees (2006) Durable solutions for Kuchi IDPs in the South of Afghanistan: options and opportunities. Report, November. Asia International Consultants, Kabul and Kandahar

UNMAP: United Nations Mining Action Program (2009) Online: http://www.un.org/apps/news/story.asp?NewsID=29515&Cr=unmaca&Cr1

Chapter 7
Constraints and Barriers to Better Land Stewardship: Analysis of PRAs in Tajikistan

Barno Kurbanova

Abstract This chapter reports on a PRA survey of 710 people in 10 separate rural villages in five *districts* of south central Tajikistan. This is a mountainous country in Central Asia, with 93% of its surface area taken up by a complex of east-west and north-south ranges. Almost half of the country is at altitudes of more than 3,000 m. Tajikistan is an agrarian country with a rural population more than 75% of the total and in which the agricultural sector accounts for 65% of employment and around 25% of GDP (averages for 1995–2009). Tajikistan remains the poorest country in Central Asia with a high level of rural poverty. Many rural people live below the poverty line ($2.15 per day).

The results of the PRA have provided a solid understanding of the situation throughout these five *districts* and of the community perception of the current problematic situation on pasture management and livestock. The main constraint relates to the uncertainties about access to grazing land and to security of land tenure. Legislation relating to issues of land use and tenure is in constant flux, and the measures to implement the existing laws are not consistent or transparent.

Keywords Land tenure • User rights • Equity and inequity • Poverty • Gender issues • Remittances • Legislation • Regulations • Implementation • Pastures • Livestock • Land degradation • Land stewardship • Barriers • Sustainability

Key Points

- Participatory rural appraisal (PRA) is a way to provide the means to bring farmers "voices" to the forefront on a range of issues surrounding pasture and livestock management and to identify barriers to livestock management as perceived by farmers themselves, as well as possible solutions or priorities for assistance.

B. Kurbanova (✉)
Donaev Management Consulting, 15 Giprozem, Dushanbe, Tajikistan
e-mail: barnojon@mail.ru

V. Squires (ed.), *Rangeland Stewardship in Central Asia: Balancing Improved Livelihoods, Biodiversity Conservation and Land Protection*,
DOI 10.1007/978-94-007-5367-9_7, © Springer Science+Business Media Dordrecht 2012

- There are many barriers to better land stewardship in Central Asia as evidenced by this study of 2,700 respondents in Tajikistan. The principal impediment is the lack of security of access to grazing lands and failings in the land tenure arrangements
- The main barrier to better land management is *poverty*. Recurrent themes from the PRA survey, across all villages, were a lack of a clear awareness of land user rights and a lack of perception of the nature, extent, and full implications of the inequitable distribution of pastureland.
- PRA participants rated the other barriers to sustainable land use as the degradation of pastureland; shortage of winter feed for livestock; limited access and lack of pastures; the shortage of water in the pasture; overgrazing; and lack of effective local institutions. The lack of local institutional support on pasture management, animal husbandry, and livestock health were especially noted.
- The results of the gender analysis demonstrate that women are involved in every aspect of livestock care and management, with the only significant underrepresentation in vaccination, slaughtering, long-medium distance shepherding, and shearing. However, women are as equally active in treating sick animals as men.

1 Introduction

1.1 Brief Outline of Participatory Rural Assessment (PRA) Approach

PRA is an approach used for many years by nongovernmental organizations (NGOs) and other agencies involved in international development. The approach aims to incorporate the knowledge and opinions of rural people in the planning and management of development projects and programs. PRA involves use of methods that involve rural people in examining their own problems, setting their own goals, and monitoring their own achievements. The PRA approach enables identification of the main constraints to improving land management from the perspective of farmers themselves while at the same time encouraging participatory identification of solutions. It involves use of semi-structured interviews conducted with rural householders (Hua, Chap. 14). PRA surveys were conducted in 10 villages from 5 pilot *districts* (Rogun, Rudaki, Faizabad, Vahdat, and Varzob) of the Region of Republican Subordination of Tajikistan. A total of 30 PRA surveys were conducted between October 2010 and April 2011.

1.2 Expected Outcomes of PRA

The *primary outcomes* of the PRA were to provide the means to bring farmers' "voices" to the sector assessment on a range of issues surrounding livestock management, according to representatives from a variety of groups who are involved with livestock ownership and grazing management practices. Barriers to livestock management were identified, as perceived by farmers themselves, as well as possible solutions or priorities for assistance. The *secondary outcomes* were to use this community-based analysis to form the basis for the demonstration activities implemented under the project to improve their likelihood of being accepted and adopted.

1.3 Characteristics of Farmers/Farm Workers and Farms Included in PRA Survey

1.3.1 Variables of Groups

In general, there are two broad categories of land users in villages (See Box 7.1):

Box 7.1 Land Reform in Tajikistan: Categories of Tenure Arrangement

***(i) Permanent heritable land use*:** Permanent heritable land use is governed by Land Code (as amended by 2009). *Dehkan* farms (literally *peasant farms*) may be established by individuals, families, or by groups (partnerships) based on shared ownership. In each case, the farm has a head, who holds the land certificate, and shareholding members who should hold share documents to a physical plot of land. The head is responsible for reporting and tax collection, but decisions on reorganization or changes to contracts between members may only be decided at a general meeting. Members may legally secede from the *dehkan* farm without permission of other members, establishing their own individual or family *dehkan* farm on their land share, with the same permanent heritable land rights. However, this is an expensive process.

Individual or family dehkan farms: When applying to establish a *dehkan* farm, former *Sovkhoz* or *Kolkhoz* workers may apply for a share of the former entity for which they worked. Areas allocated should be based on norms calculated from the area of available land and the number of former *Sovkhoz* or *Kolkhoz* members. Other Tajik citizens may apply for land from the state fund (see below).

Collective dehkan farms: Legally these are close to the group/partnership form of *dehkan* farm given in the law but were not specifically foreseen in the legislation as in fact they comprise the entire former territory of a *Sovkhoz* or *Kolkhoz*. These structures appeared in response to government targets to restructure all state farming entities by the end of 2005 and due to the high transaction costs of forming individual and family *dehkan* farms. The collective *dehkan* farm head holds the land certificate for the

(continued)

Box 7.1 (continued)

whole area, but in GBAO for example, each shareholding household farms individually and should hold a legal share document for a physical parcel. Legally, the same principles apply to that pasture which is *permanently* allocated to the former *Sovkhoz* and *Kolkhoz*, but until 2009 in GBAO this pasture continued to be used in common by all members. In 2009, the Land Registration and Cadastre System for Sustainable Agricultural Development Project (LRCSP) facilitated the conversion of a selection of collective *dehkan* farms to individual *dehkan* farms with provision of full certification for c household. This greatly improves security of tenure for those households. According to the law, each member should receive legal title both to equal shares of both arable land and the "permanent use" pastureland allocated to the now defunct collective, regardless of the number of animals owned.

(ii) Land allocated from the state land fund for long-term use: In some districts most pasture used by collective *dehkan* farms is accessed by members up to 20-year "long-term use" agreements made between farm management and the district land committee. Pasture tax is charged per hectare, but farm management usually gets around this by splitting the overall sum of tax payable so that each household pays a proportion corresponding to the number of livestock owned. Thus the *de facto* pasture management regime corresponds to common property with fixed boundaries and a fixed user group. As long as this land is not permanently allocated to collective *dehkan* farms, members do not automatically receive a share if the collective is dismantled. In the meantime, individuals may apply separately to the land committee for a permanent share of this pasture. Some officials indicated that in order to privatize pasture on long-term use land, an individual must obtain the written permission of the members of the collective *dehkan* farm to which that land was formally allocated. However, in Tajikistan, this is unlikely to constitute an effective guarantee of common user rights. It should be noted that collective *dehkan* farms may also apply to have this land transferred from long-term use to permanent use, in which case it becomes eligible for distribution to individual households upon restructuring of the collective.

***(iii) Lease of state land fund*:** Many remote pastures remain unallocated and remain part of the state fund. Any party (individual, collective *dehkan* farm, or state enterprise) Land Code (2009) allows up to 20 years, though in practice the leases often are short term and informal. As mentioned above, such land may also be privatized and incorporated into *dekhan* farms by application, at which point it ceases to be available for lease.

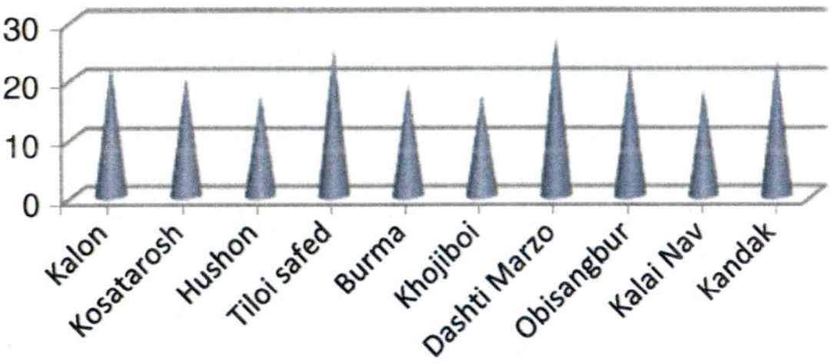

Fig. 7.1 Number of PRA participants/% of households (HHs) represented

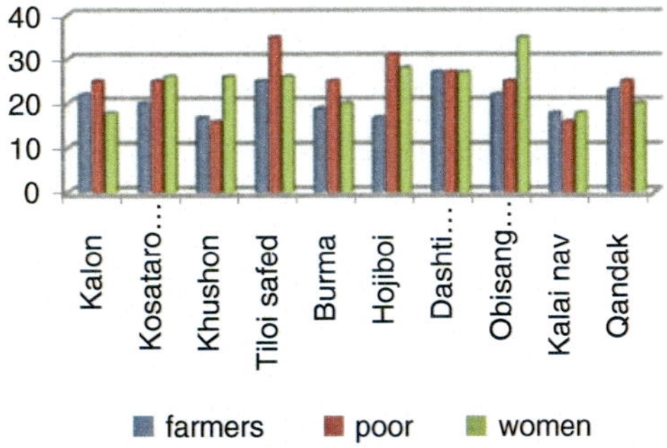

Fig. 7.2 Number of PRA participants/proportion of target groups

- Shareholders in collective farms who received lifetime heritable user rights to land after the restructuring of *Sovkhoz* or *Kolkhoz*
- Individual dehkan farms[1] and households, with their small household plots (300 m^2) and small areas of land (>1 ha) for crop, orchard, or pasture use (so-called presidential land), and who also use the land of shareholders on a rental basis

In PRA, the survey area land users can be further subdivided as dehkan farmers[2] (DF), individual DF, collective DF, cooperative farms, and family DF and landless livestock owners.

In this PRA process, three target groups were involved:

Group 1: dehkan farmer/richer farmer/shepherds/owners of larger ($n>50$) (herds of livestock)LS/male.

Group 2: nondehkan farmers, poorer farmers/owners of small LS herds ($n<50$)/mostly male.

Group 3: women farmers/mostly nondehkan farmers/owners of small LS herds ($n<50$)/some heads of households and dehkan farmers. Totally 710 participants were involved (Figs. 7.1 and 7.2).

[1] See discussion below on land reform and origin of dehkan farms in Tajikistan.

[2] Dehkan (literally peasant) farms were created after the breakup of large state farms and livestock enterprises following the collapse of the Soviet Union. They can be quite large (up to 3,000 ha) or small and may be under the control of individuals, families, cooperatives, or collectives (see Halimova, Chap. 13 for more explanation).

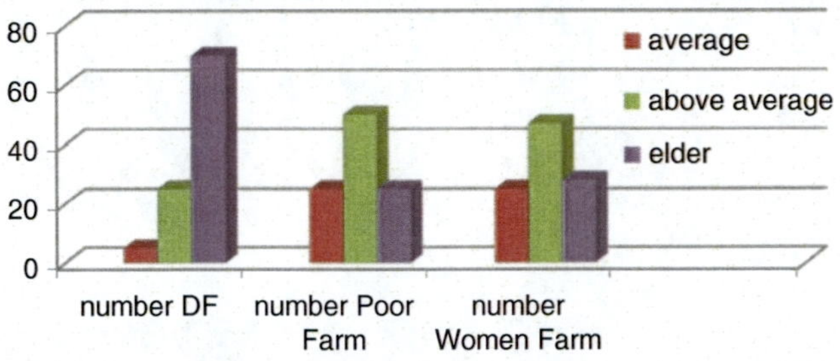

Fig. 7.3 Demographic variables in PRA groups

In some cases, the dehkan farm group included sub-certificate holders rather than private dehkan farmers (PDFs), and in some cases, the women's group included wives of wealthier dehkan farmers along with representatives of HHs with no livestock. In general however, these three groups were considered, according to local indicators, to represent the wealthier (group 1) and poorer (group 2) sections of the community. The inclusion of a separate group of women was specifically designed to allow them to participate more fully in the discussions in an informal and relaxed atmosphere to ensure they were free to voice their concerns and analyses of the issues from their own perspectives. In only one village (Khushon) did the 2nd (poorer group) include a mixture of men and women at their request. Representatives of all forms of DF accounted for 30% of all number of participants.

Demographic variables – people of middle age, older than average, as well as elderly folk – were included in the PRA groups.

As Fig. 7.3 shows, the highest participation rates (70%) were the elderly in DF group of farmers, and the lowest participation rates (5%) were people of average age. Older than average people were well represented in the other two groups (Table 7.1).

2 Knowledge of and Attitudes Toward Changes in Land Use Rights, Tenure, and Land Reform

According to the constitution, all lands in Tajikistan are the exclusive property of the government. The land tenure issue in Tajikistan is very problematical, because only 7% of the entire country is suitable for cropping and only part of the remainder is suitable for livestock raising (mostly where the elevation is

Table 7.1 PRA sites and agroclimatic zones

District	Jamoats	Village	Predominant pasture vtype	Elevation of settlement area (asl)	Elevation of predominant pasture (asl)
Varzob	Ziddee	Khalon	Summer	2,000	**>2,000**
	Luchob	Kosa Tarosh	Summer	1,820	1,850–2,500
Vahdat	Romit	Khushon	Summer	1,200–1,300	2,000
	Guliston	Tiloi Safed	Winter	957	1,200
Rudaki	Sultanabad	Burma	Spring/ autumn/winter	1,200	1,300
	Esanboy	Hojiboi	Winter	1,000	1,200
Faizabad	Kalai Dasht	Dashti Marzo	Summer	1,931	1,950
	Javonon	Obisangbur	Summer	1,200	1,600
Rogun	Kadi Ob	Kalai Nav	Summer	1,200	1,200–1,500
	Obigarm	Kandak	Spring/autumn/ summer	1,200–1,500	1,800

from 1,200 to 2,500 m (asl)). If we consider that more than 75% of the population lives in rural areas, this issue is very relevant (Lerman, Chap. 8; Robinson Chap. 11).

2.1 *Land Reform, Land Use Right, Tenure, and Pasture Management*

Land reform began at the start of the transition to the market economy by Resolution of Government 16.09.1992 № 357 "about measures on land reform in Republic of Tajikistan." The purpose of this document was to create conditions for future development of various types of methods of management, organization of a multi-sector economy, and increase agricultural production. But land legislation was amended several times in the last 15 years. Government resolution on 31.08.2004 № 349, as well as the Land Code of the Republic of Tajikistan № 326 from 13.12.1996 (on redaction law of RT № 498 from 12.12. 1997 y., № 746 from 14.5. 1999 y., № 15 from 12.5. 2001 y., № 23 from 28.2 2004 y., № 199 from 28.07.2006 y. № 357 from 5.01.2008 y. № 405 from 18.06.2008 y. (Halimova, Chap. 13)).

Land in Tajikistan is state property and provides for long-term use by dehkan farms and households. Private farms were formed on the basis of the reorganized *Sovkhoz* or *Kolkhoz*-collective farms. The legal basis for the organization and activities of dehkan farm in the Republic of Tajikistan was set down in 1992 and is regulated by law of the Republic of Tajikistan "On Dehkan (farms)." Later, a new version was introduced as № 48 from 10.05.2002; the latest version of the law was from 2009 (Halimova, Chap. 13).

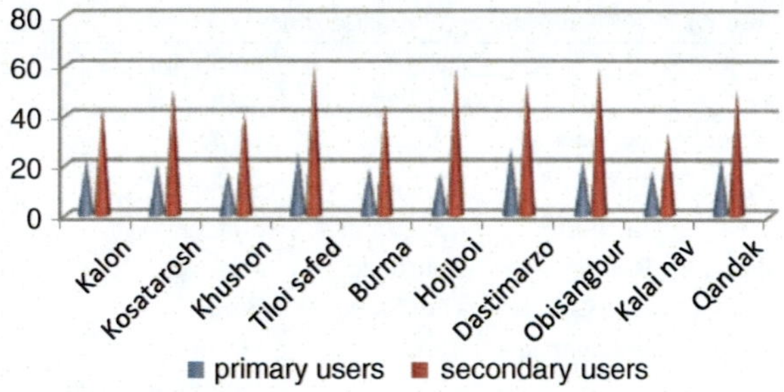

Fig. 7.4 Type of land users

2.2 *Land User Right*

Land users in Tajikistan are physical and judicial persons. Physical and judicial persons can be either (or both) primary and secondary land users. The primary users are judicial and physical individuals who use land in perpetuity (lifetime inheritable use). Secondary users are individuals and legal persons receiving land under lease agreement (Law of RT from 5.01.2008. № 357).

The primary users receive the right to obtain a certificate of land tenure on a particular land plot. Registration of tenure is through the State Committee of Land Management, Geodesy and Cartography.

The PRA process involved several types of users, primary and secondary (Box 7.1). Primary users are representatives of collective and individual dehkan farms, which have certificates. Secondary users are representatives who do not have certificates and rent land from primary users by agreement (collective and individual dehkan farms) (Fig. 7.4).

2.3 *Pasture Use and Tenure*

Pastures are the dominant component of all agricultural land. Pasture areas are used according to seasons: summer pasture, winter pasture, spring-autumn pasture, and perennial pasture. The total number of pasture hectare used (ha) was estimated for each village as follows (Fig. 7.5):

2.4 *Land Tenure*

Land tenure is regulated by the Land Code of Republic of Tajikistan. During the last 10 years, there have been amendments to better aid the development land reform process (Halimova, Chap. 13). But the current process of reform has many complications and is still not so affective.

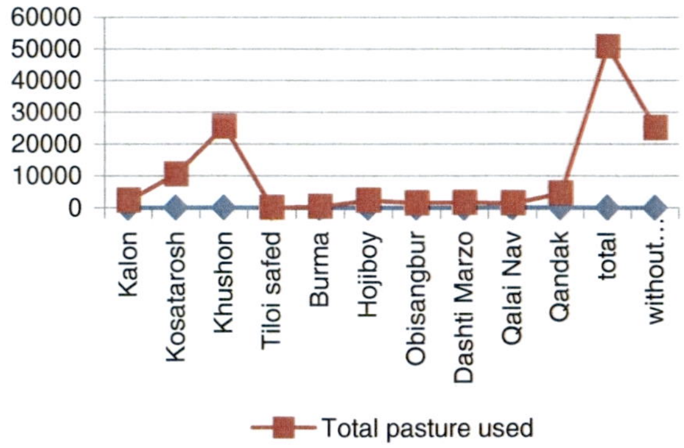

Fig. 7.5 Estimated area of pasture used by the village

Documenting the types of tenure to pasture areas utilized by the village was a complex undertaking, with often conflicting and incomplete information provided as the privatization process gathers momentum. In some instances, where people had received land certificates, the situation was straightforward. In others, people commented that they did not know their rights as former *Kolkhoz/Sovkhoz* workers and were working on the land as laborers and providing a percentage of their crop or paying rent without yet receiving sub-certificates. They were also unaware that they could create private DFs from their shares. People referred to some of the larger dehkan farms (DFs) as "collective" or "state" farms when in fact they are large private farms, allowing use of their land through various rental arrangements, often coordinated by the mahalla leader who collects the rent on behalf of the DF. Where large areas of pasture were thought to have been privatized by "influential individuals" from outside the Rayon, people often had no information at all.

In summary, the user right status of the 2,730 HHs in the project villages in relation to pastureland in particular was as follows:

- 11 collective DFs (CDFs), with a total of 372 shareholders. Of these, 225 (60%) were reported to have sub-certificates.
- A further 350 HHs were said to be "members" of a DF there, but their status was unclear with regard to certificates.
- 159 individual or private DFs with certificates are established (total number of HHs estimated at 448).
- 262 HHs are waiting for certificates to private DFs (253 of these may be entitled to sub-certificates to CDF land).
- 942 HHs are not members/owners of any type of dehkan farm (or 34% of total HHs in the PRA village). These are typically households who were not workers on the *Kolkhoz/Sovkhoz* and have no rights to "shares" (and therefore sub-certificates) or families recently established (i.e., after the *Kolkhoz/Sovkhoz* decollectivization process) and will therefore never had rights to this land.
- Khushon has no DFs at all but rents all pasture from the Forestry Department and private DFs in other areas (197 HHs).

It is important to note that dehkan farm land does not always include pasture for example (Burma: 12 DFs and Tiloi Safed: 57 DFs have no pasture), or the pasture area is too small for productive use (Obisangbur: 11 DFs share 46 ha pasture), necessitating rental agreements with larger DFs. The percentage of non-members/owners of DF HHs corresponds roughly to the number of nonlivestock HHs; however, the link between land rights and livestock ownership is not so straightforward, with many HHs that currently rent pasture. In addition, the pastureland distributed to smaller DFs was anecdotally of poorer quality than that going to the larger farms of "influential" individuals – especially hay lands.

Kalon Village, woman 43 years old: We do not know of any case where issues of ownership of pastures were settled in an equitable manner. Every year, there are more and more limited opportunities for acquisition or use of grazing land for poor people.

2.5 According to PRA Results, Knowledge About the Changes in the Law is Very Weak

Land reform in Tajikistan has been in place for nearly 20 years since independence. It directly or indirectly affects the interests of farmers, but lack of public awareness on land use issues is impeding assignment and proper registration of land rights. Laws on the dehkan farmers are defined. General provisions on the establishment of farms, definition of their powers, rights of shareholders, and so on are set out. In accordance with the law, a significant number of farms were established. This process not always followed when *Kolkhoz and Sovkhoz* were restructuring. Instead some people set up joint-stock farms and private farms, apparently in contravention of the land laws.

Since 1996, when the process of restructuring of *Kolkhoz* began, many, smaller, private farms were set up. Due to the increasing level of poverty, especially in rural areas, the government allocated land to household's so-called presidential land to help people cope with the deteriorating economic situation. These plots were covered by the provision of an inherited right to long-term use. There was no opportunity to transfer or sell this land. There is a law that authorizes and empowers dehkan farms. Indeed, many of the large farms were created by people who had access to information and to the loopholes in the laws. They immediately took action to secure their rights over large tracts of land. Owners of large farms are now major shareholders in dehkan farms. These shareholders have significant benefits, with considerable areas of land. In addition to their own agricultural enterprises and grazing their own livestock, they rent part of their land to others including those with small dehkan farms (some of which have no pastureland (see Table 7.2 above)) on an annual contract basis. The lease arrangements (often in oral form) yield significant profits. Typically, the lessee pays 30% of the harvest from the leased land (hay, livestock products), as well as payment for the number of livestock grazing on their land holdings.

Table 7.2 Estimated area of pasture used by villagers on private dehkan farms (PDF) or collective (CDF)

Village	Pasture (CDF) (ha)	Pasture (PDF) (ha)	Pasture (public) (ha)	Pasture (other) (ha)	Total pasture used
Kalon	2,250	100			2,350
Kosatarosh	10,604 (some in other District)		107 of CDF		10,604 (incl. winter pasture)
Khushon			500	25,000 owned by others/forestry	25,500
Tiloi Safed	70				70
Burma	350	100			450
Hojiboi		2,035	300	1,600 rented from other *districts*	3,935
Obisangbur	1,113	46		400	1,559
Dashti Marzo	1,177	426	(427 from CDF)	25	1,628
Kalai Nav	254		343	860 leased	1,457
Kandak	1,375			700 hay (a portion 2,565 rented from other village)	4,640 (not all 2,565 rented)
Total					50,700
Without Khushon					25,200

2.6 *Tax and Rent Paid*

The tax is paid by the primary user land, and rent is paid by the secondary users who rent land from primary users. Rate of tax or rent is not uniform.

- None – some villages mentioned that they used CDF or *mahalla* land (common land) and do not pay money
- 1 ha pastureland tax 4–6 som 50 diram/year
- 1 ha pastureland rent 15–24 som/year
- 1 cow 4–20 som/month
- 1 sheep 1–7 som 50 diram/month
- 1 cow 1 som/per day additional to rent
- 1 sheep 50 diram/per day additional to rent

Consider this, the primary user land pays tax to the state for the use of 1 ha of land at a rate, on average of 5 somoni *per year,* but receives rent of 20–25 somoni *per month* for grazing one head of cattle during the grazing season and 30% of the cost of harvest in cash or in kind. It is not difficult to calculate what benefit the large-scale dehkan farms have.

Sometimes, participants were confused with the payment of social taxes, payment of land taxes, and rent payment (Fig. 7.6).

Fig. 7.6 Kosatarosh village, Varzob District

Inoyat, 34 years old, jobless: *"For the social protection funds they are charging us 180 Somoni. We don't know whether it is right or wrong. We are not given an explanation why. But we want to know. Do we have a right to know or not?"*

Boboi Zainiddin, 80 years old: *"1 ha land is mine and 0.5 ha is from my son. For 1.5 ha, we pay 260 Somoni in one year season to the chairman of the Mahalla, and he transfers all collected funds to the head of Dehkan Farm Salim Ibrokhim but for improvement of lands no fund is ever allocated for that purpose."*

In addition, there are questions about the land area – not always a clear distinction between arable and pasture grazing land. Often, PRA participants noted that an assumption at the time of redistributing land from the former collective farms was that those people who were villagers and members of a collective farm were therefore entitled to a share. But in reality, this was not so. Many villagers who might otherwise be entitled to land missed out because of their absence at the time. Absence on the day of allotting land for any reason (labor migration to neighboring countries, moving to another village, disease, or in any other valid reasons) resulted in loss of land use rights without the possibility of transferring it to their family.

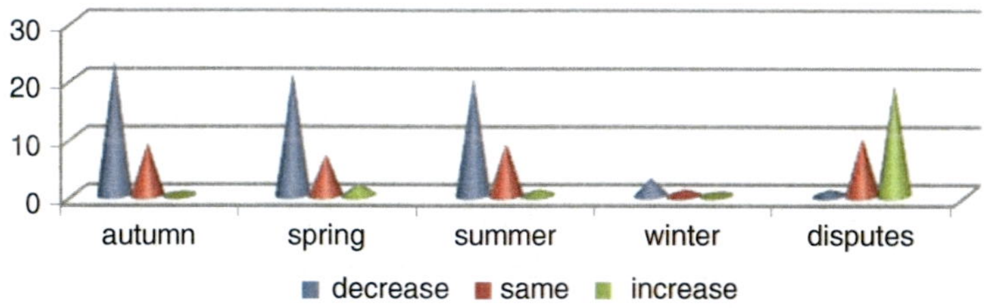

Fig. 7.7 Opinion about likely land disputes relating to access to pasture by 2020

Their land shares were redistributed among those already allotted land, that is, who were present in the village at the time of land allocation.

Often discussants noted the injustice of land distribution.

Abdukahor, 40-year-old Hojiboi: "*All lands privatized by dehkan farms. What can us – poor people do? The day will come when we find we don't have land to graze our livestock. How to establish better husbandry and increase the number of the livestock? For this reason all our options are becoming more limited day by day.*"

It was also noted that there had been cases where the shareholder pays rent to the head of the dehkan for what is his own land. This confirms that members of the farms (shareholders) often do not clearly understand the implications of the transformation from collective to dehkan farm and do not have an opportunity to use their rights.

Hojiboi village, 49-year-old male: "Many numbers of village people are counted as shareholders, but why don't they inform the people? The head of Jamoat complains that we do not pay tax, but for what kind of land should we pay tax? Who requires payment from us that we refuse to pay? If information would be provided to us about that we are shareholders then we would follow up according to that rule or information."

The discussions showed that the real-life situations in the next 10 years will tend to increase land disputes among the population and decrease access to pastureland (Fig. 7.7).

The land area allotted to the majority of shareholders is small and is located on the slopes of the foothills near the place of residence. It cannot be irrigated which limits the possibilities for increasing productivity. On the one hand, there is a law creating an investment authority to relate to dehkan farms; on the other hand, there are limitations and barriers to exploit the full potential of these small farms.

The process to receive the land user certificate is a very long and bureaucratic process. Sometimes, this process takes not only months but years, for example, in village Obisangbur, Faizabad District. This example (see below) is not unusual; the same situation can apply in other villages.

> During the Soviet era, Obisangbur village land was in the territory of Faizabad Sovkhoz. Up until 2006, it was a collective farm of Faizabad and almost 100% of the population worked there. In 2006, this collective farm was divided into 5 dehkan farms. In the territory of Obisangbur, dehkan farm "Salim Ibrohim" (farm's name) was created. For Obisangbur village, 400 ha of pasture was given. At that time, out of 46 farms, only 21 farmers had certificates. Now, out of 21 farmers, there are 11 individual dehkan farmers approved by the decision of the Chairman of Faizabad Hukumat, and only 2 farms have certificates for land tenure. The remainder is in the process of completing documentation. During the Soviet era, Faizabad Sovkhoz had winter pasture in Dangara district of Khatlon, but today these pastures are not available and they are privatized by other individuals.

Each level of PRA participants had a different attitude to changes in land use right, tenure and land reform. Several participants mentioned that unfair allocation of pasturelands and high rents is the main barrier to taking a more conservation-minded approach to land use. The fact that all higher-quality lands are already in the hands of principal shareholders and bureaucrats means that nothing will change. Other participants believe that they also have user rights for higher-quality land. The primary users believe that they will strengthen their privileges as the government policy favors "privatization." Some villagers believe that in the future, they can lose their user right, especially those who have not yet received the certificate.

2.7 *Patterns of Information Exchange and Awareness*

Discussions with villagers showed that most of the information they get is from each other. The mosque, where the men of the village make a prayer service, is important. At the end of a prayer service, in the process of dialogue and conversation, information is exchanged. Also, the process of exchange of information proceeds during celebration holiday, when people visit each other as guest, at weddings, or other traditional events (funerals).[3]

[3] Funerary practises are not only a way to pay the last respects to the deceased and to soothe the affliction of the bereaved; they are collective events during which the social identity of the living is affirmed.

Usually in rural areas from October to April each year, for many years all over the country, there is a limited provision of electricity, and villages have electricity 2 h in the morning and 2 h in the evening. This has a direct effect on the limitations in obtaining any information.

In the village, dissemination of new information on important issues is a very rapid process. Every village knows about what happened. Everybody knows who the primary users of pasture around the village are. But information about user rights and how to deal with bureaucracy to receive a certificate is not widely known.

Discussions with villages also showed their understanding that all the issues related to land use rights can only be solved through local authorities – Jamoats at the village level and Hukumats at the district level – but if those bureaucracies have a vested interest in maintaining the status quo, then they are unlikely to be too helpful.

The PRA confirmed that there is a need to spend a lot of time and patience in order to receive a certificate (even when the claim for user rights is clear). As mentioned above, many villagers are waiting to receive the land tenure certificates several years after lodging their applications (e.g., Obisangbur, and Kandak). Sometime, villagers despair and are pessimistic that they can ever receive land tenure rights, because most of the quality land has been distributed to higher bureaucrat officials already. The majority of arable and pastureland was transferred to large dehkan farms (approximately about 60–70%). Most villagers have small household plots (300 m^2) and small areas (<1 ha) of additional land for use for crops, hay, or grazing (presidential land) and access to land of primary users on a rental basis.

3 Attitude Toward and Perceptions of Land Degradation

PRA results confirmed that most villagers know about the changes – the loss of fertility of the land from year to year, significantly changed climatic conditions due to dry summers, and abundant mud flows in the spring.

Yoqub, 43 years old, driver: "There is very high snow in winter. Landslides are frequent. Recently, due to big floods, several heads of small livestock and two boys died. We couldn't do anything."

Amriddi, Kalai Nav: "Before sending livestock for grazing we are giving them two bundles of hay early morning and after coming back in the evening we are giving them two bundles of hay again. If not do that, my wife complains about the reduced quantity of milk."

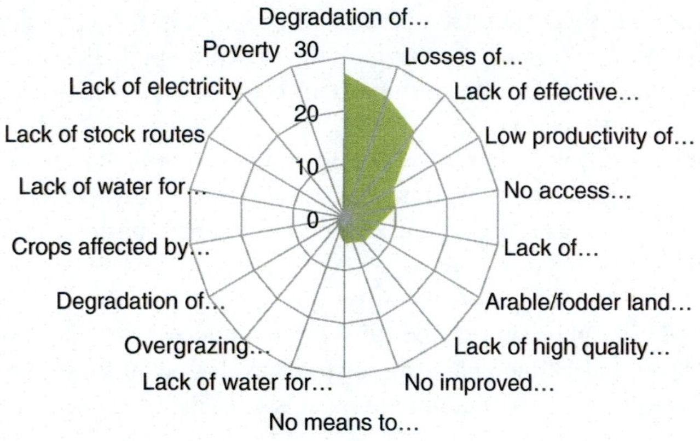

Fig. 7.8 Priority issues identified by PRA participants

Without exception, the overall condition of the pastures was believed by all groups in all villages to be in varying states of degradation, with the type and quality of pasture grasses reducing, weeds increasing, top soil being lost, rodents increasing, flooding and damage caused by droughts, wind storms increasing, and water springs drying (Dashti Marzo). Nine villages reported the disappearance or marked decrease of medicinal herbs and useful grasses and shrubs. Livestock graze all day but come back hungry because there is so little forage on offer.

Among the 18 priority issues identified in the PRA, 90% of participants indicated that the degradation of pasture and forest lands was the most significant. The reasons for degradation are marked: overgrazing (around village settlements), intensive trampling of soil along the path by herds of watering livestock during the day, limited access to winter and summer pastures, the lack of activities for the improvement of pastures, and discontinuation of the practice of rotational grazing (Fig. 7.8).

In the main, villagers and representatives of dehkan farms did not express particular concern about the real situation. During PRA discussions, some participants expressed their understanding that if they do not take into account the rapid pace of land degradation that after 10 years, the situation of the degradation will be much worse. To determine the importance of this issue, we asked leading questions, for example, the conditions of grazing lands in the Soviet Union time 20 years ago, in the period after the Civil War, at the present time, and what to expect in 10 years. The process of land degradation coincided with decreased productivity of pasture, decreased productivity of fruit trees, and perennial grasses an increase in biomass of poisonous plants resulting in worsening health of livestock through toxicity and a reduction in the yield of useful forage. Almost 100% of PRA participants noted that the condition of pastureland 20 years ago was much better. They are now considerably deteriorated, and after 10 years, the situation is likely to be even worse (Fig. 7.9).

Concerning "pasture restoration" – villagers understand that it is necessary but argue that it can only happen if there was large-scale irrigation, mechanized planting, and use of chemical fertilizers. Doubtless a legacy of Soviet-style thinking as the concept of small-scale self-help projects is not widely accepted.

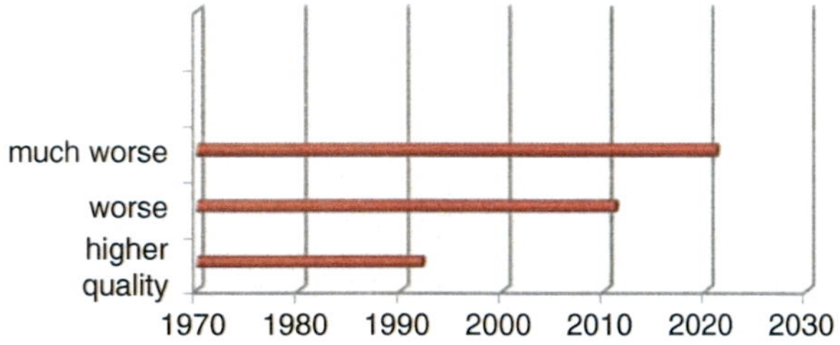

Fig. 7.9 Perceived change in pasture conditions over the past 50 years

In general, with significant benefits from the use of pastureland, users of large, private farms do not have enough experience and enough understanding to take responsibility to find ways to take action against the continuing process of land degradation. Because, not having absolute confidence in the future that they may lose rights to land, they are more interested in getting more and more benefits and profits from land today. In this regard, activities to reduce land degradation in rural areas are almost nonexistent. Many respondents referred to the fact that these restoration activities are very expensive and that people do not have such funds. Only in extreme situations is there an attempt to apply the most urgent interim measures. There is a clear understanding of reality and consequences of land degradation, but it is not possible to use expensive techniques to reverse the situation. In implementation of action, the entire community as a whole must be involved in an integrated program. Such an action requires awareness rising of the damage cost of land degradation, a clear definition of what techniques and procedures will work, and motivation of HHs to get involved. Development of the program requires commitment and a clear understanding of the importance of timely start of work in this aspect. It is unlikely to happen while primary users exploit the situation, get rich, and fail to do anything to arrest land degradation. Government intervention to improve security of land tenure (for all those with user rights for a start) would go a long way to provide the enabling environment. No one will want to spend time and money on land restoration when there is no certainty that access to the improved land will remain with those who did the repair work.

4 Livestock

Of the 2,730 households in the 10 PRA village, between 70 and 98% of HHs were reported to own at least some livestock. Those who did not own livestock were thought to be either too poor or to be newly formed household who did not have access to land. Herd sizes ranged from 1 cow to large herds of several hundred sheep, with the average herd size reported to be 2–4 cows and 5–10 sheep/goats. Most HHs have at least one donkey for transport of agricultural produce and hay from the fields to home. Horses are less common as they are expensive to keep. Although the PRA originally attempted to estimate animal numbers for each area of pasture, this proved impossible in the time allowed for the following reasons: the distribution of pastures between dehkan farms themselves varied greatly in most

Table 7.3 Estimated numbers of livestock owned by villagers

Village	Cattle	Sheep/goats	Donkeys	Horse	Total	Beehives	LS units[a]
Kalon	790	4,597	258	6	5,651	134	1,973
Kosatarosh	1,086	5,800	135	56	7,077	8	2,437
Khushon	N/A	3,600	N/A		3,600	800	720
Tiloi Safed	800	2,000	19	15	2,834	36	1,234
Burma	891	2,015	105	352	3,363		1,751
Hojiboi	547	5,300	130	50	6,027		1,787
Obisangbur	400	630	40	–	1,070		566
Dashti Marzo	1,317	4,883	6	63	6,269	57	2,362
Kalai Nav	600	300	59	–	959	905	719
Kandak	678	2,212	220	56	3,166	45	1,396
Total					40,016	1,985	14,945
Without Khushon							14,225

[a]LS unit crudely calculated as 1 cow, 1 horse, 1 donkey = 1 LS unit, 1 sheep/goat = 0.2 LS unit as used in Tajikistan

village, with many farms having a small amount of pasture and a few large farms having a lot; livestock numbers fluctuated seasonally; pastures are used jointly by herds from other villages and Jamoat within the same *district* and, in some cases, from other *district*; but most importantly, people did not always have an accurate knowledge of the situation in pastures other than the ones they used – and even then the information between groups using the same pasture was often contradictory. Nevertheless, the total number of livestock for each PRA village according to the official Jamoat statistics for 2010/2011 was recorded, and where this was unavailable, community estimates were used. Jamoat statistical data did not always coincide with date obtained from the villagers. However, it would be misleading to attempt to draw conclusions on this livestock/pasture ratio as the data on livestock distribution across the various pasture tenure arrangements and seasons is not known in detail. In addition, pasture quality is not known for each pasture area (Fig. 7.10 and Table 7.3).

4.1 Health Status

Health status of livestock depends of conditions in which livestock are housed and fed. Discussions in PRA showed that most of cattle in pens are sick during winter. During winter, animals in pens are fed a ration mainly composed of hay and various additives. Mostly they do not have enough feed, and coupled with the low nutritional value, loss of liveweight is inevitable, and this is reflected in the general health of livestock.

In spring time, a lot of livestock from the transhumance area move from other districts from winter to summer pastures through the pastures of PRA villages.

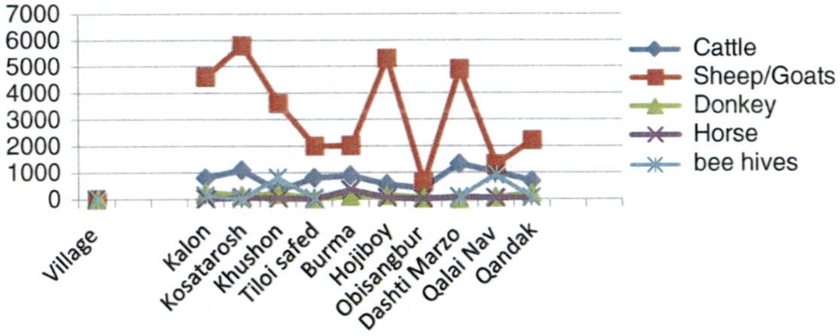

Fig. 7.10 Number of livestock in villages

In theory, transhumance livestock are meant to follow specifically designated stock routes away from villages where there are places for livestock rest and lodging for night. But usually these transhumance herds pass across village land because of (a) bad conditions of bridges and roads, (b) occupation of the transhumance place by individuals through permission of local authorities (who collect a fee), and (c) privatization of land on the former stock routes. As a result, many livestock from villages began to be infected with diseases or infested with parasites brought by the traveling stock that pass through without veterinary inspection.

"...When the village community of Obi Sangbur is going to stop the shepherd's transhumance for changing their itinerary, they immediately inform the owners of herds through mobile phone. The shepherds are telling that the owners of the livestock which we are pasturing are from "influential individuals" from outside the district. *After phone call immediately will be given permission and nobody can stop them.* ***Musoeva R. Obisangbur****: the big transhumance herds coming from Yovon, Jirgatol, Qabodiyon, Qurghonteppa, and Shahritus go by the top of our village. Last year the herds went near our house. This year, it has gone by the other side of the village".* ***Orifova:*** *"The truth will not be gained. Our words are not worth even one diram (coin). At that moment they immediately call the owners of herds and get permission. The shepherd tells us if you want, talk to them by phone. The community can't do anything against them."*

Firdavs 37 years old, teacher. *"Transhumance – this is the government service. We can't change anything. However, herds from transhumance come to our village (during move from winter to summer pastures) at night and during one day they stay here. Herds eat all young grass, trample down all around and damaging the roots of this young grass. After these herds leave our village, the land condition is terrible. How can we feed our livestock? Where we can graze? What should we do?"*

By the way, PRA participants gave examples from Soviet Union time when prevention of infectious diseases of livestock was the norm, and veterinary inspections of traveling were subject to strict regulation.

> ***Taghoeva Anor, 57 years old:*** *"Before the civil war many village women were working in the farm of Devonabegi village. There was a foot bath in the entrance of the farm which was supplied with medicines. Every day before entrance to the farm the livestock went through it and there was no hoof rot or oral infections. In advance there were vaccinations for all diseases. Now no measures will be taken against any other diseases."*

In addition to deteriorating livestock health, there are other problems relating to the conditions of grazing. Usually livestock are grazing on pasture near village on daily basis from March to April until the first frosts of autumn. There has been over severe grazing on the same pasture, far exceeding their carrying capacity. Instead of 1–2 sheep/ha there are now 15/ha. As a result, grass does not have time to grow. Additionally, there is a problem with watering the livestock. Usually for watering, the cattle return to the village, after watering and rest, they go back up to the pasture. In the evening the herd returns home not quite full and tired. All this walking and lack of feed affects the animals and contributes to poor livestock health.

4.2 Small Business Activities

According to PRA results, every household keeps poultry. During meetings with villages and visits of their household plots, we saw approximately 10–15 chickens in every garden. Twenty percent villages are farming turkey, for example, Burma and Kosatarosh.

Beekeeping is also one of the profitable sources of income from several villages in mountains area in 70% of total villages. Most income from beekeeping has come from villages like Kalai Nav and Khushon. Their bees receive nectar from flowers – from spring and summer pasture located in mountains area. Villagers of Kandak started beekeeping not long ago. Women would like to learn about beekeeping and would attend properly conducted training courses if they were available.

Feed supply. Results of PRA confirm that all 10 PRA villages are laying in of fodder for winter fattening. Laying in of fodder is collection from haymaking pastureland. But this laying in of fodder cannot provide enough fodder for winter feeding. For the rational feeding of livestock, in general without exception, all households purchase an additional concentrated feed, hay, and oats.

In general, very little fodder is planted by any of the PRA villages, and hay is cut from natural, unimproved pasture grasses. Four villages only grew hay. Others grew corn (6), alfalfa (4), and esparcet (*Onobrychis* spp.) (2) but only in very small amounts according to participants. The predominant crop residue is wheat stalks

Fig. 7.11 Future predictions for fodder production trends

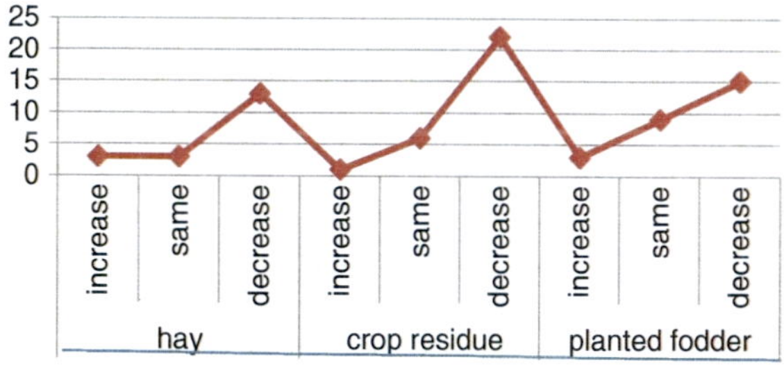

and wheat bran. Various villagers also mentioned barley (1), potato (3), flax, fruit peels, and vegetables and leftover human food. This was generally chopped, mixed together with hay, and salt was then added. Only one village (Tiloi Safed) produces silage from wheat stalks, alfalfa, and corn.

Recent experience with attempting to increase fodder production has not been profitable for the few households who have reportedly tried. Lack of irrigation and low-quality seeds were thought to be the main reasons for failure. This is not to say that there are no examples of successful fodder crop production. People generally think in terms of the large Soviet-style operations and tend to discount examples of smaller, individual areas planted, such as in Khushon's enclosed garden plantations on forestry land along the riverbed. Fodder production may therefore be underreported by communities.

> ***Jurabek.*** *"It is autumn for the shepherds that is the most important season, as well-fed animals are a guarantee of a happy winter. At this time it is necessary to make high-quality feed stocks. If now it is hard to do, what will happen in 10 years?"*

Very few groups admitted to selling any fodder but clearly they did (see Sect. 5.1.2 below). According to the PRA survey, when asked directly, only a few individual households and the big DFs were thought to sell fodder. However, in the economic survey, people in 15 of the groups (including non-DF groups) reported deriving at least some income from the sale of hay. Owners of larger areas of hay land "lease" it to shareholders or nonshareholders and charge 15–30% of the hay collected which is either provided in cash or crops and is sold to cover the cost of land tax plus profit.

Very few farms however are thought to be self-sufficient in fodder. Naturally, those with the most and best hay land and arable land come closer to producing their own fodder needs. In general, all farms buy at least some fodder, particularly for lactating cows, with hay being the most common (despite it being mostly produced on farms, the majority appear to have limited user rights to hay land and therefore have to buy at least some), followed by cotton/oil cake, bran, oats, alfalfa, silage, corn, and barley.

Figure 7.11 shows the trends of future fodder production – the opinion of most respondents was that hay production, the amount of crop residue, and the area planted to fodder will decrease.

4.3 Selling Livestock

Basically, buying, fattening, and selling cattle involved villages from Tiloi Safed. This is their constant trade and business. As reported by PRA participants during discussions, this is a good advantage. Ability to fatten cattle and sell – for this practice they learned from Soviet Union time, when they worked in a large livestock *Sovkhoz* Telman. People in other villages usually sell livestock only when there is an urgent need to counter various emergencies.

> *Zuhuriddin: "If I'm going to give my daughter to marry, or marry my son, of course I need a certain amount of money. Consulting with my wife, I decide how much we should sell livestock for in order to cover the expenses."*

5 Income and Expenditure

5.1 Income Sources

We tried to analyze the methods used to identify the allowed level of estimated economic contribution of the household. Initially the participants themselves have identified sources of income: livestock, crops, pasture/fire wood, and employment. The results showed that the group in different categories, as well as in various villages in the proportion of income, is not the same. In general, most of the population receives income from their own seasonal employment and from remittances sent from relatives working in Russia – 34%. In the villages Obisangbur and Kalai Nav, the figure was more than half of revenues – 57%. For some categories of groups, this figure in village Burma (poorer farmer group) was 80%, and in village Dashti Marzo (poorer farmer and women's groups), it was 68 and 60%, respectively.

Next importance to income generated was the income received from livestock – 33.3%, food, tree, and crop production – 25.4%; just a small part of household cash income is obtained from pastures – 7.3%.

5.1.1 Income from Livestock

Among farmers groups, the highest rate of income occurred in village Dashti Marzo (59%); even among poorer farmer groups, a higher proportion of income (51%) in Burma village was from sale of livestock or their products. Totally in 50% of villages, the proportion of income from livestock in poorer farmer group has higher rate than in farmer group. Data show that the farmers do not always derive the higher proportions of income from livestock, although the analysis should

Fig. 7.12 Images from an exercise on estimating proportions of sources of household income and mapping in the PRA survey in (**a**) Dashti Marzo, (**b** and **d**) Kandak, and (**c**) Kalon villages

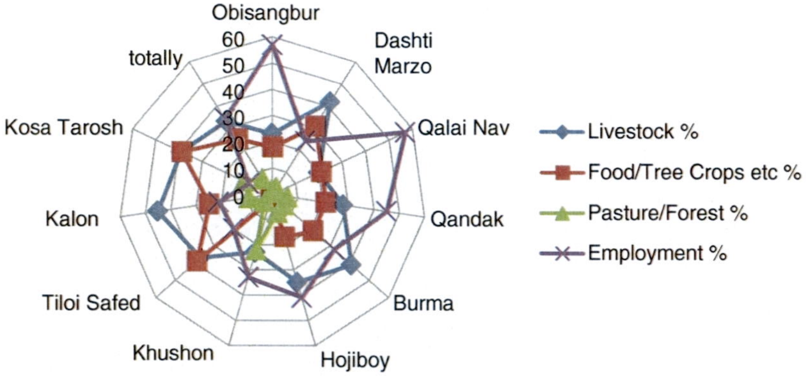

Fig. 7.13 Contribution of various economic sectors to HH income (%)

Table 7.4 Comparison of the contribution of livestock sales to income in selected PRA villages

	Proportion of household income derived from livestock %		
Name of village	Farmer group in Khushon shepherds group	Poorer households	Women group
Kalon	41	47	33
Kosatarosh	44	45	45
Khushon (Kokhu)	20	28	21
Tiloi Safed	43	47	26
Burma	39	51	32
Hojiboi	46	48	19
Obisangbur	36	17	15
Dashti Marzo	59	32	37
Kalai Nav	25	10	23
Kandak	42	18	23

take into account the fact that the proportional relationship does not mean that the nominal incomes of the poorer categories exceed the nominal income of farmers (Figs. 7.12, 7.13 and Table 7.4).

5.1.2 The Income from Pasture

The income from pasture in general was hay, wood, beekeeping, medicinal herbs, and wild rose. This proportion of income was usually very small (>8%), but in some villages like Khushon – this proportion was as high as 24%. Most incomes are from beekeeping and collection of medicinal herbs. In Khushon (Kokhu), Kalon, Kandak, and Kosatarosh, 80% of the HHs reported significant income from selling hay.

6 Constraints and Barriers to Better Land Stewardship

The results of the PRA have provided a solid understanding of the situation throughout these five *districts* and of the community perception of the current problematic situation on pasture management and livestock. The main constraint relates to the uncertainties about access to grazing land and to security of land tenure. Legislation relating to issues of land use and tenure is in constant flux, and the measures to implement the existing laws are not consistent nor transparent. As noted above, on the one hand, the government aimed at resolving land use issues, and on the other side, the process of implementing these laws, especially those relating to registering claims for user rights, constitutes significant barriers. According to the Constitution of the Republic of Tajikistan, land is the exclusive property of the government, and land use rights to it can be assigned by the government (Halimova, Chap. 13). This suggests that existing laws are not sufficiently comprehensive or consistent enough to deal with problems that arise on a day-to-day basis.

It is a fact that many large state and collective farms in the process of reorganization after independence went into the possession of a few individual and collective dehkan farms. The majority of the best pasturelands are in this category (user rights to large tracts of pastureland already assigned to a very small portion of the population). According to government statistics, over 73% of Tajikistan's population are living in rural areas. Most rural households have 300 m^2 household plots, but not many of them have the presidential land. At national level, about 44 thousand hectares, of which 25,815 ha irrigable, were granted to 409 thousand rural households as supplemental plots per 0.15 arable or 05 ha rainfed based on president's decrees in 1995 and 1997. According to statistics of State Land Committee of December 2010, by 2010, about 409,000 HHs have presidential land based on formal statistics. Many of the PRA respondents are very poor. It should be noted that the main barrier to better land stewardship is *poverty.*

The environment, economic development, and poverty – these are three concepts that are closely interrelated. The main part and especially the rural population depend on environmental factors. It is damaging water sources, natural disaster – droughts, landslides, and desertification. In addition to natural processes and phenomena which are to some extent the result of irresponsible human activities, mismanagement of land and natural resources is generally an obstacle to further development of agriculture. Cutting down of trees and perennial shrubs, overgrazing

Fig. 7.14 Barriers to better land use

on the same pastures, improper watering, and lack of maintenance of technology impedes progress. The degradation of pasturelands, the lack of watering, and diseases of animals and plants are not a complete list identified in the course of PRA, to some extent related to the environment, marked by all in the pilot area.

PRA participants rated the other barriers to sustainable land use as the degradation of pastureland – 9%, lack of effective local institutions – 7%, limited access and lack of pasture – 3%, the shortage of water in the pasture – 1.5%, and overgrazing – 1%. Recurrent themes from the PRA survey, across all villages, were:

- Lack of a clear awareness of land user rights
- Lack of perception of the nature, extent, and full implications of the inequitable distribution of pastureland
- Perception that there was a negative impact of large dehkan farms on both for pasture management and livestock productivity
- Lack of local institutional support on pasture management, animal husbandry, and livestock health (Fig. 7.14).

7 Human Welfare and Poverty Alleviation

7.1 Human Welfare

The government has identified the 12 most important priorities for achieving the Millennium Development Goals in the long term and fostering social and economic development, including poverty reduction[4] – environmental sustainability was one of these 12 priorities. So far, there is little evidence of progress in this. Results from the PRA surveys showed widespread overgrazing, inefficient use of pastureland, and limited access to drinking water for pasturing livestock, especially in summer pastures. These factors lead to changes in vegetation cover and as a consequence

[4] The most important documents of the Government and International Development Partners, aimed at structural reforms – the Government Investment Program, 2010–2012 Poverty Reduction Strategy

environmental instability. Socioeconomic reforms aimed at creating an enabling environment for human well-being and development also receive priority.[5] The degree of human development in society is determined by the Human Development Index (HDI), which reflects the life expectancy, literacy coverage of education, access to resources (clean water, land), and opportunities to live a long and healthy life, to be educated, and to have enough material prosperity. Human Development Index is a qualitative measure of the degree of the country's progress in human development – a planned growth rate of GDP and average income. HDI is also determined by the degree of poverty. The higher the score, the lower the poverty of human rights and the greater the level of literacy and opportunities to decide how to live. Reduced access to services such as education, health, and social welfare and to resources such as land and water is often associated with lower HDI scores. Despite the increase in government spending on social services, progress in macroeconomic indicators, and the significant contributions of donors, the HDI score in Tajikistan of 0.58 places it among the lowest of 112 countries with a medium level of human development.

7.2 *Poverty Alleviation*

The process of transition to a market economy has caused changes in the socioeconomic system. Under these conditions, it became inevitable that there would be a stratification of society into rich and poor.

Material well-being of families is one of the factors in the development of society. It is determined by the basic vital factors such family income, access to health care, access to education, occupation, housing conditions, and availability of the property.

The National Development Strategy (NDS) of the country until 2015, along with in-depth analysis forecast increase in living standards and poverty reduction. Despite the fact that the poverty rate fell from 80% in 2000 to 72% in 2003 to 47% in 2009, the World Bank figures[6] showed that one-third or 34% are extremely poor, and Tajikistan remains the poorest country in Central Asia. The number of people in poverty is significantly affected by the global finance crisis. Families in rural areas survive mainly on remittances sent by relatives (father, sons, husbands, brothers) who work for wages in neighboring countries. Remittances as a proportion of household income have increased, and families survive on the meager income derived from livestock, crops grown on their small plots of land, and any arable and pastureland they can use and by small business (such beekeeping) (Table 7.5).

---

[5] Millennium Development Goals, National Development Strategy, the Framework Program UNDP on the provision of development assistance to Tajikistan in 2010–2015 years.

[6] http://data.worldbank.org/country/Tajikistan Poverty headcount ratio at national poverty line % of population.

Table 7.5 The percentage of vulnerable categories of the total number of household in the PRA villages

District	Village	Total poor % (income of one person is from 60 to 100 TjS/month)	Including very poor % (income on one person is <61TjS/month)	Woman-headed household – women % (includes category of vulnerable poor)
Varzob	Kalon	98	83	2
	Kosatarosh	30	3	6
Vakhdat	Khushon	80	51	9
	Tiloi Safed	64	38	17
Rudaki	Hojiboi	54	30	5
	Burma	50	10	5
Faizabad	Obisangbur	42	13	2
	Dashti Marzo	66	37	15
Rogun	Kalai Nav	48	21	3
	Kandak	51	12	6

TJS = Tajik somoni In 2012 $1USD = 4.8 somoni

7.3 *The Vicious Circle of Poverty*

Tajikistan and other Central Asian countries suffer from a growing divide between the rich and the poor that is causing social conflicts. The income gaps between urban rural residents, different social classes, and different regions both within countries and between countries in Central Asia *sens. lat.* are all widening. There are millions of people living below the poverty line (about $2USD per person/day). So the challenge is to improve the current poverty alleviation policies (and actions) to fit the new economic conditions created by 20 years of independence and the shift to the market economy.

Governments need to pay more attention to comparative poverty, which is more serious than absolute poverty, because it is a matter of social justice. This means a change in how we define and understand poverty. Poverty is more than just low income and weak consumption ability. It is also an inability to change one's situation for the better, as it also means poor education opportunities, inadequate healthcare, unstable employment, and poor housing. In other words, poverty alleviation efforts should focus more on services and opportunities for the poor and wider social insurance (pensions and concessions). More investment is needed to create opportunities for both poor people and the impoverished regions because in Central Asia the problem of poverty is related to place – some oblasts are notoriously poorly endowed with natural resources, lack comparative advantage over other regions, and are poorly served by the state.

Therefore, in the coming years, each state should reform its poverty alleviation policy and expand its measures so that not only more people are lifted out of poverty but more people are prevented from falling into it. There should be a combination of poverty alleviation measures with national income distribution and redistribution policy. With a widening income gap, the state has no real choice. It must ensure that income distribution is fairer.

There are a great number of people in the world living slightly below the poverty line, who cannot get any government assistance, although their conditions are not much better than those living below the poverty line. There are people living on the edge of poverty, because their opportunities and personal development options are the same as those living below the poverty line.

It is responsibility of the state to prevent these people from falling into the ranks of the poverty-stricken population The experience in China and in other countries shows that poverty is vicious cycle, as children living in poverty-stricken families are more likely to be poor and deprived as adults, and their own children are likely to grow up poverty-stricken. This further curbs the mobility between different categories, which is already weak.

The various governments of Central Asia at all levels should give support to ensure that the poverty-stricken population has and can participate in social affairs, after all they are citizens and have the same right a fulfilling life as their rich compatriots.

8 Defining the Role of Women

Before any explanation of women's role, it would be appropriate to describe the social status of rural women. Social relations between male and female permeate all spheres of life. It is necessary to understand the reasons why there is inequality.

According to prevailing stereotypes, the rural woman is mostly occupied by a household duties within her plot, her traditional role, and her duty – cooking; caring for children, the elderly, and sick relatives; cleaning; washing; fetching water; gather firewood; working in the garden; sewing; and other activities. On average, women spend 6 h per day on household work. Those who do work outside the home face discrimination. From PRA information, it is clear that some women, who work in the collective dehkan farms, do so at a salary level that is lower than that of male workers doing the same job.

Typically, rural women are limited in many ways especially level. In general, they complete secondary education and the education process finishes at this time (15–17years of age). In the PRA, from the total number of women attending, 96% completed only basic school education, 3.9%, the overall secondary, and only 0.9% undertook higher education.

In everyday life, women spend time in the daily care for livestock in the stall-feeding, cleaning, milking, production of dairy products, treatment of livestock, and supervision of grazing in the pasturelands. Women's role in the process of decision-making for the care of livestock, the purchase, or sale of livestock is variable. In the majority of cases reported in the PRA, decisions were taken jointly but in some cases solely by men.

During the transition to a market economy, complicated social and economic crisis, civil war, and its consequences have changed the role of women. Women in Tajikistan more and more are engaged in search of livelihood. This is mostly due to the increasing number of women widowed by the civil war and the role of male population (between 16 and 60 years) as migrant labor in Russia and other countries to work either as permanent or seasonal labor. The absence of a male in households invariably increases the responsibilities of women for the livestock on the one hand, but on the other hand, the remittances sent back by the absent men provide funds for their families to pay for shepherding services and buy food for livestock, veterinary care, and other expenses.

More and more women are head of households and assume the role of breadwinner. Women took an active role in discussions and debates during the PRA. In general, female-headed households are risk prone and vulnerable to sliding into poverty. They are largely dependent on remittances, and they need all-round social development. Women have limited access to the productive land resources as well as to financial resources. When there was breakup of Soviet era farms, the area of land allocated to women was much less.

For the most part, this is due to traditional views, low level of awareness among women about their rights, as well as specific gender role stereotypes. The realities of life are such that women often have to bear the burden of both housekeeping and production for own consumption. It almost does not leave them much time or opportunities to address issues such as access to productive assets. These obstacles are difficult to be eliminated because they are depending with certain traditional values.

> ***Tojiniso, 45-year-old Khushon****: "Only by changing the mentality of men against women can we make a difference. Because my daughter-in-law is from the city of Dushanbe, she has a university education and began to work after marriage in the school. A man said something to my son about this, and he was forced to forbid his wife to work. She is with the higher education but sits at home, and men without higher education work in the school."*

More equitable distribution of the land use rights would not only improve the efficiency of production but could also serve as a starting point for creating their own business enterprise. Women are now dependent on their husbands for economic reasons; these options to improve their lot are limited or even impossible to realize.

According PRA each group was asked to list all of the activities undertaken by household members related to livestock husbandry. They were then asked to discuss who was responsible within the household unit for doing the work associated with each task, as well as making decisions about the task, such as when and how to do it. This joint analysis was done to encourage greater acknowledgement of women's role in livestock husbandry and ideally their greater participation in decision-making when eventually forming the proposed pasture user groups[7].

[7] One of the objectives of the rural development project and the PRAs was to facilitate the creation of pasture user groups within each village as a means of rationalizing pasture use and improve productivity and HH incomes.

The results of the gender analysis demonstrate that women are involved in every aspect of livestock care and management, with the only significant underrepresentation in vaccination, slaughtering, long-medium distance shepherding, and shearing. Vaccination may be problematic for women as it most likely means dealing with a male veterinarian, or it is done during longer distance migration of flocks and herds at remote veterinary points (if it is done at all). However, women are as equally active in treating sick animals as men. Women may be precluded from slaughtering, consistent with halal requirements, although there was an example from Kalon Village where women who took their cattle to *ailoq* also slaughtered animals on occasion.

Barfi 46, female: *"We are doing not only women's work but a man's work too. For example, there is a case when we slaughter the cow, also fence out wolves, jackals, and catch snakes. It means during this period we are converted to men, hunters, and butchers."*

The only activity done exclusively by women is milking and processing and sale of milk products. Interestingly, sale of milk products was one of the activities where five groups indicated that although women were exclusively responsible, men and women were equal in decision-making over where and when it is sold. Men dominated the sale and purchase of livestock, with only four groups indicating an equal role and another four indicating a secondary role; however, 14 indicated that although men had the exclusive role in physically buying and selling livestock, women participated equally in making decisions about when and what to buy and sell. Again, restrictions on associating with men may preclude more direct involvement of women in the market place. Five groups also indicated that men made decisions over treatment of livestock despite both men and women being involved in actual treatment. This highlights the need to provide information on treatment directly to women to ensure that they are better able to make decisions on when and how treatment is sought, paid for, and administered.

Tojiniso, 37 years, female. *"My husband a few years ago went to work in Russia. Initially he could not find work so, we managed here as best we could. Then, when he got a job, he has not had the opportunity to send us money each month. So I decided to start breeding livestock. I became more confident and solved many problems. Sale of livestock brought to my house more or less wealth. All the problems of livestock lie on my shoulders, I have a very hard time but what to do? Without a male in the house, I have to ask for help from neighboring males or my brothers when it comes to the purchase or sale or slaughter."*

The gender analysis in general indicates a fairly even contribution of men and women in day-to-day livestock management, especially around the household plot and nearby pasture areas. This would of course vary between families depending on the amount of male labor currently in Russia or working off-farm.

Additionally, every women's group mentioned that they do think about future life (their own and that of their children). They are concerned about poverty, and they would like their children to have access to higher education as a way to get qualifications as quality specialists and thus guarantee gainful employment in a worthy occupation.

9 Summary and Conclusions

Results of PRA showed that in these rural areas of the five *districts* that the quality of life of the population is dependent upon more secure access to productive land. The existing system of land use rights is not supportive of social and sustainable development. Large areas of better lands have been assigned to individuals or joint-stock companies with inheritable right to use. People from outside the Jamoat were successful in receiving rights to reserved pastureland. The perception of unequal and unfair land distribution was a recurring view expressed by all PRA groups.

The rights of the majority of small livestock owners have been violated. Violation of the principles of equality and social justice may adversely affect the access of poor rural population to land and challenge the efforts of government and aid partners (donors) poverty reduction strategies.

In the Soviet era, there were clear guidelines and a high level of awareness about what constituted effective use pastureland. Today's use of pastureland is not sustainable. Predictions about the future of the livestock sector are very pessimistic to the point that it can affect national food security. The principal reasons for the pessimism are:

- Underutilization of much productive pastureland within large privatized farms with few livestock.
- Limited access to pasture by owners of small numbers of livestock.
- High costs
- Increasing number of household (and therefore animals).
- Lack of concerted and coordinated efforts reverses land degradation in the pasturelands.
- Inadequate veterinary services and poor animal health.

At the same time, PRA villagers mentioned that crop farming is one of the good sources of income, but that arable land is limited in area and most is not irrigable.

Many poor HHs who are members of dehkan farms have access to only small areas of poor quality pasture that are often difficult to access, while large areas of better quality land were given to "influential individuals" as collective dehkan farms.

The village communities in every village surveyed have limited knowledge of their rights and entitlements and hope to receive external support. They understand that they do not have enough capacity now to independently implement better pasture and livestock management.

Finally, for regulation of all problems on land user rights especially pastureland, the following conclusions are relevant. There is a need to:

- Speed up the preparation and adoption of pasture law; this should provide the principles of sustainable land use rights of all, without exception, and acknowledge the specifics of the rural population and the existing ecological situation.
- Improve the legal awareness and transparency in the allocation land in the future in order to mitigate potential conflict as the process of land privatization occurs.
- Develop technology for pasture restoration and management.
- Develop the mechanism of pasture management through empowerment of the rural community on the use of public pastureland.
- Establish and strengthen the capacity of local public management institutions and community organization, for example, pasture user committees.

Acknowledgements The work reported here was generated during the ADB/GEF Rural Development project, and the author was engaged as a consultant by a consortium of GITEC Consult GMBH/Donaev Management Consulting under the direction and guidance of the PMU in Dushanbe. This chapter is the product of 11 months of intensive research, field work, and travel; thus, we have many individuals and institutions to thank for their advice. Thanks to the local research team, who worked under guidance of Donaev Management Consulting. I am thankful to them for their patience and goodwill during often difficult field trips. I am especially grateful to local officials in districts, Jamoats, and village levels mahalla committee for help and support. I am very grateful to Ms Helen Moriarty – international consultant – for continuous collegial and professional guidance and assistance at all times.

A special note of gratitude is extended to Dr. Victor Squires, Mr. Paul Strong, Mrs. Nargis Halimova, and other colleagues for advice and assistance during the preparation of this chapter.

Finally, our hope is that this work will contribute to the process of improving the system of land stewardship and living conditions, opening up more alternatives and opportunities for rural population of Tajikistan.

References

Halimova N (2012) Land tenure reform in Tajikistan: implications for land stewardship and social sustainability: a case study. In: Squires VR (ed) Rangeland stewardship in Central Asia. Springer, Dordrecht, pp 305–332 (Chapter 13, this volume)

Hua L. Zhang, D (2012) Engaging with land users, the first steps on a long road. In: Squires VR (ed) Rangeland stewardship in Central Asia. Springer, Dordrecht, pp 333–356 (Chapter 14, this volume)

Lerman Z (2012) Rural livelihoods in Tajikistan: what factors and policies influence the income and well-being of rural families. In: Squires VR (ed) Rangeland stewardship in Central Asia. Springer, Dordrecht, pp 165–188 (Chapter 8, this volume)

Robinson S et al (2012) Pastoral tenure in Central Asia: theme and variation in the five former Soviet republics. In: Squires VR (ed) Rangeland stewardship in Central Asia. Springer, Dordrecht, pp 239–274 (Chapter 11, this volume)

Part III
Protecting Rural Livelihoods: The Key to Land Stewardship

The three chapters deal with aspects of rural livelihoods mainly focusing on rangeland-based animal husbandry in Tajikistan. The situation in Tajikistan has many characteristics of that in other Central Asian countries.

Chapter 8 deals with rural livelihoods in Tajikistan and analyzes the factors and policies influencing the income and well-being of rural families.

Chapter 9 analyzes livestock management problems and policies in Tajikistan and draws comparison with experiences in other former Soviet republics like Azerbaijan and Georgia.

Chapter 10 provides an overview of rangeland-based livestock, a vital subsector under threat in Tajikistan.

Chapter 8
Rural Livelihoods in Tajikistan: What Factors and Policies Influence the Income and Well-Being of Rural Families?

Zvi Lerman

Abstract This chapter is an analysis of data derived from several studies of the economic effects of land reform on rural families in Tajikistan. The history of land allocation after the dissolution of the Soviet Union is briefly reviewed, and the implications for agricultural productivity and rural incomes are discussed. Enlarging family land holdings and improving productivity are the keys to raising rural family incomes, which is one of the most direct factors that mitigate vulnerability and poverty. Options for enlargement of family holdings and productivity improvement are outlined.

Family incomes today are strongly dependent on a single source, with 50–70% deriving from agriculture. The risks associated with this income strategy can be reduced through diversification of income sources, which requires strengthening nonfarm occupations, wage employment, and entrepreneurial activities. Risks associated with income variability can also be reduced by diversifying farm production between crops and livestock, on the one hand, and between a mix of different crops (cereals, vegetables, fruits), on the other.

Keywords Rural incomes • Livelihoods • Remittances • Consumption of own products • Productivity • Livestock • Land allocation • Land tenure • Vulnerability • Adaptive capacity • Rural households • Dehkan farms • Land reform • Freedom to farm • Smallholders • Horticulture • Diversification • Microfinance • Azerbaijan • Georgia

Key Points

- The typical farm in Tajikistan is small, cultivating a small land plot and taking care of a small number of animals. Agriculture suffers from low crop and livestock yields, as agricultural productivity is adversely affected by low levels of production technology.

Z. Lerman (✉)
Department of Agricultural Economics and Management,
The Hebrew University of Jerusalem, Jerusalem, Israel
e-mail: lerman@agri.huji.ac.il

V. Squires (ed.), *Rangeland Stewardship in Central Asia: Balancing Improved Livelihoods, Biodiversity Conservation and Land Protection*,
DOI 10.1007/978-94-007-5367-9_8,

- Land reform was the mechanism that distributed land to rural households and dehkan farmers after Tajikistan's independence. Land reform has had a mutually reinforcing twofold effect on rural incomes: more land leads to more income through increased production and consumption; more land also creates greater surplus and thus increases the commercial orientation of the households, generating additional sales revenue and further augmenting the income effect of land. Commercialization completes the loop between land reform and rural family incomes.
- Despite the positive income effects of land reform, Tajikistan's large rural population is still judged to be highly vulnerable to risks, including livelihood risks and food insecurity risks. By some measures, Tajikistan is the most vulnerable among all 28 countries in the World Bank's Europe and Central Asia region.
- Enlargement of land holdings, productivity improvement, and diversification of income sources are among the main factors that can be used to raise rural incomes, thus reducing vulnerability. Smallholder farms can be enlarged by redistributing inefficiently used land to more efficient users, returning unused and abandoned land to cultivation, and encouraging development of land markets as a mechanism that allows land to flow from less efficient to more efficient users. Significant productivity improvements can be achieved through a combination of hard (technology, agronomy) and soft (information, know-how) approaches, relying on the development of agricultural extension (advisory and consulting services). Diversification of income sources requires strengthening nonfarm occupations, wage employment, and entrepreneurial activities, all of which will reduce the risks associated with the currently observed reliance on agriculture as the main source of family income.
- Livestock is the most important farm resource after land and water in Tajikistan. The livestock herd is concentrated almost totally in hundreds of thousands of rural households, each with just a few animals. Sale of live animals and livestock products is an important source of rural household income, and the animal headcount is growing rapidly. Livestock productivity, however, is far from satisfactory, and milk yields in Tajikistan are the lowest among all CIS countries.
- Inadequate quantity and quality of animal feed may be one of the reasons for low livestock performance. Despite the increase in animal headcount, the area sown to feed crops declined precipitously after 1990 and the quantity of feed harvested also fell sharply (in 2007 it was merely 15–30% of the harvest in 1990). These changes are largely the outcome of government policies that impose production targets for wheat and cotton and in effect discourage or even prohibit allocation of land for feed crops.
- Pastures suffer from high stress due to the continued increase in the number of animals and general lack of sustainable pasture management practices. Reduction of areas in feed crops combined with shrinkage of pasture areas due to degradation has led to a sharp contraction of the feed base for both cattle and small ruminants. The rural households are forced to rely even more heavily than before on low-quality feed, obtained by grazing their few animals on the grassy verges of roads and canals and on postharvest stubble in the fields.

- Livestock productivity can be increased by paying more attention to availability of quality feed, by focusing on high-yielding varieties of feed crops, by adopting programs for pasture reseeding and rehabilitation, by breed improvement through artificial insemination practices, and by emphasizing all aspects of animal health and veterinary care.
- Higher livestock productivity will allow a slowing down of the rapid growth of the livestock herd without detriment to output and rural incomes. Smaller animal numbers will reduce pasture degradation and enable rehabilitation measures to be put in place. The management of common pastures and commonly herded livestock can be improved through the establishment of user associations dedicated to sustainable pasture management, including planning and monitoring of pastures and maintaining pasture infrastructure (e.g., water points, sheds, and roads). Measures designed to improve livestock productivity will increase the share of livestock production relative to crops, leading to a more balanced and less risky product mix in national agriculture.

1 Introduction

Tajikistan is judged to be highly vulnerable to risks, including livelihood risks and food insecurity risks for its large rural population (about 75% of the national total; TajStat 2011a). By some measures, it is the most vulnerable among all 28 countries in the World Bank's Europe and Central Asia (ECA) region (World Bank 2009).

An individual or a household is *vulnerable to risks* if these risks may result in a loss of well-being to a level below some threshold. The opposite of vulnerability is resilience. Vulnerability assessments usually rely on different combinations of geo-climatic and socioeconomic variables, which are always matched to the three defining dimensions of vulnerability: exposure, sensitivity, and adaptive capacity (Heltberg and Bonch-Osmolovskiy 2011). Exposure and sensitivity both act to increase vulnerability, while higher adaptive capacity mitigates vulnerability.

Exposure typically signifies the chance that assets and livelihoods will be impacted by risks or the likelihood that individuals will experience stress due to external factors – geo-climatic, environmental, or sociopolitical. Sensitivity indicates the susceptibility of assets and livelihoods when exposed to risk. Adaptive capacity signifies the ability to recover, prevent, or mitigate the effects of risks by deploying social risk management strategies (i.e., adjustments in assets, livelihoods, behavior, technologies, or policies). Both sensitivity and adaptive capacity are determined by socioeconomic variables that characterize the sustainability of agricultural production under conditions of uncertainty, the income levels, and the various endowments, including both physical and human capitals (Table 8.1). In the conventional livelihoods framework, the geo-climatic vulnerability variables loosely correspond to the natural capital, while the socioeconomic variables correspond to physical and human capitals.

Table 8.1 The effect of main socioeconomic variables on vulnerability

Indicators	Effect on vulnerability	Situation in Tajikistan
Income and well-being	–	Low but increasing
Poverty (also infant mortality, undernourishment, food insecurity)	+	High but decreasing
Debt and financial insecurity	+	Not critical
Agricultural land	–	Small holdings, land not transferable
Livestock	–	Small number of animals in each household, headcount increasing
Commercialization (share of production sold)	–	Low, underdeveloped
Population density: stress on land and water resources	+	Increasing (fast population growth)
Irrigation: stress on water resources	+	Poorly maintained, inefficient system
Water availability	–	Ample, from glaciers
Diversification of income and farm production	–	Underdeveloped
Educational attainment	–	Very high literacy levels

Source: For details and data, see Lerman (2011) (especially Chapter 5)

2 Rural Incomes in Tajikistan

Family well-being is the main determining factor for sensitivity and adaptive capacity. Well-being in turn is mainly determined by family income – both the level of income and the stability of income over time. Families with high and stable incomes are less sensitive to risk and are able to apply a wider range of coping strategies (i.e., have high adaptive capacity). Income creates wealth, and wealth provides a buffer layer that shields families from risks and adversities. Poor families are more sensitive to risk and have no resources to cope with adversity. Rural population appears to be more vulnerable than urban population due to lower per capita incomes and higher poverty rates.

Income generation requires resources. In rural families, where income largely depends on agriculture, resources are primarily land and livestock. To achieve good productivity, farming must also have access to machinery, purchased inputs (such as fertilizers and quality seeds), veterinary services, and extension information. Cash income is augmented through sale of farm output, and this in turn requires marketing channels. Adequate resources and farm services make it possible to maintain income generation at satisfactory levels, acting to reduce vulnerability and mitigate poverty. On the other hand, constraints on availability or use of resources restrict income generation and thus increase vulnerability and poverty.

Labor is another resource, which is obviously crucial for production. In Central Asia with its large families and high birthrates, labor is usually plentiful and cheap

Table 8.2 Structure of rural family income from different surveys (percent of total income for rural families)

	LSMS (2003)	LSMS (2007)	Helvetas (2011)
Wages	35	31	34
Own farm	48	49	38[b]
Remittances	9	13	16
Social transfer	7	5	4
Other[a]	1	1	9

[a]"Other" includes land rental, nonfarm business income, and subsidies/grants for education; in the Helvetas survey of dehkan farms, nonfarm entrepreneurial activity accounts for a relatively large share of 6% of total income
[b]Does not include the value of own products consumed in the household

and does not constitute a constraint. The situation may be somewhat different in Tajikistan, where many able-bodied men migrate to work outside their home village (often in Russia or other countries) and the labor force in some villages is reduced to women, youths, and pensioners. In this setting, families may experience labor shortages (especially in seasons when migrants are not at home), which in turn may lead to increased vulnerability. However, the negative effect of migration on production resources (the labor force) may be offset by the positive effect of cash remittances from migrants on rural family incomes (see Box 8.2).

Agriculture is an important source of family income in Tajikistan. Although official statistics do not publish the structure of family income by sources, they indicate that based on household surveys (TajStat 2010c, p. 112–113), the household plot – which is the small family farm cultivated by every rural household – accounts for almost 30% of per capita income for the rural population (urban household produces much less agricultural output). More detailed information on the role of agriculture in household income emerges from cross-sectional surveys conducted by various donor organizations, often in cooperation with TajStat – Tajikistan State Committee for Statistics. The World Bank's Living Standards Measurement Surveys (LSMS 2003, 2007), carried out with nationally representative samples of rural households, indicate that income from the own farm accounts for nearly 50% of total family income for rural households (Table 8.2). In a recent survey conducted by Helvetas[1] (March 2011), admittedly with a nonrepresentative sample of only 400 dehkan farmers, sale of farm products contributed nearly 40% of the family cash income. This does not allow for the value of own farm products consumed in the household, which may increase the share of agriculture to 50% or even 60% of total family income. Despite these differences, the income structure is fairly consistent across different surveys. In response to a specific question in the survey, 70% of respondents indicated that agriculture was the main (and in some instances the only) source of family income (Helvetas 2011).

[1] Helvetas – Swiss Association for International Cooperation was founded in 1955 as the first private organization for development cooperation in Switzerland. Helvetas operations in Tajikistan are sponsored by SDC – Swiss Agency for Development and Cooperation (see http://www.helvetas.tj/en/).

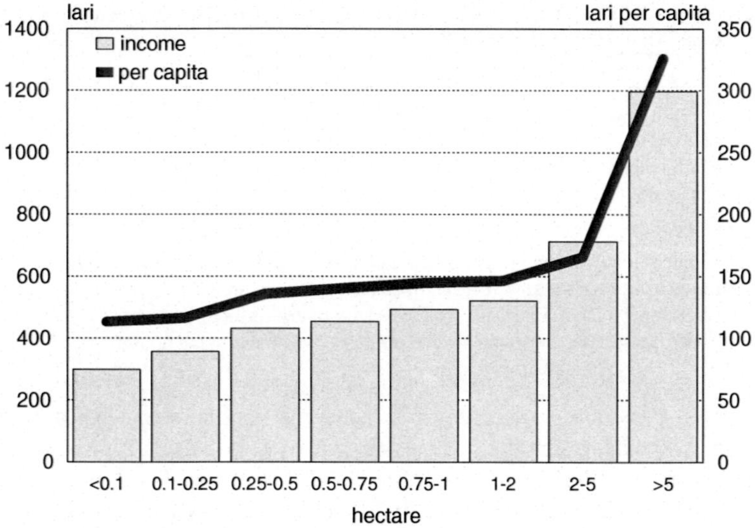

Fig. 8.1 Family income and income per capita increase with farm size (Source: Georgia Household Survey 2009 (Georgia 2009))

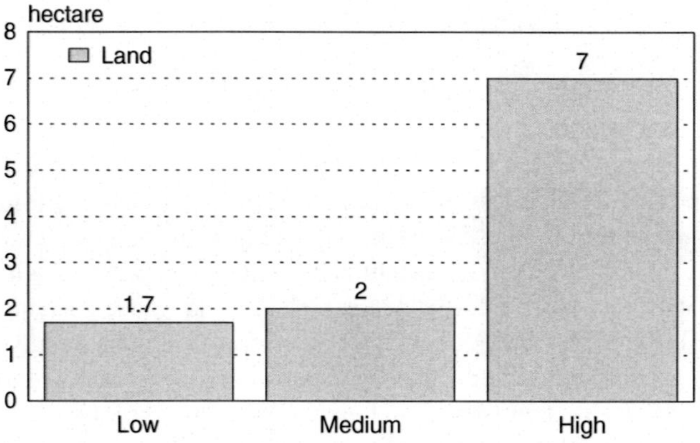

Fig. 8.2 Perceived family well-being increases with farm size (Source: Azerbaijan World Bank survey 2003 (Lerman and Sedik 2010))

3 Land and Commercialization Increase Family Incomes and Well-Being

Evidence from all CIS countries conclusively shows that per capita family incomes and family well-being increase with the increase of the land allotment in family farms. Figure 8.1 shows the effect of farm size on family income and income per capita based on the 2009 household survey in Georgia (Georgia 2009). Figure 8.2 based on a 2003 World Bank survey in Azerbaijan (Lerman and Sedik 2010) shows

Table 8.3 Farm size and share of farm sales in family income for different levels of well-being

Well-being level	Farm size (irrigated land, ha)	Share of farm sales in family income, %
High: comfortable consumption regime ($n=60$)	6.0	42
Low: able to purchase food and daily needs only ($n=46$)	3.5	30

Source: PPCR farm survey, May 2011 (Lerman and Wolfgramm 2011)

that families with more land (7 ha) are more likely to perceive their well-being as "high" compared with families that have only 1.7–2 ha. Evidence for Tajikistan from the May 2011 PPCR farm survey[2] (Lerman and Wolfgramm 2011) also demonstrates that farms with more land attain a higher level of well-being (Table 8.3, first column): respondents who report a high level of well-being (income sufficient to sustain a comfortable consumption regime) have 6 ha of irrigated land, compared with 3.5 ha for respondents reporting a low level of well-being (income sufficient to purchase food and daily necessities only). Table 8.3 incidentally focuses the attention on the importance of irrigation in Tajikistan's semiarid climate: productive farming requires water, but the engineering infrastructure in Tajikistan is not always adequate for uninterrupted delivery to the fields. A different view of the advantages of larger size in small family farms is provided by the recent Mercy Corps survey (May–June 2011) in Rasht, where households with more land (1.5 ha) were observed to be debt-free while households with smaller land endowment (0.8 ha) had to borrow (Mercy Corps 2011).

Farm size has a direct effect on family well-being by increasing food production. Part of farm output is consumed in the household, thus improving the family's food security and providing income in kind. But farm size also has an indirect effect on family income: larger farms are more likely to sell some of their output because they produce a larger surplus after satisfying the family's food needs. Sales of farm products bring in cash revenue, which increases the family's available income. Figure 8.3 based on a small survey in one of Azerbaijan's western districts (Yalcin-Heckmann 2010) indeed shows that the frequency of "commercial" farms (those producing crops for cash sales) increases markedly with the increase of holdings, while the frequency of pure subsistence farms correspondingly decreases.

The positive effect of sales on family income has been observed in several CIS countries; in particular, farm surveys in Moldova and Georgia show that households selling some of their output ("sellers") have higher income than households where the entire output is consumed in the family ("non-sellers"). The results for Moldova (Fig. 8.4) actually show that sales revenue accounts for the entire difference between the income of "sellers" and "non-sellers." The results for Georgia (Fig. 8.5) explicitly allow for the farm size dimension: "sellers" have larger farms than "non-sellers"

[2] PPCR – Pilot Program for Climate Resilience in Tajikistan (February–August 2011). The article draws on the findings of a farm-level survey conducted as part of the PPCR activities. For more details, see Lerman and Wolfgramm (2011).

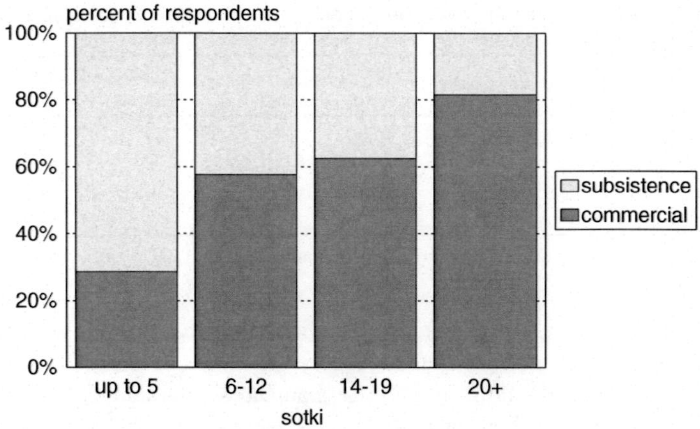

Fig. 8.3 Frequency of "commercial" farms increases with the size of holdings (sotki is a measure of land area equal to 0.10 ha) (Source: Yalcin-Heckmann 2010 (Azerbaijan, Tazakend, $n=77$ households))

Fig. 8.4 Moldova: structure of family income for "commercial" and "noncommercial" households (Source: WB/FAO baseline survey, 2000)

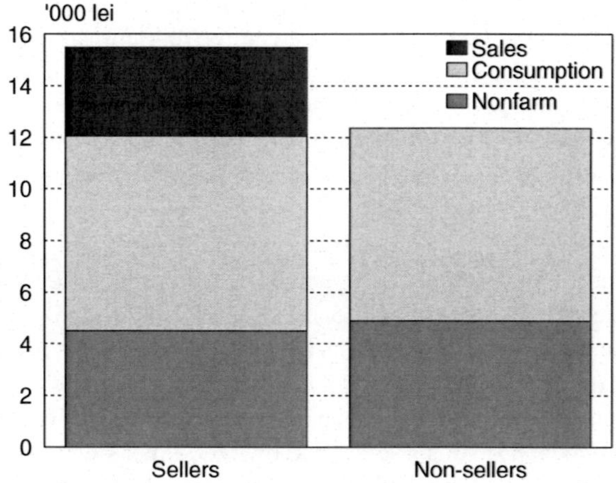

(2 ha compared with 1 ha), and their base family income is higher even before adding the sales revenue. For Georgia, we thus have a clear demonstration of the twofold effect of farm size: more income due to more production (even without sales, simply through increased consumption of own farm product) plus an additional increment due to revenue from the sale of surplus. A similar effect is observed for Azerbaijan, where both sales activity and size of holdings act to increase family well-being (Yalcin-Heckmann 2010). Figure 8.6 first shows the difference in the level of well-being between subsistence farms, i.e., "non-sellers," and farms that sell at least some of their output: among subsistence farms, 60% report their well-being as "bad" (leftmost column, dark-gray band), while among commercial farms, the corresponding percentage is substantially lower (the two right-hand columns). Furthermore, commercial farms reveal a clear size effect on family well-being: among small

Fig. 8.5 Georgia: structure of family income for "commercial" and "noncommercial" households. Average farm size: 2 ha for "commercial," 1 ha for "noncommercial" (Source: USAID/Hebrew University survey, 2003)

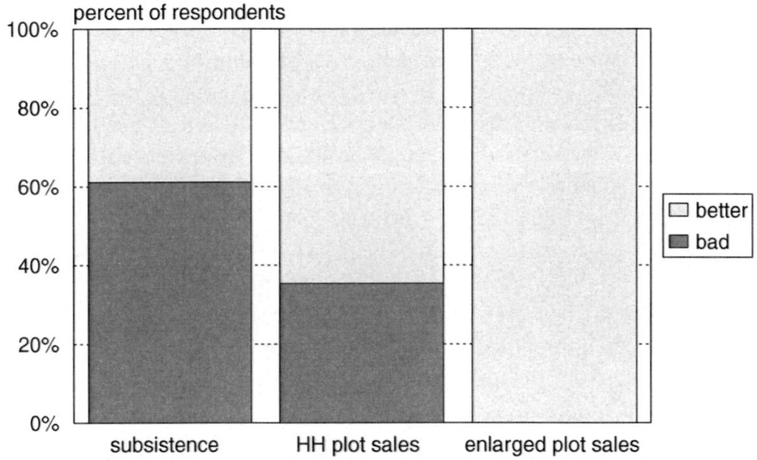

Fig. 8.6 Azerbaijan: higher level of well-being is observed more frequently for farms that sell (leftmost column for "non-sellers," the other two columns for "commercial") and for farms that have more land (farms in rightmost column larger than farms in middle column) (Source: Yalcin-Heckmann 2010 (Tazakend, $n = 77$ households))

commercial farms (selling only from the household plot), 35% report their well-being as "bad," while among larger farms (selling from both the household plot and the additional allotment received through land-share assignment), none of the respondents describes their well-being as "bad."

For Tajikistan, the May 2011 PPCR[3] farm survey (Lerman and Wolfgramm 2011) also shows that greater commercial orientation is associated with higher levels of well-being. The positive effect of commercialization on well-being is demonstrated in Table 8.3 (second column), where respondents who fall in the "high" well-being

[3] Explained in footnote 3.

category earn a higher share of their family income from farm sales than respondents in the "low" well-being category (42% compared with 30%).

Land reform in Tajikistan (and elsewhere in Central Asia) was the mechanism that distributed more land to rural households and farmers since independence (see Box 8.1). We can thus argue that land reform had a mutually reinforcing twofold effect on rural incomes: more land led to more income through increased production and consumption; more land created greater surplus and thus increased the commercial orientation of the households; commercialization created additional sales revenue, which further augmented and reinforced the income effect of land. In a sense, commercialization completes the loop between land reform and rural family incomes.

Box 8.1 Land Reform in Tajikistan with Special Reference to Arable Land

Land in Tajikistan is exclusively owned by the state, and it is given to farmers and households in use rights (legally conferred by a land use certificate). Instead of land privatization, Tajikistan has undergone individualization of agriculture – a shift to individual and family farming on state-owned land (Lerman and Sedik 2008). Prior to 1992, 95% of cultivable land was controlled by agricultural enterprises (collective and state farms), and 5% was in household plots – the smallholder family agriculture that persisted all through the Soviet era. The land reform that began in 1992 reduced the share of agricultural enterprises to just 15% of cultivable land by 2009, while the share of household plots increased dramatically to 20% (through land distribution from the state reserve in 1995 and 1997) and another 65% shifted to dehkan (or peasant) farms, a new organizational form that began to emerge after 1992. Dehkan farms and household plots combined accounted for 85% of cultivable land in 2009, up from just 5% in 1991. Most dehkan farms today are individual and family farms, as the number of originally created partnership (or collective) dehkan farms is rapidly shrinking due to the government's program reallocating land to individual farm members. The use of cultivable land in Tajikistan has been effectively individualized since 1991.

The allocation pattern for pastures is somewhat different: household plots do not have any pastures (only cultivable land); dehkan farms control about two-thirds of all pastures (roughly two million ha), and the remaining one-third is still held by agricultural enterprises (TajStat 2010a). It is impossible to say how much of the two million ha of pastures in dehkan farms has been transferred to individual and family use and how much remains in collective use.

Despite substantial expansion, household plots remain very small, averaging 0.3 ha (compared with about 5 ha on average for individual and family dehkan farms and 100–200 ha for partnership dehkan farms). There are 750,000 household plots in Tajikistan (UNICEF 2009) and only 50,000 dehkan-farms (TajStat 2010a). The increase of land resources in household plots has inevitably led to an increase of their share in agricultural production.

(continued)

Box 8.1 (continued)

While the share of agricultural enterprises in gross agricultural output (GAO) dropped from 65% in 1995 to less than 10% in 2009, the share of household plots soared from 35 to 65%. The remaining 25% comes from dehkan farms – the second component of the individual farm sector that started to contribute after 1997. Agricultural production, like land use, is now fully individualized in Tajikistan. Since household plots produce 65% of agricultural output on 20% of cultivable land, they are obviously much more productive than other farm types. Due to their high productivity, household plots are the engine of agricultural growth: they are responsible for the recovery of Tajikistan's agriculture, with GAO more than doubling between 1998 and 2009, despite the sharp decline in the output of agricultural enterprises.
(See also Halimova, Chap. 13)

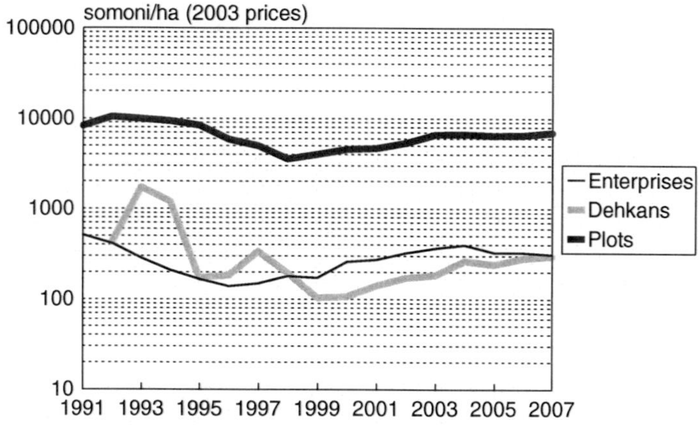

Fig. 8.7 Productivity of land by farm type, 1991–2007 (GAO per hectare of agricultural land, by type of farm, somoni per ha in constant 2003 prices, log scale) (Source: Lerman and Sedik 2008)

The small household plots, despite their high vulnerability due to limited land holdings, have managed to demonstrate exceptional adaptive capacity over time by achieving productivity levels that are orders of magnitude above those achieved by the larger dehkan farms (Fig. 8.7; see also Box 8.1, where small household farms are shown to produce much more than their share of cultivable land). They are also the main driver for agricultural growth: while Tajikistan's gross agricultural output doubled between 1997 and 2008, the output produced by household plots increased by a factor of 2.5, offsetting (together with dehkan farms) the shrinking production of agricultural enterprises (Lerman and Sedik 2008). Similar results are observed for all Central Asian countries and the rest of CIS (Lerman 2010). These achievements may be attributed to the well-known advantages of the family form of farm organization with its strong internal cohesion and accountability.[4]

4 Livestock

Livestock is the most important farm resource after land and water. In Tajikistan, the livestock herd is concentrated almost totally in rural households (Table 8.4): very few dehkan farms keep animals, and in aggregate they account for less than 3% of the value of livestock production in the country (compared with 41% of crop production) (see TajStat 2011b). The livestock herd is dispersed among a very large number of rural households, each with one to two animals. In the May 2011 PPCR farm survey (Lerman and Wolfgramm 2011), close to 50% of households fall in the category that WFP classifies as "vulnerable" by animal headcount (up to two heads of cattle). The small numbers of animals per household are naturally reflected in low levels of livestock production (milk, meat, fattened live animals) and low levels of livestock-related wealth, increasing the vulnerability of the rural households in this dimension.

The animal headcount increased sharply after 1998, rising from 1.3 million cow equivalents in 1997–1998 to 2.1 million cow equivalents in 2007 (Fig. 8.8, black curve). The household herd continues increasing because livestock is an important source of both food and income for the rural households. There is a ready cash market for live animals, while milk is easily sold to dairies or directly to consumers. All households periodically sell some of their live animals in the livestock bazaar or to intermediaries, but they always treat their herd as a store of value, carefully replenishing the stock to ensure continued growth of the headcount. Livestock sales represent 56%

Table 8.4 Livestock in rural households 2009

	Headcount in rural households	% of national headcount	Average per household[a]
Cattle	1,676.3	92	2.2
Cows	909.7	96	1.2
Sheep and goats	3,456.9	82	4.6

Source: TajStat (2010a)
[a]Based on 757,608 rural households (UNICEF 2009)

[4]Despite the exceptional productivity of household plots compared to other farm types, the average yields of cultivated crops in Tajikistan are generally below the CIS averages (CISSTAT 2010): for cereals 2.1 tons/ha compared with 2.5 tons/ha in other CIS countries and for raw cotton 1.6 tons/ha compared with 2.2 ton/ha in CIS. Tajikistan performs relatively well only in horticultural crops (potatoes and vegetables): 21 tons/ha for potatoes compared with 15 tons/ha in CIS. Tajikistan's predominantly smallholder agriculture is apparently better suited for cultivation of labor-intensive horticultural crops rather than broad-scale cash crops.

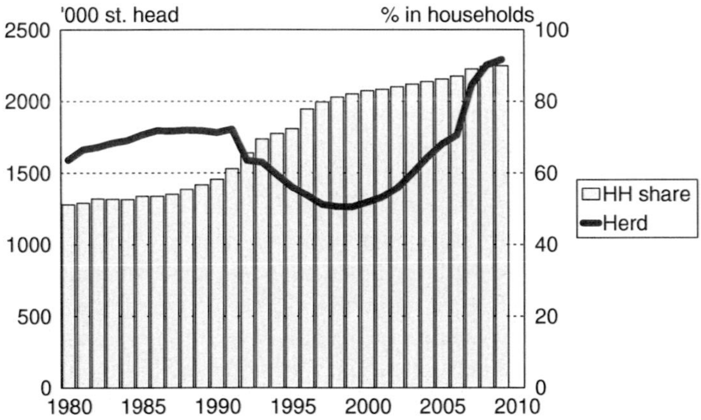

Fig. 8.8 Livestock herd, 1980–2009 ('000 standard head) (Source: pre-1990 from CISSTAT 2010; after 1990 from TajStat 2010a and earlier years)

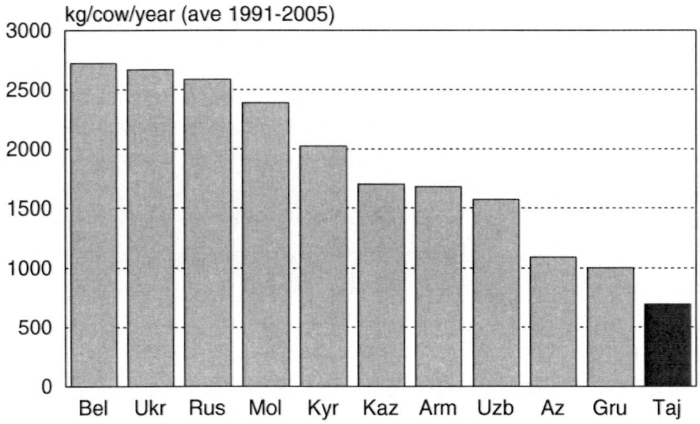

Fig. 8.9 Milk yields in Tajikistan and other CIS countries (averages for 1991–2005) (Source: CISSTAT 2010)

of total sales revenue from household plots, most of it (36%) from sale of live animals and the rest mainly from milk sales (Lerman and Wolfgramm 2011).

The performance of the livestock sector is far from satisfactory. Milk yields in Tajikistan are the lowest among all CIS countries, averaging 800 kg per cow per year (Fig. 8.9). Inadequate supply of animal feed may be one of the reasons for low livestock performance. Despite the increase in animal headcount, the area sown to feed crops declined precipitously after 1990, and the quantity of feed harvested also fell sharply (in 2007 it was merely 15–30% of the harvest in 1990). These changes are largely the outcome of government policies that impose production targets for wheat and cotton and in effect discourage or even prohibit allocation of land for feed crops. In principle, decrease in feed crops should have been compensated by increased grazing, but in reality the area of pastures has decreased by 300,000 ha (about 10%) since 1997 (TajStat 2010a). This was another outcome of government policies, which ignored efficient pasture management, contributing to unsustainable use and degradation of pastures. Pastures in Tajikistan suffer from high stress due to the continued increase in the number of animals and general lack of sustainable pasture management practices.

Reduction of areas in feed crops combined with pasture shrinkage indicates a sharp contraction of the feed base for both cattle and small ruminants. The rural households are forced to rely even more heavily than before on low-quality feed, obtained by grazing their few animals on the grassy verges of roads and canals and on postharvest stubble in the fields. Inadequate quantity and quality of animal feed have a negative effect on livestock performance. Lack of systematic breed selection and artificial insemination programs is another obvious reason for low milk yields due to poor genetic profiles of livestock in Tajikistan.

5 Policy Measures to Improve Rural Incomes and Livelihoods

The typical farm in Tajikistan is small, cultivating a small land plot and taking care of a small number of animals. Moreover, agriculture in Tajikistan suffers from low crop and livestock yields: agricultural productivity is adversely affected by low levels of production technology. The analysis of rural incomes and well-being in the previous section suggests that enlargement of land holdings and increased commercialization is among the main factors that can be used to raise incomes and improve livelihoods. Productivity improvements through a combination of hard (technology, agronomy) and soft (information, know-how) approaches should also make a significant contribution to better livelihoods. The following measures counteracting the effects of smallness can fundamentally improve rural livelihoods:

- Implement policies that will lead to enlargement of household plots and small individual and family dehkan farms – the most vulnerable segment in rural Tajikistan.
- Implement policies to increase agricultural productivity (technology, information).
- Implement policies that will improve the access of small farms to market services:
  - Services for marketing farm products (to facilitate commercialization)
  - Channels for purchase of farm inputs (including quality seeds)
  - Farm machinery services (rental and maintenance)
  - Veterinary and artificial insemination services
  - Extension services to raise the level of technology and know-how among small farmers
  - Credit services for small farms
- Implement policies to encourage greater diversification at the farm level.

6 How to Get More Land to Small Family Farms?

A clear policy prescription for increasing rural incomes and improving livelihoods is to enlarge the highly productive household plots with the object of increasing their contribution to Tajikistan's agricultural growth and productivity (see Box 8.1). In addition to supporting household plots, it is also necessary to enable individual and family dehkan farms to enlarge their land holdings, as world experience definitely proves that individual and family forms of organization have higher productivity than collective and corporate farms. Enlargement of smallholder farms can be accomplished in several ways.

6.1 Distribution of Inefficiently Used Land

The state land reserve in Tajikistan is less than 1% of arable land (land balance data for January 2010), which rules out another wave of land distribution for expansion of the small farms. However, 15% of cultivable land (nearly 130,000 ha) is still managed by agricultural enterprises (TajStat 2010a), which achieve relatively low productivity levels. In addition, a substantial area of cultivable land is held in collective ("partnership") dehkan farms, which are not more productive than the former Soviet-era *kolkhozes* that they succeeded. Land in agricultural enterprises and collective dehkan farms is a large hidden reserve that may be as high as 30% of Tajikistan's 1.1 million ha of cultivable land (Lerman and Sedik 2008). Making at least part of this land available for distribution to small family-based farms could further increase the productivity of the agricultural sector and increase rural incomes. Government initiatives promoting transformation of partnership dehkan farms into family farms since 2007 have already produced noticeable increases in the average size of small farms (TajStat 2010a). These and similar efforts for expansion of small farms should be broadened and intensified.

6.2 Development of Land Markets

Since the options for additional land distribution are inherently limited, it would be important to enable farmers to adjust the size of their holdings through land market transactions. Land markets allow land to flow from less efficient or inactive users to more efficient and productive ones, and the development of land markets will allow enterprising farmers to increase the size of their farms and achieve higher incomes. Given that all agricultural land in Tajikistan is owned by the state and thus cannot be bought or sold, the only feasible way for land markets to develop today is by allowing transferability of land use certificates – either temporarily (through leasing) or permanently (through selling). This option is included as one of the proposed amendments in the new Land Code approved by the government of Tajikistan as recently as June 2012 (Halimova, Chap. 13; Robinson, Chap. 11). Safeguards ensuring that distressed smallholders cannot be pressured into giving up their land to more powerful land users will have to be introduced as land markets develop. These safeguards may restrict transactions in agricultural land to bona fide farmers, thus precluding land accumulation in the hands of rich investors; they may ensure that no household remains landless through bankruptcy or forced sales by allowing distressed families to keep at least their household plots.

6.3 Return of Unused Land to Cultivation

Another option involves identification of currently unproductive or unused lands that have a potential for being upgraded to productive use. To implement this option, unproductive and unused lands should be fully inventoried at the village level and earmarked for distribution to small farmers for productive cultivation. Such practices have already been implemented in Tajikistan: examples are listed in the online database of World Overview of Conservation Approaches and Technologies – WOCAT (www.wocat.net).

The process may require overcoming certain legal obstacles, such as permissions to convert pastures into orchards. Farmers willing to invest in rehabilitating degraded land should benefit from incentives, such as tax credits, exemption from the higher tax applicable to orchards, and access to low-interest microloans or grants for investment in conversion.

7 How to Improve Livestock Productivity?

As noted, livestock accounts for 56% of total sales from the household plot (Lerman and Wolfgramm 2011). Given the importance of livestock production for rural incomes, it is essential to achieve higher animal yields than so far. Measures for improving livestock productivity may include the following:

- Greater attention to feed sufficiency, including development of high-yield varieties of feed crops and rehabilitation of pastures (reseeding, fencing, adoption of pasture rotation schemes, gully rehabilitation)
- Provision of more watering points for animals in grazing areas
- Attention to animal health through modernization of veterinary services
- Improvement of animal breeds through artificial insemination, including breed selection for both higher yields and greater tolerance to local climate

Higher livestock productivity will enable to slow down the rapid growth of the livestock herd without detriment to rural incomes. Smaller animal numbers will reduce pasture degradation and enable rehabilitation measures to be put in place. It is necessary to improve the management of commonly used pastures and commonly herded livestock through the establishment of user associations engaged in sustainable pasture management, including planning and monitoring of pastures and maintaining pasture infrastructure (e.g., water points, sheds, and roads).

8 How to Improve Farm Services?

While larger farms almost automatically have greater commercial orientation, the willingness to sell should be supported by ensuring access to functioning market services, especially channels and mechanisms for selling farm products. Improvements in other farm services – input supply, machinery (rental and maintenance), extension, and credit – will lead to more efficient production and generate higher incomes. Renewed focus should be placed on extension services: this is the key to knowledge transfer and human capital development. As such, extension services are central for increasing crop and livestock yields and thus raising agricultural productivity.

Best-practice world experience suggests that farmers' service cooperatives provide the most effective way of improving the access of small farmers to market services. Such cooperatives can cover the whole field-to-market value chain, including joint purchase of farm inputs, organization of machinery pools for field work, establishment of sorting and packing facilities, transport of farm products to markets, processing, etc. Service cooperatives do not rule out private initiative: private trade intermediaries, integrators, and service providers should be allowed to coexist with service cooperatives and continue their currently developing operations. The main issue here is the attitude of the government. It has to undergo a radical change from neglect and disdain of household plots and small farmers to full recognition of the huge role that small farms play in Tajikistan's agriculture. Government officials and decision makers have to acknowledge the contribution and importance of small farms, abandon the traditional preference for large farms, and focus on policies that ensure a supportive market environment for successful operation of the small-farm sector. This change of attitude will require a comprehensive "reeducation" effort in all ministries and should probably be guided from the very top.

9 Diversification of Income and Farm Production

More land, improved livestock productivity, and greater commercialization will allow farmers to achieve higher levels of income and better livelihoods. Yet income flows are always prone to variability and therefore risky. Diversification is a standard risk-reducing tool in economic theory and financial practice.

Diversification should be practiced on two levels: (a) diversification of income sources to reduce income risks and (b) diversification of the agricultural product mix to reduce production risks. Farmers in Tajikistan diversify both their income sources and their production. However, in both dimensions diversification is still not sufficient.

Table 8.5 Impact of remittances from migrant labor on household income

Wealth indicator	Families without migrants	Families with migrants	1 migrant	2 migrants	3 and more migrants
Has a car, % of households	29	68[a]	63	83	83
Expenditure on fuel, somoni/year	1,390	2,240[a]	2,000[b]	2,700	3,000[b]
Number of animals, st. head	1.7	2.3	2.1	2.8	3.2

[a]Difference between families with and without migrants statistically significant at $p=0.1$
[b]Difference between families with 1 migrant and 3 or more migrants statistically significant at $p=0.1$

Box 8.2 Remittances from Migrants Help Build Household Wealth

Shahtuti Bolo is a village of 569 inhabitants in Hakimi Jamoat, Nurabad District, RRP,[5] located at the end of a single mud track some 5 km from the Jamoat center, 40 min by a mudslide-prone track from the main communication route between Dushanbe and Garm. A small survey designed to collect data on energy usage prior to implementation of a rural energy efficiency program was conducted in spring 2011 among all 58 households in the village. The survey unexpectedly provided interesting insights into the role of remittances from migrants working abroad to the well-being of the village population.

Three-quarters of the households in the village (44 out of 58) have someone working abroad (basically in Russia). The number of migrant workers is generally 1–2 per household, but 10% of the households in the village report from 3 to 5 migrants. A total of 67 people work abroad, which constitutes 23% of the working-age population in the village (aged 16–60). According to informal interviews in the village, the migrants usually work abroad from March to October and return home for the winter months.

Family incomes today are strongly dependent on a single source, with 50–70% deriving from agriculture (Lerman and Wolfgramm 2011). Diversification of income sources is generally achieved by accepting wage employment outside agriculture and by expanding entrepreneurial activities. Wage employment (including remittances from family members working abroad as migrants) constitutes already now a significant component of family income (see Table 8.5, Box 8.2). Income from entrepreneurial activities, however, is so far negligible. Policy measures should be

[5]RRP stands for Raiony Respublikan skogo Podchineniya – Districts of Republican Subordination, a region consisting of 13 districts (raions) that extends from the border with Uzbekistan in the west to the Pamir region (GBAO) in the east. Unlike the other three major administrative divisions in Tajikistan (Sughd, Khatlon, and GBAO), RRP does not have a regional capital or a regional governor of its own: its constituent districts answer directly to the central government in Dushanbe.

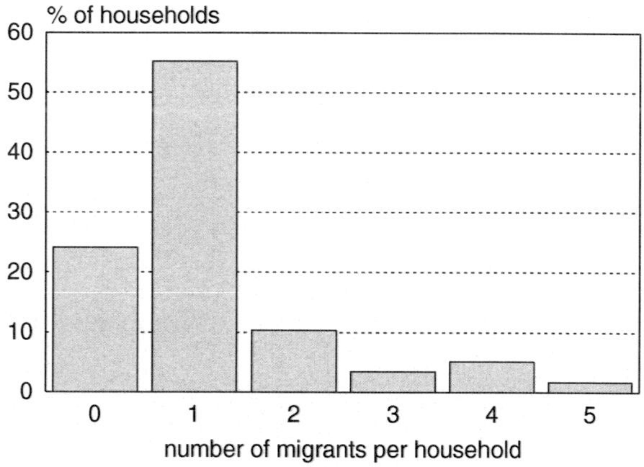

Fig. 8.10 Shahtuti Bolo: distribution of households by number of migrants working abroad

put in place to encourage development of off-farm activities in rural areas, including small-scale processing (dairy and meat, dried fruits, jams), cottage industries (carpet weaving, traditional arts and crafts), and small business initiatives (transport, trade, intermediation). Encouragement of off-farm activities requires an intelligent public awareness and education campaign, with information and support materials developed and provided by the government and NGOs; it may also require microfinancing with reasonably priced loans or grants, establishment of alternative credit mechanisms (e.g., credit unions as an alternative to commercial banks), and innovative tax measures to provide additional incentives.[6]

In terms of product diversification, Tajikistan's agriculture is 70% crops and only 30% livestock (TajStat 2010a). Only one-quarter of small dehkan farms engage in both crop and livestock production (Helvetas 2011; Lerman and Wolfgramm 2011); the rest produce crops, but no livestock. Tajikistan's agriculture thus bears an unbalanced risk due to the dominance of crop enterprises with their exposure to weather and climate change risks. Measures designed to improve livestock productivity (see above) will inevitably result in higher output and increase the share of livestock production in national agriculture, leading to a more balanced and less risky product mix.

There is a clear relationship between various household wealth indicators and the fact that at least one of the family members works abroad. Families with migrants are more likely to have a car; families with migrants can afford to spend more on

[6] There is no clear information on the borrowing capacity of rural households in Tajikistan. While survey results published by various NGOs (WFP, IPC, Save the Children, Mercy Corps) paint a picture of considerable financial stress in rural households, official statistics based on a representative sample of 11,600 rural households indicate that just 2.4% of respondents reported any borrowing (TajStat 2011c). Among dehkan farms, debt is mainly concentrated in relatively large cotton-growing collective dehkan farms, while the smaller individual and family farms are without significant indebtedness. Overall, it seems that rural indebtedness in Tajikistan is not a critical factor that should seriously constrain future borrowing for purposes of diversification and productivity improvement.

Fig. 8.11 New houses being built by returning migrants from Russia. Jamoat Saied, near Shaartuz (Photographed by Zvi Lerman, 8 April 2011)

coal and on fuel in general; families with migrants have more livestock. Furthermore, all three indicators increase as the number of migrants in the household increases. The differences are statistically significant, except for livestock.

Wealth creation in Shahtuti Bolo is thus facilitated by the earnings of migrant workers abroad.[7] This is a positive effect of labor migration, but it is apparently achieved at a huge human cost, not least because of the discrimination and violence that Tajik migrants suffer in Russia and other countries where they go to work. It is interesting to note that, unlike the model of a Turkish "gastarbeiter" in Europe, Tajik migrants work abroad mainly during the summer months and return home in the winter. Many of them return permanently after a number of years, once they have accumulated sufficient wealth. Returning migrants contribute in various ways to the rejuvenation of their community, e.g., by building new modern homes, as is seen in the photograph from Jamoat Sayed near Shaartuz (Fig. 8.11).

[7] Yalcin-Heckmann (2010, p. 192) reaches a similar conclusion regarding the economic contribution of migrant remittances to rural development in western Azerbaijan (the village of Tazakend). In addition to supporting livelihoods, remittances also provide funds for investment, such as the construction of a regional wholesale market near Tazakend (2007), development of trade links in Russia for local herbs and vegetables, and promotion of domestic livestock sales.

Tajikistan never became a cotton monoculture in the Soviet era. Both cotton and cereals (mainly wheat) were always present in Tajikistan's crop mix, occupying in varying proportions up to 70–80% of total sown area (TajStat 2010a). The remainder was split between feed crops and horticulture (potatoes, vegetables, melons, fruits, and grapes), with area under feed crops shrinking significantly since 1980 and the area devoted to horticultural crops increasing with the progress of farm reforms after 1995. Here again, as with the crop/livestock mix, we witness basic diversification of crop production, but the diversification is not particularly pronounced: cotton and wheat dominate the cultivated area.

Labor-intensive horticultural crops are ideally suited for small farms, with their abundance of relatively cheap labor, which may in fact explain the growth in horticulture since 1998. Vegetables account for a much larger share of the cultivated area in household plots than in either dehkan farms or agricultural enterprises.

The specialization in cotton and wheat is an inherited feature of the government policies that prevailed until about 2008. During most of the period since independence, authorities in effect dictated the allocation of land to "strategic" crops and set production targets for cotton and wheat. Dehkan farmers were not free to decide what to grow, and failure to meet the targets could trigger the ultimate sanction leading to confiscation of one's land. To encourage further diversification of dehkan farms away from cotton and wheat, while strengthening livestock production and improving pasture management, the government should ensure strict compliance – at all levels – with the full intent of the "freedom to farm" provisions adopted in several rounds since 2007. These provisions release farms from production targets on cotton and wheat, eliminate administrative intervention in production and land allocation decisions, and allow farmers to decide where and how to sell their output (Lerman 2011). Implementation of the "freedom to farm" principles should enable small farms to maximize their relative advantage by specializing to a greater extent in labor-intensive horticultural crops, which are ideally suited for small farms with their abundance of relatively cheap labor. Relaxation of production constraints should also lead to allocation of more land to feed crops, ensuring that livestock is supplied with enough feed to maintain reasonable milk yields.

10 Summary and Conclusions

The rural population in Tajikistan is judged to be highly vulnerable to risk due to low incomes and high poverty levels. Empirical evidence demonstrates that land and commercialization increase family well-being and thus reduce vulnerability. This evidence suggests four policy recommendations for increasing family income and mitigating rural poverty: (1) enlargement of family land holdings, (2) improving livestock productivity, (3) increasing commercialization through improvement of farm services, and (4) diversification of income and farm production.

More land, improved livestock productivity, and greater commercialization will allow farmers to achieve higher levels of income and better livelihoods. Yet income flows are always prone to variability and therefore risky. Diversification is a standard risk-reducing tool in economic theory and financial practice.

Diversification should be practiced on two levels: (a) diversification of income sources to reduce income risks and (b) diversification of the agricultural product mix to reduce production risks. Farmers in Tajikistan diversify both their income sources and their production. However, in both dimensions diversification is still not sufficient.

Acknowledgements This chapter is based in part on research carried out in the framework of the Pilot Program for Climate Resilience (PPCR) in Tajikistan, Phase 1, Component A5, Agriculture and Sustainable Land Management (SLM), February–August 2011. The author acknowledges helpful comments from the PPCR team in Dushanbe and Bern – Bettina Wolfgramm, Shane Stevenson, Julie Zaehringer, and Hanspeter Liniger.

References

CISSTAT (2010) Official statistics of the countries of the commonwealth of independent states. Statistical Committee of the CIS, Moscow

Georgia (2009) Integrated household survey 2009. Main statistics – standard of living. National Statistics Office of Georgia, Tbilisi. http://www.geostat.ge/index.php?action=page&p_id=176&lang=eng

Heltberg R, Bonch-Osmolovskiy M (2011) Mapping vulnerability to climate change. Policy research working paper 5554. World Bank, Washington DC

Helvetas (2011) Assessment of legal issues of farmers in Tajikistan and knowledge of their risks. SDC-Helvetas, Dushanbe

Lerman Z (2010) Agricultural recovery and individual land tenure: evidence from Central Asia. In: Ferto I, Forgacs C, Jambor A (eds) Changing landscape of European agriculture: essays in honour of professor Csaba Csaki. Agroinform, Budapest

Lerman Z (2011) Tajikistan's vulnerability to climate change. Discussion paper 7.11. Center for Agricultural Economic Research, The Hebrew University, Rehovot, Israel. http://departments.agri.huji.ac.il/economics/en/publications/discussion_papers/2011/index.htm

Lerman Z, Sedik D (2008) The economic effects of land reform in Tajikistan. Policy studies on rural transition 2008–1. FAO Regional Office for Europe and Central Asia. http://www.fao.org/fileadmin/user_upload/Europe/documents/Publications/Policy_Stdies/Tajikistan_en.pdf

Lerman Z, Sedik D (2010) Rural transition in Azerbaijan. Lexington Books, Lanham

Lerman, Z, Wolfgramm D (2011) Vulnerability to risk among small farmers in Tajikistan: results of a 2011 survey. Discussion paper 8.11. Center for Agricultural Economic Research, The Hebrew University, Rehovot, Israel. http://departments.agri.huji.ac.il/economics/en/publications/discussion_papers/2011/index.htm

LSMS (2003, 2007) Tajikistan Living Standards Measurement Survey. World Bank, Washington DC. http://go.worldbank.org/SPQ7R4D6S0

Mercy Corps (2011) Rasht valley food security, nutrition, and livelihood assessment. May-June 2011, Mercy Corps Tajikistan (December)

Robinson S, Wiedemann C, Michel S, Zhumabayev Y, Singh N (2012) Pastoral Tenure in Central Asia: theme and variation in the five former Soviet republics. In: Squires V (ed) Rangeland stewardship in Central Asia. Springer, Dordrecht Chapter 11, this volume

TajStat (2010a) Agriculture in Tajikistan 2010. Tajikistan State Committee of Statistics, Dushanbe

TajStat (2010b) Analysis of poverty in Tajikistan [in Russian]. Tajikistan State Committee of Statistics, Dushanbe

TajStat (2010c) Statistical yearbook of Tajikistan 2010. Tajikistan State Committee of Statistics, Dushanbe

TajStat (2011a) Demographic yearbook of Tajikistan 2010. Tajikistan State Committee of Statistics, Dushanbe

TajStat (2011b) Gender indicators of Dehkan farms 2011. Tajikistan State Statistical Committee, Dushanbe

TajStat (2011c) Survey of household plots in rural areas (November-December 2010). Tajikistan State Statistical Committee, Dushanbe

UNICEF (2009) Tajikistan standard of living survey 2007. Tajikistan State Statistical Committee and UNICEF, Dushanbe

World Bank (2009) Adapting to climate change in Europe and Central Asia. World Bank, Washington DC

Yalcin-Heckmann L (2010) The return of private property: rural life after Agrarian reform in the republic of Azerbaijan, vol 24, Halle studies in the anthropology of Eurasia. Lit Verlag, Berlin

Chapter 9
The Feed-Livestock Nexus: Livestock Development Policy in Tajikistan

David Sedik

Abstract This chapter discusses livestock development policy in Tajikistan in the context of the transition from intensive to extensive livestock husbandry since the late 1980s. The structure of feed demand and supply in Tajikistan in the postindependence period is discussed in order to understand the driving factors behind feed imbalances. Measures to address feed shortages with projections to indicate anticipated effects are analyzed. A comprehensive livestock development strategy for Tajikistan could employ these measures as cornerstones of a program for improving the feed-livestock nexus for Tajikistan.

Keywords Feed balance • Silage • Hay • Barn feeding • Dehkan farm • Feeding • Feed sources • Feed concentrates • Feed units • State farms • Collectives • Privatization • Animal productivity [and indicators] • Milk yield • Grain yield • Grain self-sufficiency • Crop rotations • Livestock inventories • Lucerne • Irrigated land • Yaks • Donkeys • Horses • Poultry • Feed requirements • Pasture Trust • Freedom to farm • *Jayits* • Pasture user groups • Russia • Kazakhstan • Kyrgyzstan • Azerbaijan • Moldova • Georgia

Key Points

- A three-tier system of animal feeding was the norm in Central Asia in Soviet times, and it rested on three pillars: (i) an elaborate organization for procuring animal feed for winter feeding based on intensively cultivated feed crops raised in large-scale state and collective farms, (ii) sizable imports of concentrates, and (iii) an organized system of pasture management and utilization, including pasture maintenance, transportation along established routes, clear assignment

D. Sedik (✉)
FAO, Food and Agriculture Organization of the UN, Rome, Italy
e-mail: David.Sedik@fao.org

V. Squires (ed.), *Rangeland Stewardship in Central Asia: Balancing Improved Livelihoods, Biodiversity Conservation and Land Protection*,
DOI 10.1007/978-94-007-5367-9_9, © Springer Science+Business Media Dordrecht 2012

of pasture rights, animal veterinary points, shepherd supply facilities along the routes, and an established schedule of transhumance pasturing. This system was dismantled after the break up of the Soviet Union, leading to the present situation of low productivity and dire feed shortages.
- Breaking the downward spiral of animal yields and poverty requires the gradual implementation of policy measures to address the feed shortage in the country. The first step toward formulating these policies is to understand the root cause of the imbalance between the supply and demand for feed.
- The main constraint on the development of the livestock sector in Tajikistan is an extreme imbalance between the supply and the demand of cultivated feed and concentrates for dairy and beef cows. First, demand for all-year winter and fall-spring pastures is much too high to be sustainable. Clearly, demand for these pastures needs to be limited in order to ensure sustainable use of these resources. The main users of these pastures are dairy and beef cows. In order to limit grazing of cows and cattle on these fields, the supply of cultivated feed and concentrates must be increased and made available to farmers. This will allow farmers to keep animals in barns for feeding rather than letting them feed in nearby fields.
- The mismatch between feed and animals is such that household farms raise only 6% of required feed on farm and must utilize the pastures of agricultural enterprises and dehkan farms or purchase cultivated feed from enterprises and dehkan farms. Agricultural enterprises and dehkan farms, however, have nearly five times the feed resources required to support their livestock inventories.
- The feed-livestock nexus is only one of a number of issues that should be addressed under a sustainable livestock development policy. Other issues, including the establishment of a viable plan for supplying livestock advisory and health services, a forward-looking livestock breeding policy, development of marketing channels for livestock products, livestock product safety, and environmental impacts of livestock, should also be part of such a policy.

1 Introduction

The end of the 1980s and the beginning of the 1990s signaled the deterioration of the socialist system of livestock production in all CIS countries. The changes that transpired during those years transformed the livestock husbandry system in many CIS countries, and certainly in Central Asia, from one based on intensive livestock farming to one based on extensive livestock husbandry. Intensive farming is an agricultural production system characterized by high inputs of capital, labor, or heavy usage of technologies such as pesticides and chemical fertilizers relative to land area. In the Soviet Union, including Soviet Central Asia, intensive livestock farming was conducted in large dairy and meat complexes where cultivated feed and purchased concentrates were fed to dairy cows, poultry, and hogs. The livestock production system that has emerged since 1991 relies primarily on grazing of livestock with limited feeding of cultivated feed and concentrates.

The transition from an intensive to extensive system of livestock production is not necessarily a bad thing. However, like any system, it needs to be managed sustainably. When the number of animals allowed to graze on pastures is not controlled, extensive livestock systems can lead to overgrazing and to a vicious cycle of inadequate feed and ever-lower animal yields, contributing to ever-lower returns from livestock husbandry. Because of the risk of a persistent decline in yields and hence rural incomes, the transition from an intensive to an extensive livestock production system carries a significant danger of pervasive and continuing rural poverty.

Breaking the downward spiral of animal yields and poverty requires the gradual implementation of policy measures to address the feed shortage in the country. The first step toward formulating these policies is to understand the root of the imbalance between the supply and demand for feed. Since animal feed derives from two main sources—cultivated feed for barn animals and pasture feed for grazing animals—the analysis of the feed demand overhang should address supply and issues for both types of feed. The heart of the feed-livestock nexus is in understanding the reasons for the excess demand for feed.

The feed-livestock nexus is only one of a number of issues that should be addressed under a sustainable livestock development policy. Other issues, including the establishment of a viable plan for supplying livestock advisory and health services, a forward-looking livestock breeding policy, development of marketing channels for livestock products, livestock product safety, and environmental impacts of livestock, should also be part of such a policy. However, the first key challenge of livestock policy is ensuring adequate and accessible supplies of feed for livestock. Feed adequacy is the first-level constraint on income from livestock husbandry. Previous studies have concluded that livestock product marketing and safety, advisory and health services, and environmental impact issues do not represent first-level constraints on farm incomes (Bravo 2005; World Bank 2007). Instead, most studies (e.g., Nolan 2005, 2006; O'Mara 2006; Bravo 2005; FAO 2009) conclude that the feed-livestock nexus is the most immediate problem for sustainable livestock development, though little has been written on this problem.

This chapter discusses livestock development policy in Tajikistan in the context of the transition from intensive to extensive livestock husbandry since the late 1980s. The structure of feed demand and supply in Tajikistan in the postindependence period is discussed in order to understand the driving factors behind feed imbalances. Measures to address feed shortages with projections to indicate anticipated effects are analyzed. A comprehensive livestock development strategy for Tajikistan could employ these measures as cornerstones of a program for improving the feed-livestock nexus for Tajikistan.

2 The Transition from Intensive to Extensive Livestock Husbandry in Tajikistan, 1991–2007

The socialist intensive livestock husbandry system in Tajikistan relied on three separate subsystems for support of livestock (FAO 2009, pp. 29–32). The first consisted of livestock inventories (predominantly milking herds) in large-scale enclosures on state and collective farms or complexes attached to industrial concerns. Along with these large-scale holdings went the livestock of employees (predominantly dairy cows) on individual subsidiary plots. The animals in this subsystem fed on hay, mixed feed, and cut feed all year.

The second subsystem consisted of livestock that spent the winter-spring period in enclosures and the summer-fall period in pastures. This was predominantly beef cows, beef cattle, animals of certain alpine regions without winter pastures, and animals in the majority of northern regions. This subsystem required 210 days of cultivated forage for feeding in large enclosures. These first two subsystems included the large-scale industrial livestock complexes common in the former Soviet Union for dairy, hogs, and poultry.

The third subsystem was entirely pasture-based, with transhumance grazing of livestock in summer, spring-fall, and winter pastures throughout the year. This system covered all sheep, goat, and horse inventories of the absolute majority of regions in the south, Khatlon oblast and the Regions of Republican Subordination (RRP). For this subsystem, it was necessary to hold only an emergency stock of cut feed. Intensive livestock husbandry in Tajikistan was based on use of chemical fertilizers and irrigation for grains, resulting in significantly increased yields. Higher grain yields freed up area for planting feed crops, which were also fertilized and irrigated. Central Asia as a whole, including Tajikistan, was also a net importer of feed and food grains. Mixed feed imports assisted in filling the winter feed gap.

In addition to these "technological" aspects of intensive agriculture, the governments of the Soviet Socialist republics also made great efforts to properly manage pasture maintenance, utilization, and transportation, and to supplement pasture feeding with adequate cultivated fodder. This involved matching feed demand and supply through feed balances which took account of use of summer, spring-fall, and winter pastures and the procurement of sufficient fodder to fill the winter feed gap.

The Soviet three-tier system of animal feeding in Central Asia rested on three pillars: (1) an elaborate organization for procuring animal feed for winter feeding based on intensively cultivated feed crops raised in large-scale state and collective farms, (2) sizable imports of concentrates, and (3) an organized system of pasture management and utilization, including pasture maintenance, transportation along established routes, clear assignment of pasture rights, animal veterinary points, shepherd supply facilities along the routes, and an established schedule of transhumance pasturing. Box 9.1 explains the elements of the livestock feed base in Tajikistan.

Table 9.1 illustrates the collapse of the first two pillars supporting the socialist intensive livestock husbandry system in Tajikistan, showing falls in the availability of cultivated feed crops and concentrates. All figures are shown in standard feed units, allowing aggregation. The fall in area and yields of feed crops caused a sharp decrease in the production of cultivated feed crops, the first pillar supporting the socialist intensive livestock husbandry system (Table 9.1, line 1a). Between 1991 and 2000, the total cultivated feed available to livestock in Tajikistan fell by 79%. The second pillar of support for intensive livestock husbandry—imported concentrates—was nearly completely eliminated between 1991 and 2000 (Table 9.1, line 1c). The availability of domestic concentrates also fell by half in this period

Box 9.1 The Livestock Feed Base in Tajikistan

Feed	Definition
Cultivated feed crops	Crops raised specifically for feeding domesticated livestock, which include: (1) Dry forage (perennial grasses, harvested as hay, haylage (from alfalfa), and straw) (2) Green chop (lucerne, a legume), annual grasses, corn and other silage (fermented, high-moisture fodder that can be fed to ruminants, such as cattle and sheep. Usually made from grass crops, including corn, sorghum, or other cereals, using the entire green plant (not just the grain)) (3) Succulents without silage (feed roots and melons, sugar beets for feed)
Domestic and imported concentrates	(1) Coarse grains such as corn, barley, and oats (2) Bran (the hard outer layer of grain, a by-product of milling in the production of flour) (3) Oil meals (in Tajikistan, cotton meal) (4) Mixed feed, feed additives, and other concentrated feed mixtures (grass flour, etc.)
Pasture	Pasture is land with low-growing vegetation cover used for grazing of livestock. Pasture growth can consist of grasses, legumes, other forbs (such as clover or milkweed), shrubs, or a mixture

In the Soviet period, area under *cultivated feed crops* in Tajikistan grew from 7 to 30% (1940–1985) of total sown area, allowing for the rapid growth of the livestock sector. The primary feed crops raised in irrigated fields of collective and state farms were lucerne, corn, sorghum, and sugar beets. Feed crops were raised through a variety of multiple cropping techniques in order to maximally utilize the long vegetative period in Tajikistan. These techniques included planting two harvests of silage per year, adding feed roots to land sown with corn, planting of lucerne together with feed grains, and other methods. In the Soviet period, there were 14 specialized seed farms for supplying lucerne seeds for rotation with cotton. Tajik farms practiced rotation of lucerne with cotton in order to raise cotton yields, secure ample supplies of fodder, and guard against verticillium wilt in cotton-growing areas of Tajikistan.

Concentrates refer to feed that has a higher concentration of energy than a forage diet. These are coarse grains, wheat, oil meals, and feed mixtures. A concentrate diet is the primary basis of intensive livestock production in developed countries.

In addition to cultivated feed crops, Tajikistan has ample *pasture* land used for grazing livestock. In mountainous countries as Tajikistan, pastures are classified according to their season of use depending predominantly on their altitude. Summer pastures in Tajikistan are located from 2,200 to 3,400 m above sea level and are used between June and August. Spring-fall pastures are usually located between 900 and 1,500 m above sea level and are used from March to May and September to November. Winter pastures are used between November and March and are located 500–1,200 m above sea level. All-year pastures are located at the same level as winter pastures but used all year round.

Sources: "Agriculture" in Academy of Sciences of Tajik SSR (1974) and FAO (2009)

Table 9.1 The collapse of available cultivated feed and concentrates in Tajikistan (in tons of standard feed units), 1991–2007

No.	Feed source	1991	2000	2007	Percent change, 1991–2000	Percent change, 1991–2007
1	*Total from cultivated feeds and concentrates (feed units, tons)*	2,196,062	458,131	738,744	−79	−66
a	Total cultivated feed crops	1,500,404	274,858	386,748	−82	−74
b	Domestic concentrates	368,658	182,954	344,439	−50	−7
c	Imported concentrates	327,000	319	7,557	−100	−98
2	*Feed availability per animal (feed units/head)*[a]					
a	Feed per standard animal head[b]	13	4	5	−69	−73
b	Feed per cow[c]	38	8	9	−78	−77

Sources: FAO (2009), p. 22. *Sel'skoe khoziaistvo respubliki Tadzhikistan: statisticheskii sbornik (2007)*

[a]This measure is incomplete because it does not include feed consumed through grazing in pastures

[b]Includes all animals measured in cow equivalent units

[c]Only cows

(Table 9.1, line 1b). Thus, the first two pillars supporting 1.6 million standard head of animals in 1991 were eliminated nearly entirely within 9 years and probably by 1995. Overall, though there was some recovery in the availability of feed in Tajikistan after 2000, it is today a mere 44% of what it was in 1991 (Table 9.1, line 1).

Feed availability per head of livestock also fell after 1991 and has not recovered (Table 9.1, line 2). Total feed availability from cultivated feed and concentrates per standard head fell by 73% between 1991 and 2007, and total feed per cow (the primary consumer of cultivated feed and concentrates) fell by 77%.

The deterioration of the feed base in Tajikistan was largely responsible for an unprecedented initial fall in livestock inventories. In the period 1991–1998, livestock inventories in Tajikistan fell by 30%. The initial fall was nearly exclusively due to liquidation of livestock inventories in agricultural enterprises, as evidenced in Fig. 9.1. Inventories on household plots, not directly supported by the socialist industrial feeding system, remained predominantly untouched by this initial downturn.

The disintegration of the Soviet intensive agricultural system and the resulting fall in both crop and livestock production led to the decision to partially privatize agriculture. The first legal acts on land reform and farm restructuring in Tajikistan were issued in 1992, but land reform began in earnest only in 1995, with a presidential decree allocating additional land to household plots. In parallel (1995–1996), Tajikistan moved to reorganize the traditional collective and state farms into new corporate forms in the hope that restructuring would improve productivity in a notoriously inefficient sector. When this largely cosmetic restructuring failed to produce efficiency gains, the government switched the focus of its attention to dehkan (peasant) farms as a model of family farming. Since 1999, dehkan farms have

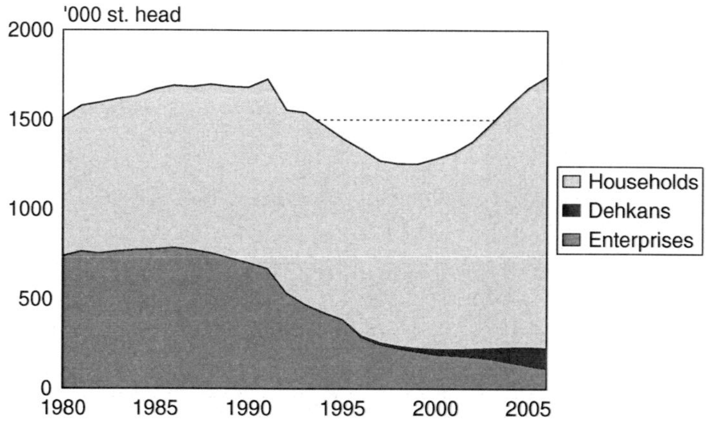

Fig. 9.1 Livestock inventories by farm type, 1980–2007 ('000 standard cow head) (Source: CISSTAT 2008)

largely supplanted the corporate farms—limited liability companies, leaseholding enterprises, joint-stock companies, and agricultural cooperatives—as the main agricultural land users.

The above reforms led to two crucial changes for the livestock production system in Tajikistan: (1) the virtual complete individualization of livestock inventories and (2) an initial fall and then rapid growth of livestock numbers. The individual sector in Tajikistan controlled most livestock even back in the Soviet era. In 1990, 62% of livestock was held outside of corporate farms (Fig. 9.1). But by 2007, the share of household plots in livestock had risen to over 90% (measured in standard head), so that the household farm sector now dominates livestock production, while enterprises and dehkan farms remain minor players. This situation is not unique to Tajikistan: A similarly extreme concentration of livestock production in household plots is also observed in Uzbekistan.

The rapid individualization of livestock herds and the end of hostilities in Tajikistan ushered in a new era of rapid growth in livestock inventories based on household farms. Overall livestock inventories in Tajikistan increased by 82% from 1998 to 2007, nearly exclusively as a result of growth in household farms (Fig. 9.1). The rapid recovery of livestock inventories after 1998 meant that (using official published statistics) by 2007 total livestock inventories were 16% higher than in 1991. The rapid expansion of livestock inventories despite the fall in feed availability has kept feed availability per animal (Table 9.1) extremely low.

Taken together, the above changes signify no less than the transformation of the livestock husbandry system in Tajikistan from one based on intensive livestock farming to one based on extensive livestock husbandry. The livestock production system existing today in Tajikistan relies primarily on grazing of livestock with limited feeding of cultivated feed and concentrates.

The hallmark characteristic of an intensive farming system is relatively high output per unit of input. In livestock, this meant that meat production per animal and

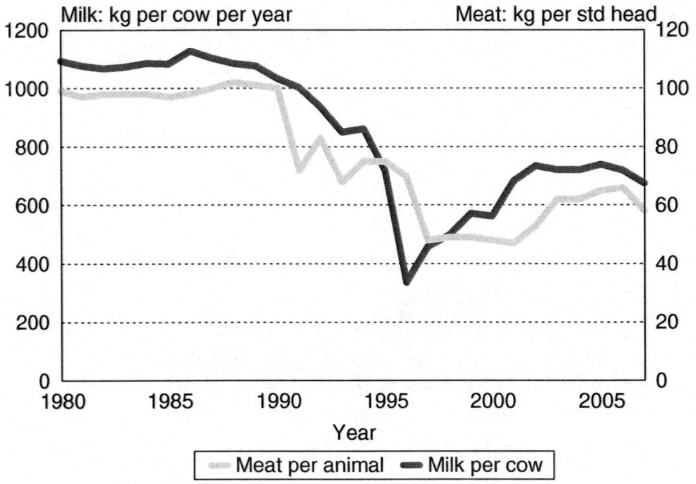

Fig. 9.2 Animal productivity in Tajikistan, 1980–2007 (Source: CISSTAT 2008)

milk produced per cow in Tajikistan were at all-time highs in the 1980s (Fig. 9.2). The decline began in the end of the 1980s when milk per cow and meat per animal started declining gradually. After 1990, this gradual decline turned into a free fall that lasted through 1997, after which both indicators began to rise and level off. Today both productivity indicators have recovered somewhat from the severe decline of the early 1990s but have stopped rising since 2003.

Despite recent increases, there is a generally low level of livestock productivity in Tajikistan. Milk yields are representative of the problem. Though milk yields recovered and have remained fairly constant since 2002, their recovery and stabilization does not appear to be directly linked to improvements in animal nutrition. The availability of feed crops per cow declined sharply from 1991 to 2000 and then stabilized (Table 9.1). At less than 700 kg per cow per year, milk yields in Tajikistan are far below yields in Western countries and rock bottom in the CIS (Fig. 9.3). Even during the heyday of Soviet Tajik agriculture, milk yields were far below those of the other 15 republics.

3 Crop Policies Limiting Feed Resources in Tajikistan

The fall in available feed crops per animal and the rapid expansion of livestock inventories after 1998 raise the issue of an expansion of the feed base in Tajikistan through increasing feed crop production. However, feed issues in Tajikistan should not be considered in isolation from other crop policies which impact on feed availability in the country. The three principle crops grown in Tajikistan have historically been cotton, grain, and feed crops, and the area in feed crops has always been linked to the area sown to the other two major crops. Figure 9.4 illustrates the complementary relationship between the three crop groups from 1940 to 2007. The postwar

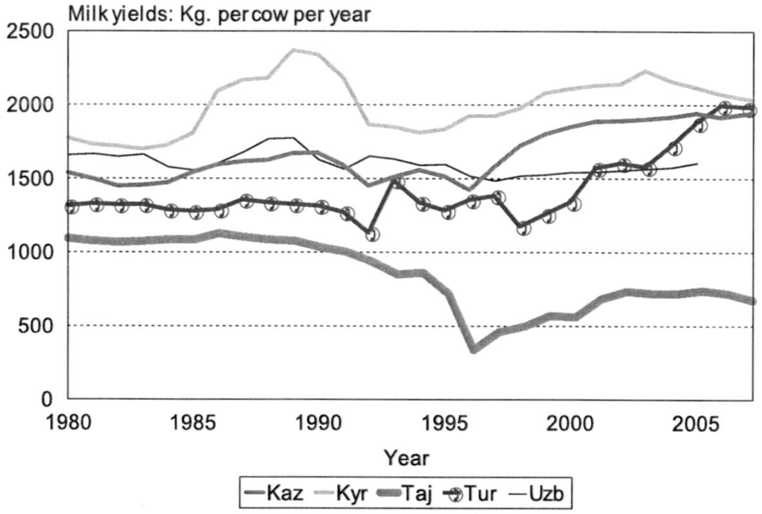

Fig. 9.3 Milk yields for Tajikistan and other Central Asian countries, 1980–20007 (Source: CISSTAT 2008)

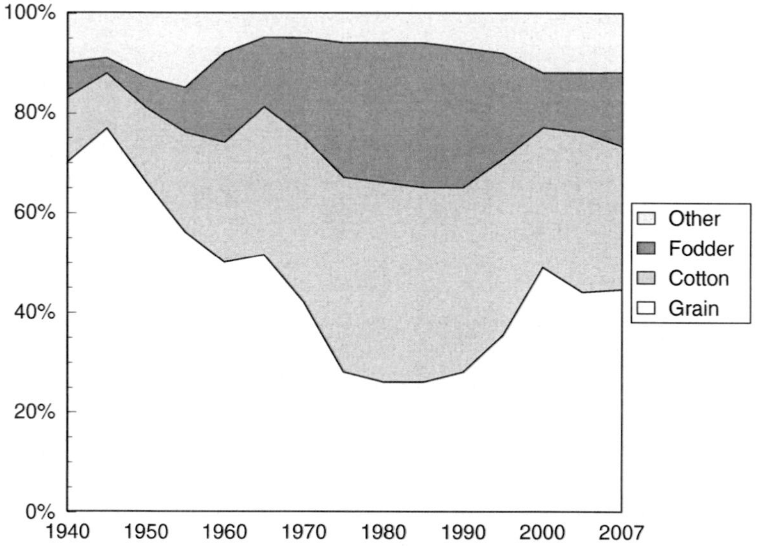

Fig. 9.4 Tajikistan crop areas, 1940–2007 (in percent of total sown area) (Sources: *Sel'skoe khoziaistvo respubliki Tadzhikistan: statisticheskii sbornik* (2001, 2002, 2003, 2004, 2005, 2006, 2007); *Narodnoe khoziaistvo Tadzhikskoi SSR: statisticheskii ezhegodnik* (1961, 1965, 1971, 1972, 1976, 1977, 1978, 1979, 1980, 1981, 1984, 1985, 1988))

Soviet period was characterized by shrinking area in grains as the area under cotton and feed crops increased through 1985. Falling grain area had no adverse effect on production volumes, because grain yields were growing quite rapidly due to increased applications of fertilizers, herbicides, and insecticides, as well as improvements

Table 9.2 Grain area, production, and imports in Tajikistan, 1988–2007

Year	Cereal production (without beer) (1,000 tons)	Cereal imports (1,000 tons)	Cereal availability[a] (1,000 tons)	Imports (% of availability)	Food use (1,000 tons)	Per capita food use (kg/cap)
1988	303	1,320	1,623	81	915	185
1989	322	1,300	1,622	80	915	180
1990	252	1,350	1,602	84	890	170
1991	286	1,250	1,536	81	700	131
1992	265	1,771	1,135	156	980	178
1993	259	1,503	1,089	138	978	175
1994	209	1,011	1,028	98	934	164
1995	235	627	1,134	55	1,048	182
1996	380	271	1,037	26	926	158
1997	545	389	1,148	34	1,063	179
1998	477	516	1,225	42	1,162	194
1999	459	439	1,163	38	1,100	181
2000	518	410	1,056	39	1,024	166
2001	464	355	1,064	33	1,030	165
2002	671	473	1,144	41	1,024	162
2003	846	419	1,265	33	1,129	177
2004	843	492	1,335	37	1,238	192
2005	882	789	1,671	47	1,250	191
2006	877	829	1,706	49	1,286	194
2007	889	970	1,859	52	1,234	183

Sources:
1988–1991: USDA, Economic Research Service, Former Soviet Union Commodity Balances
1992–2005: FAOSTAT Supply and Use Tables (http://faostat.fao.org/site/368/default.aspx)
2006–2007: *Sel'skoe khoziaistvo respubliki Tadzhikistan:statisticheskii sbornik (2008)*
[a]Production plus imports

in agronomic practices. In addition, starting in the early 1960s, the Soviet Union began to import grain, and Central Asia became a net importer of grain within the country.

When yields fell after 1985, it was natural that grain area would grow at the expense of fodder and cotton area. However, by 2005, though Tajik grain yields had surpassed peak levels during the Soviet period, grain area remained high. In fact, grain production and production per person in Tajikistan were at all-time highs in 2007. Yet grain sown area remained at nearly 50% of total sown area rather than at 26% as it had been in 1980 and 1985.

Grain area in Tajikistan remains at levels not seen since the 1950s because of state policies that limit areas sown to cultivated feed or rather support areas sown to grain and cotton. The first policy is the encouragement of grain self-sufficiency. The calculation of grain availability in Table 9.2 illustrates the issue. Tajikistan today imports about one-third to one-half of the grain imported annually during the late Soviet period. Because of significantly lower grain imports, the domestic production of grain in Tajikistan required to attain Soviet levels of food grain availability is far greater than under the USSR. The second policy limiting sown area to cultivated feed relates to cotton. Even in the food insecure years during the war, cotton area in Tajikistan remained at over 30% of sown land.

It is not clear that Tajikistan actually requires so much area in cereals. According to FAOSTAT, *per capita food use since 2004* has been above or at its level in the late Soviet years (Table 9.2). The cereal supply and utilization balances for Tajikistan in Table 9.2 are rough estimates at best. However, they do raise the question of whether current policies aimed at supporting grain area are really necessary. In considering ways and means to increase the availability of feed crops in Tajikistan, we will concentrate on limiting demand for and raising production of feed crops, leaving the important issues of cotton and grain areas alone. However, the wisdom of these two policies should not be taken for granted, and the issue deserves further study.

In addition to the limitations on fodder crops imposed through the wheat self-sufficiency policy, the absence of crop rotation on soils used for cotton cultivation also limits the area available for fodder. Crop rotation is a planned order of specific crops planted on the same field. Crop rotation also means that succeeding crops are of a different genus, species, subspecies, or variety than the previous crop. Examples would be barley after wheat, row crops after small grains, grain crops after legumes, etc. The planned rotation sequence may be for a 2- or 3-year or longer period. Some of the general purposes of rotations are to improve or maintain soil fertility, reduce erosion, reduce the build-up of pests, spread the workload, reduce risk of weather damage, reduce reliance on agricultural chemicals, and increase net profits.

4 Demand and Supply of Feed Resources in Tajikistan, 2007

In order to understand the specifics of livestock feed inadequacy in Tajikistan, the first step is to understand the nature of demand for and supply of feed resources.

4.1 *Feed Demand*

Feed demand may be calculated using standard feed units and standard head allowing one to aggregate demand over species and over feed types. Table 9.3 illustrates the calculation of total feed demand of animal inventories in Tajikistan. Livestock inventories by species are first converted into *standard head* by comparing the total feed a given species consumes in relation to the standard animal, in this case beef cattle (Table 9.3, column 4). For instance, one sheep requires on average only about 15% of the total oat unit equivalents required by beef cattle per year to lead a healthy and normal life. This means that one sheep is equal to only 15% of a beef cow in terms of *standard head*. The feed demand of standard head can be calculated using the information that one beef cow requires 2.12 tons per year of oat unit equivalents to lead a healthy and normal life. Feed units are calculated in a common *feed unit equivalent* based on the nutrient value of oat feed. The last column of Table 9.3 gives the portion of total feed demand in Tajikistan by species.

Table 9.3 illustrates an important fact about demand for feed in Tajikistan: Nearly 70% of demand originates from cows (dairy and beef), whereas sheep and goats, though they are numerous in Tajikistan, are responsible for a mere 20% of overall

Table 9.3 Feed demand based on animal inventories in Tajikistan, 2007

Livestock	Livestock inventories (1,000 s)	Beef cattle equivalent per head	Total beef cattle equivalents (1,000 s)	Feed units required per year (tons)	Percent of total demand
Beef cattle	838.2	1.0	838.2	1,776,984	33
Dairy cows	864.3	1.0	864.3	1,832,316	34
Hogs	0.6	0.4	0.2	445	0
Sheep and goats	3,798.4	0.1	531.8	1,127,365	21
Poultry	3,280.4	0.0	65.6	139,089	3
Horses	78.5	1.0	78.5	166,420	3
Donkeys	155.0	1.0	155.0	328,600	6
Yaks	15.2	1.0	15.2	32,224	1
Total demand			2,393.8	5,074,843	100

Source: *Sel'skoe khoziaistvo respubliki Tadzhikistan: statisticheskii sbornik (2007)*
Note: Beef cattle equivalents per head are Soviet era coefficients still used in Tajikistan to calculate standard head in beef cattle units. Required (oat) feed units are based on 2.12 tons of oat units required for feeding cattle per year

feed demand. This is important because cows, and particularly dairy cows, spend most of their time near the village grazing in local pastures or eating forage and concentrates.

A second important issue of feed demand is that it must adapt to the seasonality of pasture use over the course of the year. That is, because of seasonal variation, each species eats different feeds at different times of the year. It is of little use to aggregate all feed demand and all feed supply and compare the two. Both feed demand and feed supply must be compared for different categories of pasture, forage crops, and concentrates.

Pastures in Tajikistan are divided into those utilized in winter, spring-fall, summer, and year round. Table 9.4 illustrates the various types of pastures and their characteristics.

Different animals spend their time feeding from different sources during the course of the year. Small ruminants, such as sheep and goats, graze in pastures for a long period during the year (often in quite distant alpine pastures), while milk cows spend their time eating forage and concentrates and grazing in nearby pastures. Table 9.5 illustrates these differences by animal species in Tajikistan.

By distributing the feed requirements of each animal over pasture and forage resources according to Table 9.5, the total feed requirements can be estimated for each animal species by source of feed (Table 9.6). The resulting calculations illustrate an important mismatch in Tajikistan between demand and supply of pasture feed. Though Tajikistan has ample summer pastures (over 50% of total pasture area, see Table 9.4), most demand pressure is on all-year and fall-spring pastures, which account for only 28% of total pasture area. All-year pastures, with only 10% of pasture land, serve 41% of total pasture feeding needs. This mismatch between demand for feed and availability of pasture resources is a serious difficulty for livestock development.

Table 9.4 Pasture types in Tajikistan

Pastures	Winter	Spring-fall	Summer	All-year
Altitude (meters above sea level)	500–1,200	900–1,500	2,200–3,400	500 to 1,000–1,200
Use months	Nov–Mar	Mar–May, Sep–Nov	June–Aug	Jan–Dec
Use days	120–150	90–110	80-90	300–330
Total area (1.1.08) (1,000 ha)	699.0	675.9	2,081.3	400.0[a]
Percent of total pasture area (%)	18	18	54	10
Of which, in farm units (1,000 ha)	625.0	598.6	1,334.6[b]	360.0
Yield average (tons/ha of edible dry mass)	0.35	1.15	2.25	0.29
Distance from villages (km)	0.8–1.4 to 4–5	1.2–1.8 to 30	200–600[c]	less than 1 km

Source: Safaraliev (2009)
[a]85–90% degraded
[b]76.2% of area in dehkan farms
[c]6–8 weeks per year are spent traveling from winter to summer to winter pastures by animals using summer pastures per year

Table 9.5 Animal feeding throughout the year, by animal species and feed source (percent)

	Percent of time through year by feed source (%)					
Animal types	Summer pasture	Fall-spring pasture	Winter pasture	All-year pasture	Cultivated feed and concen-trates	Total
Beef cows	17	17	7	21	38	100
Cows	0	16	4	22	58	100
Hogs	0	0	0	0	100	100
Sheep and goats	22	18	12	24	24	100
Poultry	0	5	0	41	54	100
Horses	13	14	11	32	31	100
Donkeys	0	14	10	45	31	100
Yaks	34	2	38	26	0	100

Source: Safaraliev (2009)
Note: This table is distilled from a larger table of feed days in Tajikistan by region by animal

4.2 *Feed Supply*

Feed supply, just as feed demand, is differentiated by type, consisting of forage crops, concentrates, and various pasture types. The supply of feed available in Tajikistan is calculated in Table 9.7. The total tons of feed units derived from forage and concentrates is taken from Table 9.1. The supply of pasture feed is a function of the area of pastures by type and the yield of edible dry matter (DM) obtained. Dry matter available per year is then converted into tons of oat feed units.

Table 9.6 Feed demand in Tajikistan, by animal species and source, 2007

Feed demand (tons of feed units)	Pasture feed					Cultivated feed and concentrates
	Total pasture	Summer	Fall-spring	Winter	All-year	
Beef cows	1,072,952	303,919	288,191	121,203	359,639	669,522
Cows	768,389	0	293,258	72,782	402,350	1,061,275
Hogs	0	0	0	0	0	436
Sheep and goats	856,990	246,057	203,692	137,864	269,377	268,788
Poultry	64,221	0	6,992	0	57,229	74,670
Horses	115,336	20,996	22,989	17,563	53,788	50,908
Donkeys	249,934	0	50,560	34,884	164,490	112,671
Yaks	32,136	11,021	514	12,266	8,334	0
Total feed demand	5,398,226	581,993	866,196	396,563	1,315,206	2,238,268
Structure of total feed demand, %	*59*	*11*	*16*	*7*	*24*	*41*

Source: Table 9.5

Table 9.7 Feed supply and feed adequacy in Tajikistan, by source, 2007

	Feed supply	Pasture feed					Cultivated feed and concentrates
		Total pasture	Summer	Fall-spring	Winter	All-year	
1	Pasture area in 2007 (ha)	3,856,246	2,081,287	675,909	699,003	400,047	n/a
2	Yield of edible DM (t/ha)	1.53	2.27	1.15	0.36	0.39	n/a
3	Total edible DM (t/year)	5,910,608	4,723,750	780,246	248,759	157,853	n/a
4	Total feed supply (tons of feed units)	2,955,304	2,361,875	390,123	124,379	78,927	738,744
5	*Total feed availability (percent)*	*80*	*64*	*11*	*3*	*2*	*20*
6	*Feed adequacy coefficient (ratio of supply to demand, %)*	*94*	*406*	*45*	*31*	*6*	*33*

Source: Estimates based on Safaraliev (2009)
Note: This table is derived from a larger table of pasture area and yield by region
DM dry matter

Line 6 of Table 9.7 indicates feed adequacy in Tajikistan by category of feed. This is calculated by comparing total feed demand in Table 9.6 by type with total feed supply by type in Table 9.7. The feed adequacy results in Table 9.7 illustrate that, though pasture resources in Tajikistan may be adequate in total, fulfilling 94% of total demand, this calculation carries very little meaning as it hides extreme differences in feed adequacy of different pasture types. While summer pastures are in excess supply, pastures grazed during other seasons meet a very small share of demand. Moreover, demand for forage crops is satisfied by only one-third.

The information in Tables 9.6 and 9.7 leads to the conclusion that the main constraint on the development of the livestock sector in Tajikistan is an extreme imbalance between the supply and the demand of cultivated feed and concentrates for dairy and beef cows. First, demand for all-year winter and fall-spring pastures is much too high to be sustainable. Clearly, demand for these pastures needs to be limited in order to ensure sustainable use of these resources. The main users of these pastures are dairy and beef cows. In order to limit grazing of cows and cattle on these fields, the supply of cultivated feed and concentrates must be increased and made available to farmers. This will allow farmers to keep animals in barns for feeding rather than letting them feed in nearby fields. Only after such limitations on use are introduced can the pastures themselves be rehabilitated, thus increasing total pasture feed available in Tajikistan. Second, demand for cultivated feed and concentrates far outstrips supply. Clearly, there is an immediate need to make cultivated feed and concentrates for cows more available and accessible to farmers.

The two above imbalances are complicated by yet another imbalance, an extremely unequal distribution of livestock inventories and feed resources across farms of different types in Tajikistan. Table 9.8 illustrates the concentration of feed resources in agricultural enterprises and dehkan farms in contrast to the extreme concentration of livestock inventories in household farms. Only 36% of cultivated feed resources are raised in household farms, while 90% of animal inventories are in their ownership.

The mismatch between feed and animals is illustrated in Table 9.9, which shows that household farms raise only 6% of required feed on farm and must utilize the pastures of agricultural enterprises and dehkan farms or purchase cultivated feed from enterprises and dehkan farms. Agricultural enterprises and dehkan farms, however, have nearly five times the feed resources required to support their livestock inventories.

5 Policies to Support a Return to a Sustainable Livestock-Feed Balance in Tajikistan

This chapter has analyzed the issues surrounding what has been described as the most important immediate constraint on livestock yields and rural incomes from livestock husbandry, animal nutrition. It was argued in the previous section that the main impediment to the reestablishment of proper animal nutrition in Tajikistan is

Table 9.8 Estimated distribution of cultivated feed and concentrates in Tajikistan, 2007

Cultivated feed in Tajikistan	Household farms (percent)	Agricultural enterprises and dehkan farms (percent)
Total cultivated feed	*36*	*64*
Dry forage		
Hay	26	74
Straw	42	58
Haylage	n.d.	n.d.
Green chop[a]	*10*	*90*
Succulents without silage[a]	*10*	*90*
Concentrated feed		
Corn	75	25
Barley and oats	36	65
Bran	42	58
Cotton and other meals	40	60
Imported concentrated feed	0	100

Source: FAO (2009), p. 22

[a]There are no statistical data on these feeds. It is assumed that 10% of green chop and succulents are raised in household farms

Table 9.9 Distribution of total feed resources by farm type in Tajikistan, 2007

		Total	Household farms	Agricultural enterprises and dehkan farms
1	Cultivated feed and concentrate availability, 2007 (tons of feed units)[a]	738,744	265,247	473,497
2	Pasture production, 2007 (tons of feed units)[b]	2,955,304	0	2,955,304
3	Total available feed resources, 2007 (tons of feed units)	3,694,048	265,247	3,428,801
4	Standard head (beef cattle units) of animals in Tajikistan (Jan 1, 2008)[c]	2,393,794	2,154,415	239,379
5	Feed units per std head per year (tons, 3/4)	1.54	0.12	14.32
6	Feed units required per std head per year (tons)	2.12	2.12	2.12
7	Deficit (surplus) (6–5) (tons of feed units/std head/year)	0.58	2.00	(12.2)
8	Percent of requirement met (5/6, percent)	*73*	*6*	*675*

Sources: Tables 9.1 and 9.7

[a]From Table 9.1, production and imports in 2007

[b]Uses estimates from Table 9.7

[c]Inventories from Table 9.3 converted to standard head in beef cattle equivalents

increasing the availability of cultivated feed and concentrates primarily for dairy and beef cows. It would be a mistake, however, to understand the problem as merely one of excess demand for feed. There is a clear need to limit the number of animals allowed to feed in fall-spring, winter, and all-year pastures. This can only be accomplished by giving farmers themselves a stake in the sustainable management of pastures (Box 9.2).

Box 9.2 Limiting Pasture Use Through Pasture Users' Associations

Goal	Policies	Short run results	Longer run results
Raise milk and meat yields through increased supply of cultivated feed and concentrates	Make pasture management the responsibility of users through a pasture users association	Clearly defined rules of pasture use that will establish access to pastures on a sustainable basis. Sustainable use of pastures will include limitations on pasture use enforced by pasture user associations	Pasture rehabilitation possible due to controlled use

The pasture management system in Tajikistan remains largely unchanged since Soviet times with the exception that the lowest rung in the management system (agricultural enterprises) no longer has adequate resources for pasture maintenance and management. In some other CIS countries (e.g., Azerbaijan), pasture lands have been transferred into a separate category of municipal lands with a municipal management structure. However, in Tajikistan, pastures are under state ownership, just as all agricultural land, but are mainly held by agricultural enterprises and dehkan farms. Farms of all types—agricultural enterprises, dehkan farms, and household farms—use pasture land for their animals.

Legislation on pasture management has not been adapted to the post-land reform farming structure (Halimova, Chap. 13). According to existing legislation on pasture management, the following institutions are responsible for the rational utilization of pasture resources in Tajikistan:

- Local executive organs of the state at the regional level
- Regional and municipal land-surveying organizations
- Local (village-level) authorities
- The state committee on environment
- The Pasture Trust of the Ministry of Agriculture

The existing system of pasture management responsibilities in Tajikistan is ill-defined and not designed to involve the end-user in the management of pasture resources. The existing system is not well adapted to managing a public good such as pastures in the post-socialist period when over 90% of animals are held in household farms, not in large-scale enterprises. In other countries, pasture land is state-owned, just as in Tajikistan, but the management of pasture land, as for other public goods, such as irrigation works, involves significant involvement of and financial contributions from users.

One institution ensuring that pasture management incorporates the needs of pasture users is the pasture users association. Kyrgyzstan has recently adopted pasture legislation that changes the system of pasture management to one which may be better suited to the environment of smallholder agriculture. Table 9.10 compares pasture legislation in Tajikistan and Kyrgyzstan. In the Kyrgyz system, pastures are state-owned public goods, just as in Tajikistan. However, the management, including permission to use pastures, pasture rehabilitation, fee assessment, and collection, has been decentralized to the level of the pasture users association. The idea of decentralizing such decisions puts the users themselves in control of the public good they require to graze their animals. Thus, it could be expected that pasture users would have an intrinsic interest in better husbandry of pasture resources (Robinson, Chap. 11) (Box 9.3).

There are many policies that can potentially raise the availability of livestock feed. A first area of policy is the introduction of better agronomic practices such as crop rotation for cotton and wheat. A second issue under good agronomic practices is to increase funding for agronomic research aimed at raising feed crop yields. Certainly, feed crop yields could be raised to levels of 1991 and greater.

A second general area of policy concern is so-called freedom to farm. Elimination of informal controls on cotton sowing area would enable farms to increase feed crop production as demand increases. According to a USAID and World Bank survey of dehkan farms, cotton growers in Tajikistan have much less freedom of decision than other dehkan farms (Lerman and Sedik 2008). Hukumat intervention is quite pervasive for cotton growers and virtually nonexistent for other farms. Among cotton-growing dehkan farms, only 14% have freedom of decision, whereas in 56% of the farms, the decision is made by the manager, and in a staggering 28% of the cotton growing farms, the local authorities (the hukumat) directly intervene in planting decisions. This is in a striking contrast with the decision-making process in other dehkan farms, where 60% make the decisions themselves and the hukumat intervenes in only 5% of the cases.

A third area for policy attention is to raise forage and mixed feed availability for farmers by making land and business permits easily available for mixed feed production facilities and forage and mixed feed sales points operated by farmer cooperatives in rural areas. This may necessitate new investment for construction of mixed feed storage or production plants. Uzbekistan, for example, has a program to improve access of rural households to concentrate feed by an expansion of storage facilities and sales outlets in rural areas (Presidential Decree of the Republic of Uzbekistan PP-308, 23 March 2006; Resolution of the Cabinet of Ministers of the Republic of Uzbekistan no. 67, 21 April 2006). Under this program, the state-controlled Uzkhlebprodukt system has been required to establish feed storage facilities and sales outlets in rural areas. The program envisages a sevenfold increase in the number of sales outlets for concentrated feed across the country, from 113 in 2005 to 773 in 2010.

Table 9.10 Description of legal responsibilities for pasture management in Tajikistan and Kyrgyzstan

Tajikistan	Kyrgyzstan
Ownership	
All pastures owned by the state	All pastures owned by the state
Land users	
Physical or juridical persons of Tajikistan	Physical or juridical persons of Kyrgyzstan or foreign country (by international agreement or intergovernmental agreement)
Pasture border demarcation	
Majlisi Oli (parliament)	Carried out by local commission appointed by the local state administration The government of Kyrgyzstan establishes a commission to settle disputes
Pasture management	
A multitude of agencies and government at various levels are responsible for pasture management. The Tajik national government is responsible for the organization, development, and realization of government and intergovernmental programs for the rational use of pastures, raising productivity and soil fertility, and for environmental protection. It also has responsibility for general management of pasture resources, establishment of the plan for cadastre works, and monitoring pasture resources. *Oblast level government organs (goskomzem, giprozem, and their local organs)* are responsible for cadastre, monitoring the state of pastures, improvements in management within existing legislation, and the development of legislation for state management of pastures. *The Jamoat (municipality), the local representative organ,* is to control the utilization of pasture and protect pasture lands. *Local regional government administrations within the administrative boundaries of their region* are to propose pasture use plans to the local government, control the use of pasture, and protect pasture lands. They are also responsible for the establishment of agricultural land, state forest land, and other lands for pasture use. *Pasture users are responsible for the protection of pastures as well*. They are responsible for the rational use of pastures, rehabilitation of soil fertility, protection of pastures from water and wind erosion, other soil damage, and protection from weeds, all to prevent degradation. *The Pasture Trust of the Ministry of Agriculture is* responsible for state control of pasture use and protection	The *local self-government body* (pasture users association) has responsibility for management of pasture resources except for the right of disposal. The interference of state organs and local state administrations in the work of local self-government bodies and the pasture users associations in the area of pasture utilization is forbidden, except in cases foreseen in legislation (Article 4) The *pasture users association* represents the interests of the users from within a certain territory. This association draws up a community pasture management plan on an annual basis. *The executive committee of the pasture users association is the Jayit* which consists of representatives of pasture users, deputies of local self-government bodies, and heads of executive local self-government bodies. Pasture users elect their members to the Jayit. The authority of the Jayit is to develop community pasture use plans, annual pasture use plans, implement these plans, monitor pasture conditions, issue pasture use tickets according to the plans, fix and collect pasture fees, resolve disputes, and manage pasture revenue for pasture improvement *Community pasture use plan* includes pasture utilization, maps for boundaries, stock routes, protected areas, watering places, pasture infrastructure, and pasture conditions and quality; also contains a map of carrying capacity of various pastures, plans for the development of pastures and maintenance operations, and plans for the reconstruction of infrastructure. These plans are updated annually and are approved by local self-government bodies. *The annual pasture use plan* includes a list of pasture users holding tickets for the year and an inventory of their livestock, a list of animal health measures that users must carry out in order to exercise their grazing right, a map of seasonal grazing routes, a pasture rotation plan, herd movement routes, cattle pens, etc.

(continued)

Table 9.10 (continued)

Tajikistan	Kyrgyzstan
Establishment of a list of pasture users	
A multitude of agencies and government at various levels are responsible for allowing or not allowing pasture use. The Tajik government has the right to grant and confiscate pasture plots for government purposes in Tajikistan in agreement with local executive authorities. *Oblast level government organs (goskomzem, giprozem, and their local organs)* state registration of right to use pasture and issue of land certificates for pasture use. *Local government (oblast and regional)* grants pasture plots for agricultural production and grants and confiscates plots when used not according to their prescribed use, including pasture land use rights. *The Jamoat, the local representative organ,* makes proposals to local governments on granting pasture use rights on the establishment and approval of tax norms within the bounds established by tax legislation. The Jamoat also keeps a list of pasture users and pasture lease agreements and regulates the use of additional pasture land established from non-pasture resources. *Local regional government administrations* propose pasture use plans to local governments and establish the uses of agricultural land and state forest land and other lands within the administrative boundaries of the region. *The Pasture Trust of the Ministry of Agriculture is* responsible for state control of pasture use and protection. *Jamoats* are supposed to grant pasture use rights to citizens with livestock from reserve land, forest land, urban land, and agricultural land, though need to gain consent of various organizations that are responsible for that land	*The Jayit* establishes a list of pasture users based on pasture use plans
Pasture use fees	
Payment for pasture use is made yearly in the form of a flat land tax. The proceeds of the land tax are distributed according to the tax laws of the country. Tax proceeds for use of pastures and hayfields are used for the protection and rehabilitation of pastures, for maintenance of soil fertility, for monitoring of pastures, etc.	*The Jayit* fixes and collects pasture use fees and manages pasture revenue for pasture improvement. A portion of pasture use fees are transferred to local budgets

Sources: Tajikistan: *Land Code of the Republic of Tajikistan* (2008), Kyrgyzstan: *Law on Pastures* (2008)

Box 9.3 Policies to Increase Availability of Livestock Feed

Goal	Policies	Short run results	Longer run results
Raise milk and meat yields through increased supply of cultivated feed and concentrates	*Good agronomic practice policies*		
	Reintroduce cotton-lucerne crop rotation cycle in Tajikistan widely following proper agronomic practices	Supply of forage crops increased, lowering prices for livestock farmers. Cotton yields increase due to decrease of cotton wilt and better soil management	More sustainable production of cotton and increased supply of forage crops
	Raise forage crop yields to 1991 levels	Increase in domestic production of forage crops	Demand for forage crops better satisfied
	Increase domestically raised wheat yields through crop rotation following good agronomic practices	Increased supply of wheat	Allows increased area for forage and cotton
	Freedom to farm policies		
	Eliminate informal controls on cotton sown area	Farmers are free to make their own decisions on what crops to produce	As feed demand increases, farmers will be able to produce feed crops to meet demand
	Marketing and investment policies to improve feed access		
	Raise forage and mixed feed availability by making land and business permits easily available for mixed feed storage and production plants and for forage and mixed feed sales points operated by farmer cooperatives in rural areas	Farmers have easy access to forage and mixed feed for their animals	Increased demand for forage crops from farmers. Decreased demand pressure on nearby pastures, allowing them to be rehabilitated
	Trade policies		
	Eliminate any barriers on wheat flour imports from Russia and Kazakhstan	Increases availability of domestic soft wheat for feed use	Higher milk and meat yields from feeding increasing rural incomes

Table 9.11 Effects of implementation of policies for a sustainable livestock-feed balance in Tajikistan

	2007 base	Scenario after 10 years of policy change
1. Milk production, cow inventories, and milk yield		
Total milk production (1,000 tons)	584	1,402
Total cow inventories (1,000)	864.3	864.3
Milk yield (liters/cow/year)	675	1,622
2. Feed adequacy (%)		
Summer pastures	406	406
Fall-spring pastures	45	73
Winter pastures	31	51
All-year pastures	6	10
Forage crops and concentrates	33	70

Sources: Calculations based on input-output tables underlying previous tables

Feed mills in the state-controlled Uzkhlebprodukt system are to be allowed to purchase grain directly from peasant farmers ("fermerskie khozyaistva") as a raw material for concentrated feed production.

A last area of policy is trade barriers on wheat flour imports which can influence availability of domestic soft wheat for feed use. Higher milk and meat yields can come from better feeding, and this leads to increasing rural incomes.

6 Effects of Policies to Support a Sustainable Livestock-Feed Balance in Tajikistan

A sustainable livestock development policy should aim toward implementing all the policies suggested here: (1) introduce an institutionally viable pasture management system to limit the number of animals authorized to graze in fall-spring, winter, and all-year pastures; (2) raise cultivated feed crop yields through crop research; (3) expand the area of cultivated feed crops through introduction of proper crop rotation for cotton and wheat; (4) eliminate informal controls on cotton sown area; (5) raise forage and mixed feed availability by making land and business permits easily available for mixed feed production facilities and forage and mixed feed sales points operated by farmer cooperatives in rural areas; and (6) eliminate trade barriers for imported wheat flour from Russia and Kazakhstan, thus making more domestically grown soft wheat available for feed use.

Table 9.11 illustrates the combined effects that could be expected if these policies were enacted over 10 years. The specific assumptions of the scenario are:

1. Increases in pasture yields each year for 10 years at 5% per year
2. Raising cultivated feed crop yields to their 1991 level
3. Expanding cultivated feed crop area by 10% through increased rotation with cotton crops

4. Raising wheat flour imports by 2.5% per year
5. Raising Tajik wheat yields by 10%

Even under these quite conservative assumptions, there are dramatic improvements in livestock feeding and milk yields in Tajikistan over baseline levels of 2007 (Table 9.11). It should be emphasized that Table 9.11 illustrates *only* the effects of the implementation of the policies discussed under the assumptions indicated independent of growth in livestock inventories and productivity increases due to other causes.

The most direct and largest effects on milk yields in Tajikistan are to be gained by improving cultivated feed crop yields, increasing their area, and raising flour imports and wheat yields. Pasture rehabilitation has considerably less impact on milk yields because cows feed predominantly on mixed feed and forage crops.

7 Conclusions

The imbalance between feed demand and supply is perhaps the most important limiting factor on the sustainable development of the livestock sector in Tajikistan. The purpose of this study is to analyze the feed supply-demand balance and to suggest policies that may support a return to a sustainable livestock-feed balance in Tajikistan. The study described the transformation of the livestock husbandry system in Tajikistan after independence from one based on intensive livestock farming to one based on extensive livestock husbandry. It then went on to describe some of the crop policies in Tajikistan that limit feed resources in Tajikistan, followed by a calculation of feed demand and supply in Tajikistan by feed type. The study then suggested a number of policies to support a return to a sustainable livestock-feed balance in Tajikistan, followed by an estimate of the possible effects on the supply-demand balance for livestock feed in the country.

The study concludes that the policies suggested could, if implemented, substantially reduce the pressure on some pastures if consistently implemented over the next 10 years. Moreover, livestock production and productivity could be significantly increased. A vital part of the policies suggested is the management of pastures through an improved management incentive system of pasture user associations.

The feed-livestock nexus is only one of a number of issues that should be addressed under a sustainable livestock development policy. Other issues such as the establishment of a viable plan for supplying livestock advisory and health services and a livestock breeding policy should also be part of such a policy. However, this study has concentrated on a first-level constraint on rural incomes that, unfortunately, has not received the attention it deserves. It is hoped that this study has shed some light on this issue and provided some basis for beginning a dialogue between the government of Tajikistan and donors on a sustainable livestock strategy for the country.

References

Academy of Sciences of Tajik SSR (1974) Encyclopedia of the Tajik Soviet Socialist Republic (in Russian). Academy of Sciences of Tajik SSR, Dushanbe

"Agriculture" in Academy of Sciences of Tajik SSR (1974)

Bravo H (2005) Livestock value chains in Tajikistan. Consultant report (World Bank, processed)

FAO (2009) Strategicheskoe issledovanie sektora zhivotnovodstva (Policy study on the livestock sector). Prepared by Fund Biotechnology, Tajikistan (processed)

Halimova N (2012) Land tenure reform in Tajikistan: implications for land stewardship and social sustainability: case study In: Squires VR (ed) Rangeland stewardship in Central Asia. Springer, Dordrecht, pp 239–274 (Chapter 11, this volume)

Land Code of the Republic of Tajikistan (2008) (Zemel'nyi kodeks Republiki Tadzhikistan, Dushanbe)

Law on Pastures (2008) (Zakon Kyrgyzskoi respubliki, "O pastbishchakh," 18 Dec 2008)

Lerman Z, Sedik D (2008) The economic effects of land reform in Tajikistan. FAO Regional Office for Europe and Central Asia. Policy studies on rural transition no. 2008-1

Narodnoe khoziaistvo Tadzhikskoi SSR: statisticheskii ezhegodnik (Dushanbe: Gosudarstvennyi komitet Tadzhikskoi SSR po statistike, 1961, 1965, 1971, 1972, 1976, 1977, 1978, 1979, 1980, 1981, 1984, 1985, 1988)

Nolan T (2005) The animal/vegetation relationships component of Rangeland management. Consultant report for Republic of Tajikistan Livestock Sector Review (processed)

O'Mara F (2006) Animal nutrition and livestock productivity in Tajikistan. Consultant report for Republic of Tajikistan Livestock Sector Review (processed)

Official Statistics of the Commonwealth of Independent States (2008) (CISSTAT) CD-ROM 2008-13. Statistical Committee of the CIS, Moscow

Presidential Decree of the Republic of Uzbekistan PP-308 (2006) On measures for increasing cattle inventories in household, dehkan and family farms, 23 March 2006

Resolution of the Cabinet of Ministers of the Republic of Uzbekistan no. 67 (2006). On measures for implementation of the cattle inventories development program in household, dehkan and family farms for the period 2006-10, 21 April 2006

Safaraliev (2009) September 2009 personal correspondence

Sel'skoe khoziaistvo respubliki Tadzhikistan: statisticheskii sbornik (Dushanbe: Gosudarstvennyi komitet statistiki respubliki Tadzhikistan, 2001, 2002, 2003, 2004, 2005, 2006, 2007)

World Bank (2007) Republic of Tajikistan: priorities for sustainable growth: a strategy for agriculture sector development in Tajikistan. World Bank, Washington DC

Chapter 10
Rangeland-Based Livestock: A Vital Subsector Under Threat in Tajikistan

Paul J.H. Strong and Victor Squires

Abstract This chapter is a detailed case study of Tajikistan, one of the Central Asian countries, and has three main sections: (1) a macroeconomic analysis of the rangeland/livestock subsector of Tajikistan's agricultural sector, (2) an assessment of challenges and opportunities, and (3) strategic recommendations for long-term and operational priorities for development of the rangeland-based livestock industry.

The importance of the livestock economy to the rangelands is clear. The current and potential importance of the pasturelands/rangelands to the current and future Tajikistan economy cannot be overstated. Specifically the rangelands, through livestock mainly, are a source of direct rural income support for over four million people as well as providing much of the nation's meat and milk requirements. It will be productive developments within the livestock sector that will largely influence and determine whether sustainable development will be possible in Tajikistan and several neighboring countries (Afghanistan, Kyrgyzstan, parts of Kazakhstan, Uzbekistan, and northwest China).

The assured future development of the rangelands requires a clear and strategic vision of their role within the overall economic fabric of Tajikistan. Because raising livestock on rangelands is the prime agricultural system and source of outputs for the rangelands that impacts directly and indirectly on almost all of the Tajik population, the restoration and maintenance of the rangeland resource should be the key focus of implementation scenarios. Livestock productivity can be increased by

P.J.H. Strong
GITEC Consulting GMBH, Cologne, Germany

V. Squires (✉)
College of Grassland Science, Gansu Agricultural University, Lanzhou, China

GITEC Consulting GMBH, Cologne, Germany
e-mail: dryland1812@internode.on.net

V. Squires (ed.), *Rangeland Stewardship in Central Asia: Balancing Improved Livelihoods, Biodiversity Conservation and Land Protection*,
DOI 10.1007/978-94-007-5367-9_10, © Springer Science+Business Media Dordrecht 2012

almost 40% by modest improvements in husbandry practices, resulting in reduced mortality and improved calving intervals and growth rates. In view of the importance of the rangeland sector, it is considered essential that the status of all development efforts must be elevated to a very high level. Implementation partners will include state and local government agencies, NGOs, and rural communities especially through community-based organizations (CBOs).

Keywords Fuelwood • Soil erosion • Soil creation rate • Wind erosion • Gullies • Sheet wash • Community-based management • Forest • Biodiversity • Income sources • Ecological health • Land tenure • Privatization • Export income • Cropping • Tax regime • Land market • Female-headed households • Farm gate prices • Offtake • Shepherds • Forage supply

Key Points

- The rangelands are a valuable national resource that provide, and will continue to provide, a source of income and a livelihood to the largest proportion of the rural population.
- The main income-generating activity of the rangelands now and in the foreseeable future will be livestock production. Livestock are a major contributor to the rural economy, but numbers have risen to unsustainable levels relative to the feed available, and consequently their output is low. They are often the only easily redeemable asset for millions of rural poor, many of whom are highly food insecure. For livestock-owning families, milk and milk products provide a vital source of protein, and sales of surplus products aid cash flow for poor households.
- A sustainable livestock industry is a crucial component to the future development of a balanced and prosperous Tajikistan economy (in view of the current widespread livestock ownership and socioeconomic implications).
- The primary basis for the sustainable maintenance and development of the rangelands' natural resources is the fostering of a sustainable and profitable livestock industry. In turn, this will require a broad range of interventions that will directly and indirectly support and sustain the rangeland's ecological health with the associated environmental overspills.
- The rangelands require certain specific legal considerations (including appropriate land tenure and a tax regime that is conducive). These will assure guidance, checks and balances, and incentives for rangeland protection. In addition, it will encourage good principles of land stewardship (e.g., preservation of biodiversity) to ensure the rangelands remain a productive income source for future generations.
- Land degradation in much of Tajikistan is characterized by significant recent loss of natural forest, severe and continuing forest degradation, and widespread soil erosion on sloping land. Most farming communities rely on remnant forests and scattered trees for fuelwood supplies and on degraded mountain lands for livestock grazing.
- The local maintenance and development of the rangelands should be placed under the direct control of organizations comprising users of the rangelands. This will require the greater development of democratic community-based management groups more conducive to land stewardship.

Fig. 10.1 Tajikistan is a small (143,100 km^2) mountainous, landlocked country in Central Asia

- In Tajikistan and other developing countries in Central Asia, the self-actualization, fundamental to many of the ideas embodied in land stewardship, is an impossible dream for all but the very few, buried as they are under the realities of a daily struggle to survive.

1 Overview of the Current Status of the Pastureland/ Rangeland Sector

1.1 Macroeconomic and Socioeconomic Setting

Tajikistan is a small landlocked country bordered by 5 countries (Fig. 10.1) and one of the poorest countries in the Commonwealth of Independent States (CIS). After a civil conflict (1992–1997), it is in the process of strengthening its democracy and

Fig. 10.2 Many of Tajikistan's rivers arise in extensive glacier fields

making the transition to a free market economy. The country suffers from high unemployment and widespread poverty, with a significant proportion of the population living below the poverty line. In rural Tajikistan, women are the major workforce – forming more than 70% of the total agricultural workforce.

Tajikistan may be seen as a small open economy with a narrow economic base that specializes in the export of labor, alumina, electricity, and cotton. The mountain ranges of northern Tajikistan and their branches create several hydrographic areas, forming two main river systems: the Amu Darya and Syr Darya. The Amu Darya river arises in the glacier fields and flows to the Aral Sea. The rivers of Tajikistan (Fig. 10.2) are not only a source of water for irrigation, without which the area of crops would be extremely limited, but also represent the highest hydropower potential among all the CIS countries. Tajikistan ranks second after Russia.

The largest export is labor with remittances estimated at 36% of GDP. This cash inflow has proven to be a stable cash income source particularly during the 2009 financial crisis. Other key exports are alumina and cotton. The country, however, has a large food deficit which *per se* is not a problem, but low average export earnings as a result perhaps of high exporting costs (costs of production) combined with relatively high importing costs due to its geographic position, together with difficult geopolitics, make Tajikistan vulnerable. The country would become more secure if a national strategic grain reserve were established. It is noted that the donors especially World Food Programme (WFP) are providing technical and financial support to help achieve this but progress so far has been slow. The estimated per capita income (2009) of US$700[1] though above the world average for low-income countries (US$512) is much lower than Europe/Central Asia (US$6793) and the lowest of any CIS country.

The economy is therefore fragile which is exacerbated by factors such as weak governance, high unemployment, seasonal power shortages, and external debt. Development is also constrained by low levels of literacy and life expectancy at birth (67 years) as reflected in the Human Development Index (HDI) value of 0.58 (ranked 112th in the world). This is mainly due to poverty as well as deteriorating health and education services which are commonly considered as key contributors required for sustainable development. In rural Tajikistan, women are the major workforce – forming more than 70% of the total agricultural workforce (Fig. 10.3).

[1] World Bank.

Fig. 10.3 Tajik women make up a large part of the rural workforce, and female-headed households are the norm as many able-bodied men work outside Tajikistan as migrant labor

Table 10.1 GDP by sector 2005–2010 (constant 2010 prices in TJS millions)

Sector	2005	2006	2007	2008	2009	2010	% Annual growth 2005–2010
GDP	17,939	19,194	20,691	22,326	23,197	24,705	6.6
Industry	4,090	4,088	3,899	3,171	3,317	3,113	–5.6
Agriculture	3,203	4,165	4,014	4,420	4,291	4.620	7.6
Services	7,229	7,793	8,815	9,757	10,879	11,809	10.3

Source: GoT, GITEC consultants
Note: Services clearly have risen and agriculture has really remained flat (the jump in 2005 to 2006 was probably due to price rises if we examine the figures). Industry understandably has fallen
TJS Tajik Somoni $1 USD=4.5 somoni in 2010

Agriculture was a key contributor in 2005–2010 to GDP growth (Table 10.1) mainly as a result of external remittances (boosting local demand and prices) and high export prices. Cotton fiber, the main agricultural export, in value terms contributed 16% to the value of total exports and 90% to agricultural exports. Other agricultural exports included fresh and processed fruits, vegetables, raw animal hides, and silk products. Domestic industry is limited and in the same period was lagging, i.e., it made a negative contribution to GDP growth (Table 10.1).

Agriculture accounts on average for 26% of export earnings and 39% of Government of Tajikistan (GoT) tax revenue despite limited arable land and virtually no public investment in infrastructure, farm machinery, and equipment. Agriculture also employs about 66% of the labor force but accounts for only 19% of official GDP implying low average returns to labor.

The national development (NDS) and poverty reduction (PRS) strategies of the GoT-NDS (2007–2015) and PRS (2007–2009) endeavor to secure sustainable economic development including improvement in the living standards in Tajikistan. The ongoing land distribution and land registration reforms (Halimova, Chap. 13) aim at strengthening the private sector in agriculture, developing entrepreneurial activity, and ensuring equal rights and security for land tenure. These are among the key priorities for driving rural economic growth, improving household food security, and enhancing livelihood. The land reform is as yet incomplete: modern land registration is still nascent, and land markets are blocked due to unresolved issues with sales of land rights and security of land tenure – revealing some internal opposition to reform and insufficient political will.

Fig. 10.4 Milk and other dairy products from ruminant livestock provide the basis for subsistence living

Most livestock[2] are owned by the rural households, the majority of which are poor. Livestock, therefore, are important in the coping strategy[3] of these households despite low livestock productivity. Sustainable livestock development faces many constraints especially winter feed shortage which is caused by factors such as high animal numbers and degradation of, or lack of access to, pastures. One result is low-quality underfed animals which are more susceptible to disease, less fertile, and die more easily. This in turn leads to low animal yields and income, thereby causing or contributing to poverty. Poverty itself is also a cause of low animal yield and income since farmers have no spare cash to invest in better feeding and health care.

The number of poor in Tajikistan is estimated[4] at just fewer than 50% of the total population equivalent to approximately 3.5m people, while the number of extreme poor is estimated at about 17%, equivalent to about 1.2m. Most of the poor are (i) in Sughd and Khatlon regions,[5] (ii) rural, (iii) involved in crops and livestock, and (iv) characterized by having few animals, small land areas, limited or no access to finance, and limited education. Many of the poor are also in female-headed households[6] partly due to labor migration (Kurbanova, Chap. 7).

Many of the poor, however, are reluctant to invest in land due to land tenure insecurity (Halimova, Chap. 13), which may have contributed to the rapid increase in livestock numbers, thereby demonstrating livestock's importance to the poor in terms of (i) livelihood and (ii) perhaps more importantly a survival coping mechanism as mentioned earlier.

Small ruminants (Fig. 10.4), for example, provide the majority of Tajikistan's rural households with nutrition and income as well as serving as "bank" or buffer in times of social, natural, or financial emergency situations.

---

[2] PRA work indicates 82% GoT statistics, say, up to 90% (see Kurbanova, Chap. 7, this volume).

[3] Also called social adaptive strategy (see Squires, Chap. 2, this volume).

[4] UNICEF, World Bank, LSMS 2007.

[5] UNICEF (2007).

[6] Tajikistan has the highest number of female-headed households in Central Asia.

Livestock are the main economic activity of the rangelands, and it will be developments within this sector that will influence their potential sustainability. Using 2010 national statistics, there are approximately 1.9m cattle and 4.4m sheep and goats. The current gross output is estimated at US$670m using 2011 farm gate prices. This is equivalent to a per capita income of US$160 based on a livestock-owning rural population of 4.2m.[7] Broken down into meat and milk terms, this is equivalent to an average annual per capita income of US$107 from meat and US$53 from milk or a daily income of about US$2.6 per household and $US 0.44 per capita assuming an average household size of six. This simple financial estimate excludes other benefits such as the financial security derived from livestock ownership, nutrition benefits, as well as the value of dried dung as a fuel.

Typical gross margins from sheep, goat, and cattle when all fodder (excluding grazing) is taken into account are, at best, at break-even point. This reflects the present relatively high price of fodder compared to output values. Where a farmer must buy most of his fodder needs and livestock and some of the output is home consumed, livestock will represent a cash drain. Where most of the roughage is home produced, there will be a higher perception of profitability. Current annual gross margins are negative, for example, per breeding, sheep are (−) TJS 1, goat (−) TJS 42, and cow (−) TJS 355. Despite the negative figures, the potential for livestock and pastureland development is considered to be high once some of the much needed structural changes (tenure, pasture user groups) are enacted. This will then facilitate many standard husbandry improvements (mainly quality fodder production, better nutrition, improved animal health, etc.) to be introduced that, cumulatively, will have an enormous impact on the current low levels of performance. At the same time, all opportunities for value-adding within the livestock sector should be sought. Nevertheless, a key to the sustainable development of the rangeland economy, and the population dependent upon it, will be seeking alternate employment opportunities and diversification in the enterprises (Lerman, Chap. 8). This will remove employment pressures from the rangeland areas, thus making it possible for those remaining in the livestock sector to realize real income increases in the future.

2 The Natural Resource Base for the Livestock Industry

2.1 Rangeland and Forage Resources

The total area of rangelands/pasturelands[8] is 3.9m hectares and represents over 80% of the country's agricultural land. The largest areas of rangeland are found in Khatlon and the Districts of Republican Subordination (DRS) and represent 60% of

[7] Based on a national resident (i.e., excludes migrant workers) human population of 7 m, 74% rural population of which 82% own livestock (PRA survey work).

[8] The terms rangelands and pasturelands (often abandoned cropland) are used interchangeably.

Table 10.2 Classification of pasturelands in Tajikistan based on altitude and season of use

Pasture type[a]	Altitude (meters asl[b])	Season of use	Remarks
Winter	350–1,000	November–March	Very severely degraded
Autumn/spring	900–1,600	March–May September–November	Degraded
Summer pastures	1700–3,500	June–August	Fair to good
Year round	All areas below snow line	All year	Severely degraded

[a]Moderated locally by such factors as aspect and slope and presence/absence of shelter
[b]Above sea level. Hay is typically taken from pastures that are relatively close to villages, i.e., from the autumn/spring categories or land set aside for the purpose

the total national rangeland area. Table 10.2 provides an overall breakdown of land areas and pasture in the country. The pasture area includes some 360,000 ha of grazing land included in state forest reserves which are generally used by livestock owners on short-term leases. The balance of the agricultural land is 0.8m hectares of which 70% is irrigated. The remaining lands are dominated by rocky and snow-covered mountains with little or no vegetative cover.

The rangelands/pasturelands are traditionally categorized by their season of use, summer, spring/autumn, and winter pastures, and thus are a reflection of altitude (Table 10.3). Livestock owners typically with more than 25 sheep will use seasonal pastures (usually collectively, employing shepherds). Owners with small numbers particularly cattle increasingly use pastures close to villages, and as a result, many of these near-village pastures show the worst signs of overgrazing and degradation.

Of the total 4.7m hectares of agriculture land, 702,000 ha (14.9%) are arable,[9] 110,000 ha (2.3%) are shrubs and trees, and approximately 3.9m hectares (about 82%) are rangelands/pasturelands. The 3.9m hectares are categorized according to seasonal use; more than half comprise summer pastures. Only about 2% of all pastures are provided with water (Fig. 10.5).

The state of the natural rangelands/pasturelands has markedly deteriorated in the 20 years ending 2010. The reasons are many, and it is the cumulative effect of these that has led to the widespread deterioration of this natural resource:

(a) There is no official separation of rangelands/pasturelands from arable land in the Tajikistan Land Code, both being grouped into the land-use category "agricultural land" (or farming land) where only "irrigated or rainfed cropland" is further designated.
(b) The Pasture Trust (PT) of the Ministry of Agriculture is responsible for state control of pasture use and protection. The PT, however, has very few resources and lacks a strategic plan and vision for the development of the natural resource base of the country. Its limited budget allows only a small area of rehabilitation each year.

[9] This includes an estimated 49,000 ha of saline land. In addition, there are further 200,000 ha comprising household plots and presidential plots (168,000 ha) and 32,000 ha of "abandoned land" – the total official area of arable is thus 902,000 ha.

Table 10.3 Pasture area ('000 ha) by Oblast in Tajikistan

Category	Tajikistan total	DRS	Khatlon	GBAO	Sughd
Total land area	14,256 (100%)	2,861 (20%)	2,468 (17%)	6,408 (45%)	2,519 (18%)
Total area of agricultural lands	4,719 (33%/100%)	1,222 (26%)	1,622 (35%)	771 (16%)	1,104 (23%)
Area of pasturelands (includes forestry grassland areas of 360,000 ha)	3,855 (27%/100%)	1,083 (28%)	1,227 (32%)	751 (19%)	794 (21%)

Source: "The Report on Land Areas by Land Categories and Land Usage in the Republic of Tajikistan," State Committee of Land Management and Geodesy of the Republic of Tajikistan – as of 1 January 2011

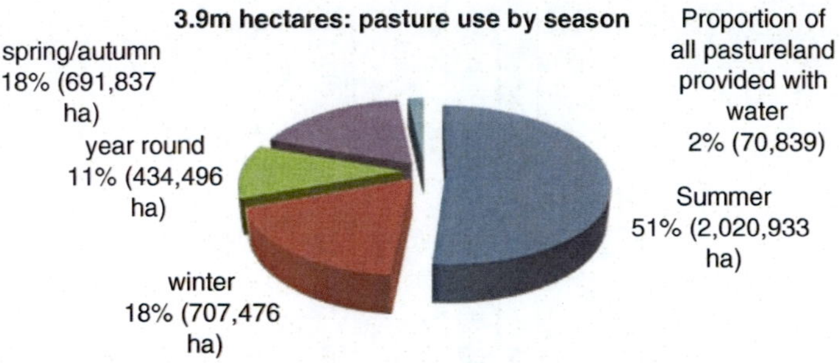

Fig. 10.5 Classification of pastures per season of use (Excludes small areas of pastures outside the main national boundaries)

(c) There has been no organized uptake of rangeland revegetation or land improvement by villages and *raions* and very little by individual farmers. The Soil Science Research Institute (SSRI) in the Academy of Agrarian Sciences is the nominated lead institution for soil conservation activities. Though valuable demonstrations can be seen on their experimental stations, work has rarely "passed the fence" for widespread uptake.

(d) The management of the rangelands has changed in the past 20 years, from well-organized pasture maintenance, established stock routes, clear assignment of pasture rights, animal veterinary points, and an established schedule of transhumance pasturing. Only emergency stocks of cut feed were required to cover feed deficiencies caused by such events as colder-than-usual winters. Now, the rangeland system is based on extensive livestock grazing with limited supplementary feeding of cultivated fodder and concentrates.

(e) There is now a greater proportion of livestock owners, generally of smaller herds, who no longer send their animals to winter pastures, rather they allow them to graze within a 3–5-km radius of their village as they cannot afford to buy hay or feed concentrates for stall feeding. This has led to intensive, year-round overgrazing of traditional spring/autumn pastures. These lands are grazed by village livestock in wintertime as well.

(f) Changed too is the "seasonality of use" of pastures toward reliance on year-round grazing. For instance, in five raions within the DRS, there has been a shift in the way/how the pasture is used – with year-round grazing increasing substantially (Kurbanova, Chap. 7).

The production of forage crops for animal feed has fallen by 45% over the 20 years ending 2010. In Soviet times, large collective farms were engaged in animal husbandry and consequently specialized in producing large amounts of fodder crops and grain. Currently there are now only 88,000 ha of fodder land, whereas the number of livestock (cattle, sheep, and goats) in the 8 years to 2010 has increased by 67%. Such data only confirms the clear observation that livestock in Tajikistan are underprovided with fodder and are in a state of undernutrition, particularly in the later winter months and spring. A principal cause for the decline in fodder production is a structural one as a result of changes in farm size and the consequent change in farming patterns combined with the lack of knowledge about farming and animal husbandry on the part of many of the first-time farmers. Many had been employed in Soviet times as technicians, drivers, mechanics, etc.

2.2 *Land Degradation and Loss of Forage*

The vegetative deterioration of the Tajikistan rangelands has wider ecological and environmental impacts on regional water supplies, biodiversity, and climatic variations. Tajikistan is one of the major water tanks of the world, with 840,000 ha (6% of the total land area) comprising glaciers that feed some 50% of the rivers in Central Asia. Their continued deterioration, therefore, has severe implications across the entire region. Preserving the vegetative cover of Tajikistan's extensive pasturelands is critical to achieving short- to midterm sustainable water supply to the region (and Tajikistan). This is a major goal of rural development since it is recognized that it will be the pasture-based livestock sector that will ultimately provide the economic stimulus for national development. Hence, it is the stabilization and future growth of this sector that must become a national priority.

Land degradation in much of Tajikistan is characterized by significant recent loss of natural forest, severe and continuing forest degradation, and widespread soil erosion on sloping land. Most farming communities rely on remnant forests and scattered trees for fuelwood supplies and on degraded mountain lands for livestock grazing. Natural forests have long been relied upon for non-wood forest products, such as naturally growing fruits, to meet household needs and for sale. A sharp deterioration in the area, ecological structure, and quality of forests in Tajikistan since independence contrasts sharply with the situation throughout most of the twentieth century. The illegal collection of fuelwood, mostly near villages, is the principal cause of forestland degradation, leading not only to widespread erosion but also loss in natural biodiversity of indigenous fruit and nut trees, many of which are considered to be endangered and of global horticultural importance. Fuelwood is the primary household energy source for cooking and heating, used by 70% of the country's population living in rural households. Nationally, the demand for fuelwood is of the order of 20 million m^3 annually and is probably increasing. Non-wood forest products (important products of the rangelands) are essential to the daily lives of many Tajik people. Not only do they contribute to their subsistence throughout the year, but non-wood forest products can also be traded for cash. Fruits are particularly important, especially apples and apricots, also various nuts, honey, and medicinal herbs.

Soil erosion is widespread and severe across the rangelands of Tajikistan. FAO has published figures of "up to 150 tons/ha/year" and a general, rangeland-wide figure of "30 tons/ha/year." Each of these falls within the universally accepted class of "severe," particularly considering that a yield of 6–11 tons/ha/year is considered to be the tolerable erosion limit and the average rates of natural soil formation are in the order of 0.5–1.0 tons/ha/year. According to Tajikistan's National Action Plan to Combat Desertification, about 90% of the country's agricultural lands are suffering from erosion, compared to 68% in 1973 with mudslides and gully erosion especially strong in the foothills as a consequence of greatly increased human and animal densities in nearby towns over the past 20 years with concurrent increased demand for meat and grain production.

A major contributor to this continuing degraded state is the lack of national or local conservation programs to counteract the degradation, let alone improve land quality for the sustained development of the rangelands. Degradation of the water supply system to the rangeland/pastureland area commenced at the end of the Soviet era when all upland irrigation systems, both pumping stations and concrete channel supply systems, inexorably fell into ruin. It is unlikely that it will ever be feasible to rehabilitate them. The alternate solution is likely to be low-cost, gravity-fed irrigation systems from natural springs to support small-scale fodder and orchard production, to ensure better productive capacity through dry seasons.

Soil erosion from water is the most widespread form of land degradation, evident as gullies, slumps, sheet wash, and mudflows. Wind erosion is found in many mountainous areas of the country. The more visible effect of soil erosion is the physical removal of soil; however, the removal of the fertile rich topsoil is an additional major concern. Topsoil can take hundreds of thousands of years to form but can be lost within minutes with acute erosion events. The presence of widespread soil erosion shows a strong imbalance of land management practices and the resilience of the soil resource. The aim with sustainable farming systems is the achievement of a natural balance of demand and supply, where land usage necessary to maintain economic and human development does not destroy the environment required to assure this growth. The Tajikistan rangelands demonstrate the opposite of sustainable management, and the most evident outcome is widespread and severe soil erosion. Eroded land has far lower productive capacity whether for grazing or fodder production. The challenge is to find and adopt widely applicable, practicable systems of land management that not only ensure the long-term socioeconomic development of the sector but concurrently protect and enhance the resource base that underpins and facilitates its growth and prosperity.

There are several factors that act both individually and together to lead to the widespread degradation of Tajikistan's pasturelands:

(a) There has been a disproportional increase in the rural population relative to Tajikistan as a whole; whereas the numbers in the irrigated cropland area have remained fairly constant since 1980, the rural population has more than doubled.
(b) Winter and spring-autumn pastures are under relatively greater pressure. Though they are less extensive than summer pastures, they are year-round accessible and close to the villages where livestock come from.
(c) Unregulated grazing causes further pressure on already degraded pasturelands, particularly large herds/flocks that tend to remain in one location beyond its carrying capacity. Tajikistan's natural rock and soil formations cause them to be

most vulnerable to soil erosion with 93% being hill and mountain lands where soils have developed on loose-packed loess. Loess soils rapidly erode and wash away, once the topsoil is devoid of vegetation.

(d) The nation's rainfall patterns are such that 50% falls in spring. Together with spring snowmelts, this causes a short period with intense water flows across the rangeland landscapes at the very time of least surface vegetation. The cumulative impact is widespread soil erosion.

(e) Erosion causes far more damage than the loss of topsoil. Off-site negative impacts include blockage of hydro engineering structures and sediment-full waters that are rich in nutrients and deleterious farm chemicals, leading to polluted waterways and storages.

(f) Poor land management directly leads to land and vegetation degradation. In Tajikistan, a series of events has greatly exacerbated the unsustainable demands on the rangelands. From 1991, grain imports dropped markedly, while at the same time some 40,000 ha of sloping rangelands were converted to arable production with disastrous outcomes in terms of low-yielding grain crops and widespread soil erosion. At the same time, the failure of electricity production and gas imports caused Tajiks to deforest the hill lands on a massive scale and utilize animal dung for cooking and heat, further exacerbating soil erosion and desertification. It has been estimated that degradation from erosion fostered by overgrazing, plowing (arable land), and deforestation now extends to 3 Mha, or about 85% of the total Tajikistan rangelands.

(g) Despite the above recognition of the widespread and severe degradation of the pasturelands, there seems, currently, to be most limited map-based (cartographic) information to describe and delineate areas of degraded land and its nature.

In conclusion, the widespread and severe land degradation of Tajikistan's pasturelands strongly and negatively impacts on rangeland productivity as well as its capacity for off-site, clean water supply. Poor land management practices on a most fragile resource are the main drivers of accelerated land degradation, and little seems to be done to redress the problem, far less to improve the productivity of these lands.

2.3 Biodiversity

The biodiversity of the flora of Tajikistan is the richest in Central Asia and comprises 5,000 species of higher plants and over 3,000 species of lower plants. Both categories display a high level of endemism (being unique to a defined geographic location). The fauna of Tajikistan is also very diverse. There are 84 species and subspecies of mammal, 385 species of bird, 46 species of reptile, 52 species of fish, 2 species of amphibian, and more than 10,000 species of invertebrate. A few rare and endangered species of animals can also be listed, such as screw-horned goat, argali, urial, Bukhara red deer, snow leopard, Central Asian cobra, desert monitor, peregrine, snow cock, and others.

Such diversity arises from the specific geographical location of Tajikistan inside the Eurasian continent with its heterogeneity of habitats, ranging from the hot lowland deserts of southern Tajikistan to the high mountains of western and eastern Pamir. Vertical zonation, geomorphologic contrasts, and the many and varied landforms with their different aspects and slopes create many habitats – thus favoring great diversity of the microclimatic conditions. The unique and high-value biodiversity of Tajikistan's flora and fauna is under serious threat, a result of unsustainable agricultural practices; socioeconomic pressures; overuse and illegal uses of forest, rangeland, and wildlife; insufficient integration of biodiversity considerations into local land-use plans; weak institutional and human capacity; and limited public awareness (Sedik, Chap. 9).

Tajikistan has ratified a number of international conventions including the UN Convention on Biodiversity, the UN Convention to Combat Desertification, the Ramsar Convention about conservation of wetlands, and others relating to wildlife conservation, trade in endangered species, etc. At the institutional level, the State Committee for Environmental Protection, local authorities, and administration of reserves, protected areas, and national parks are engaged in protection of ecosystems. Within the framework of realization of the State Ecological Program, measures on restoration of vulnerable landscape complexes on separate ecological areas of Tajikistan are stipulated. In development of the National Strategy and Action Plan on preservation and sustainable use of biodiversity, the state has attempted to apply an integrated approach. In order to administer a wildlife management policy, with a view to increasing the level of public awareness, the GoT has developed and approved a State Ecological Program in 1996 and the State Program of Ecological Education of the population in 1997.

3 Structure of Livestock Farming

Tajikistan has a total of 6.3m livestock, comprising 1.9m cattle and 4.4m sheep and goats (Table 10.5). They are spread among the four main types of farm found in the country, and the first two of them are the most numerous[10]:

1. *Landless households* are strictly speaking not farmers as they do not have a land certificate[11] but often rent arable and pastureland from dehkan farmers. A typical family owns one to two cows, a few sheep and goats, and possibly a donkey (Kurbanova, Chap. 7). Output of the animals is low, and the milk they produce is the main supplier of protein and calcium for the family diet. Householders will typically sell an animal when there is a need for cash rather than selling at a more optimum time as regards livestock rearing (Table 10.4).
2. *Dehkan farms* include two subtypes – family/individual and joint dehkan farms (see Halimova, Chap. 13 for more discussion on the various types of dehkan farm).

[10] More details about these types of farms and livestock will be found in Halimova, Chap. 13, and Kurbanova, Chap. 7.

[11] Land-use certificates are a form of land entitlement associated with pasturelands (see Halimova, Chap. 13).

Table 10.4 Livestock populations in Tajikistan as a whole and for specific regions (2010)

	Cattle		Sheep/goats	
No's '000	No.	%	No.	%
Tajikistan	1,897	100	4,394	100
Sughd	511	27	1,192	27
Khatlon	795	42	1,827	42
GBAO	104	5	313	7
DRS	488	26	1,062	24

Source: TAJSTAT (2011)

Fig. 10.6 A joint venture feedlot complex that is situated in the transition zone between the irrigated lowland and the more extensive rangelands. The feedlot operators can source fodder and grain from the lowlands and livestock from the rangelands

The size of the farm can vary considerably, from less than a hectare of arable land and some pastureland up to several hundred hectares of arable land and several thousand hectares of pastureland. The herd size on farms also varies considerably, and some of them acquired large land holdings and several hundred or more sheep and goats which give them considerable influence in a village.

3. *Joint stock companies and limited liability companies* are large private farms such as the cattle farm (feedlot) near the village Tiloi Safed in Rudaki *raion* in Fig. 10.6.
4. *State farms:* These are similar to the former *kolkhoz*[12] and *sovkhoz,* but today very few exist.

[12] See glossary of terms.

3.1 Current Production from the Livestock Subsector

National statistics for 2010 indicate that the approximate livestock populations are 1.9m cattle and 4.4m sheep and goats. The estimated current gross output from these animals is US$670m, using 2011 farm gate[13] prices. This is equivalent to a per capita income of US$160 based on a livestock-owning rural population[14] of 4.2 million. Broken down into meat and milk terms, this is equivalent to an average annual per capita income of US$107 from meat and US$53 from milk or a daily income of about US$2.6 per household and US$0.44 per capita assuming an average household size of six. This simple financial estimate excludes other benefits such as those derived from livestock as a form of financial security and nutrition benefits as well as the value of dried dung as a fuel source.

Between 70 and 95% of the Tajik rural population either own or have some financial interest in ownership of livestock. The rangeland-based livestock sector therefore has tremendous importance to the income provision and related livelihoods for the whole rural sector. In-kind benefits are derived from milk, home slaughtered animals, and dried dung. Performance is acknowledged as being low with the current gross annual estimated red meat production being worth about US$448m. Applying better nutrition and health care that raises livestock fertility and growth rates and lowers mortality would raise offtake[15] of all livestock by 45% to $652m. This higher offtake could be achieved after 10 years while keeping livestock populations constant, an important goal to avoid additional pressure on the already overutilized rangelands.

After red meat, milk is the largest contributor of the livestock sector to agricultural output and is the main supplier of protein and calcium for many rural families. Its gross contribution is estimated at $222m using certain assumptions[16]. When this is added to the value of the national red meat offtake of US$448m, this gives a total gross output value of US$670m/year. For livestock-owning rural households, this is equivalent to an average annual per capita income of US$107 from meat and US$53 from milk or a daily income of about US$2.6 per household and US$0.44 per capita[17] assuming an average household size of six (Table 10.5).

There are many pressures on the livestock sector in Tajikistan:

(a) Table 10.6 shows that though the total cattle and sheep/goat numbers have risen by 52 and 41%, respectively, from 1992 to 2010, the fodder crop area is reported to have fallen by 56% and actual fodder production by 74% during the same

[13] Prices received from buyers at point of sale in the village.

[14] Based on a national resident (i.e., it excludes migrant workers) human population of 7 m, 74% rural population of which 82% own livestock.

[15] Offtake refers to the annual sale of livestock surplus to those required for breeding or to their salable products, milk, cheese, wool, hides, skins, etc.

[16] For 985,000 cows (TAJSTAT) yielding 1,000 kg with a calving interval of 540 days, 20% of the milk is consumed by the calf, and the average milk price is TJS 2/kg.

[17] Here we are talking about the gross value of livestock products/capita; this is not the same as GDP which will reflect the spending power of external remittances.

Table 10.5 Calculation of national red meat and milk production

Current livestock population (m)		Carcase/milk weight (tons)/year	Farm gate value/year	
			TJS (m)	US$ (m)
Cattle (meat)	1.9	71,400	1,428	297
Sheep (meat)	2.7	19,900	476	99
Goats (meat)	1.7	9,500	248	52
Total (meat)		100,800	2,152	448
Milk (cows)	1.0	532,800	1,066	222
Total gross value of meat and milk			3,218	670

Table 10.6 Trends in livestock and fodder production in Tajikistan 1992–2010

	1992	2002	2010
Cattle ('000)	1,246	1,136	1,897
Sheep ('000)	2,237	1,591	2,729
Goats ('000)	871	842	1,666
Horses ('000)	53	73	76
Fodder crop area ('000 ha)	201	99	88
Total fodder production ('000 tons dry weight)	870	230	480

Source: TAJSTAT (2011)

period. In 2010, the fodder crop area only accounted for about 10% of the arable area.

(b) Small farmers and landless households own up to 90% of cattle and about 80% of small ruminants. They seldom have arable land and struggle to feed their livestock and as a consequence usually the least productive (Kurbanova, Chap. 7). These households are typically the poorest in the village unless there is income support from remittances (see above).

(c) There has been a significant change from the large collective herds during the Soviet era to family ownership of livestock, and the rangelands have become fragmented, rented to groups of private individuals. The consequence is modification of traditional seasonal grazing patterns based on transhumance to year-round grazing at one location.

(d) Transhumance has been a traditional practice for many decades. The practice continues today, but increasingly difficulties are faced along stock routes,[18] and the summer pastures are being allocated to new owners.

[18] Not only are there actual land barriers due to "privatized" land, there are access problems with roads and bridges in a poor state of repair. Often these can be rectified with small sums of money. By allocating small sums with minimum of administration through *jamoats* or *mahallas* to undertake these activities (such matters are a main activity of the *mahallas*), much could be achieved at little cost. *Jamoats* are local administrative regions within a *raion* (like a county), and *mahallas* are local village councils.

Table 10.7 Summary of livestock budgets

	Unit	Dairy beef	Sheep	Goat
Current situation (1)				
Gross income per animal	TJS	2,533	247	208
Net income per animal	TJS	−355	−1	−42
Net income excl. roughage	TJS	1,358	111	73
Improved situation (2)				
Gross income per animal	TJS	3,385	389	338
Net income per animal	TJS	−268	125	65
Net income excl. roughage	TJS	1,951	244	198
(2) – (1)				
Net incremental income per animal	TJS	87	126	107
Net incremental income/animal excl. forage	TJS	593	133	125

(e) The main failure of efforts to modernize agropastoral resource management is the limited capacity of farms to produce adequate quantities of food for the household and enough feed for penned livestock. In the case of feed, quality is probably the most limiting factor rather than absolute bulk.

(f) Village-based livestock owners in the uplands seldom send their animals away. Thus, village pastures are seriously overgrazed. Many spring/autumn rangelands have been converted to year-round pastures and now are never rested since they are also grazed by village livestock in winter.

(g) Livestock health is very poor principally due to the lack of quality feed. Farmers involved in recent PRA survey (Kurbanova, Chap. 7) highlighted disease as one of the three priority problems that their animals face.

The feed surfeit in summer, a shortage in spring and autumn, and a deficit in winter result in alternate periods of weight gain and loss and, coupled with disease, result in low fertility, long calving intervals, poor milk yields, and slow growth rates. For many families, livestock are a part of their more subsistence way of life and will often be subsidized indirectly through remittances and other sources of income. Table 10.7 summarizes their net incomes per animal and net incomes per hectare.

The budgets reflect a commercial situation and illustrate that with current levels of performance with all the (relatively expensive) hay and roughages taken into account, negative figures result. In a strictly commercial sense, the results are very poor reflecting the high cost of maintaining animals which have a very low productivity. Many farmers who produce their own roughage will not perceive this as a cost, and thus, in simple cash terms, the livestock produce positive margins. In the case of cattle, milk is invariably all home consumed and thus does not represent any cash income. If a farmer produces his own roughage for winter feed, he arguably can sell it instead of feeding to his own stock – that is why it is included. These results would also be the case if a farmer had to buy all his winter roughage requirements. Margins are also shown excluding forage costs which then transform the negative margins into what may be perceived as high returns.

Landless households and small dehkan farmers sell animals when they need cash which could be at any time of the year. However, the income from livestock sales is low, only about TJS 900 per year per household. These households and dehkan famers seldom invest in good quality feeds to add value to their animals at sale because they either cannot afford to buy feed or they are unwilling to borrow cash to buy it because of the risks of not covering their costs at sale. Farmers in more favorable areas derive the benefit of the value addition at the expense of those in the uplands. Thus, there are villages in the lowlands where livestock are fattened, particularly on the irrigated plains since feed availability is better and markets are closer. The livestock for fattening either belong to local producers or are purchased from upland farmers. Unless they live close to a market, small producers sell to local traders who come to the village, and the traders transport the animals to a large market such as close to Dushanbe and sell them to buyers. It is often said the sellers are paid a low price but they avoid the high transport costs of taking one or two animals to a distant market. The buyers in the main livestock markets might be butchers who slaughter finished animals or they feed them for few months to add finish. Buyers might also be specialized farmers who fatten and finish livestock.

4 Livestock: Possible Development Actions

Livestock are a major contributor to the rural economy, but numbers have risen to unsustainable levels relative to the feed available, and consequently their output is low. They are often the only easily redeemable asset for millions of rural poor, many of whom are highly food insecure. For livestock-owning families, milk and milk products provide a vital source of protein, and sales of surplus products are often low. Moreover, demand for meat and milk will continue to rise due to a rapidly increasing human population and because red meat in particular has a high income elasticity of demand.[19]

In view of the steadily increasing livestock numbers, policies and husbandry practices are needed that slow and eventually stop this growth, but they should be accompanied by increases in productivity.[20] A policy environment that is conducive to these practices being applied should be put in place. The following section looks at a number of actions that need to be taken to develop the livestock sector in Tajikistan.

4.1 Possible Actions

From a nutritional perspective, the two contrasting scenarios shown in Table 10.8 should be considered when attempts are made to develop the livestock sector. The options are briefly discussed below.

[19] As real incomes rise, demand for meat increases at a faster rate.

[20] Shorter calving intervals, lower mortality, better nutrition, faster growth rates, and better feed utilization.

Table 10.8 Possible livestock development scenarios

Scenario	Options	Possible actions
1. Match livestock numbers to feed supply	Selling surplus animals	Encourage more timely selling and trust in rural finance
		Stimulate more lowland fatteners, thus drawing some young stock off the rangelands for fattening – rural finance plays a role here
	Better-performing animals	Adjust parturition period to match nutrient needs to feed supply and milk production to demand
		Improve veterinary services – high mortality and morbidity waste scarce and costly feed
		Disciplined pasture management through group herding schemes which apply optimal stocking rates and raise livestock productivity
		Judicious use of improved breeds, timely parturition (see above)
2. Match feed supply to livestock numbers	Improve forage output	Management of rangelands (pasture groups, deferred grazing, modest use of fertilizer), rural finance needed
		Where profitable optimize growing of forage crops for sale/use in arable rotations – scope for cotton growers to introduce legumes – rural finance needed
		Balanced diets to improve overall feed utilization (use of balancing supplements and low levels of concentrates), rural finance needed
	Home rearing/ penned stock	Where appropriate promote "cut and carry" systems to improve feed utilization

As Table 10.8 and accompanying commentary shows, there is a need for more investment. Access to rural credit lies at the heart of efforts to develop an efficient livestock sector. Short-term credit can be used profitably for such things as supplementary feed for livestock and growing/finishing livestock, fertilizers, improved seeds, and materials for better animal housing. Larger and longer-term loans are needed for the purchase of cows, milk processing equipment, and the materials needed to construct better livestock sheds. To assist the farmers, there is need to improve their knowledge and transfer technologies related to animal health and nutrition. Capacity building at the village level is an essential part of efforts to improve livestock productivity. In remote locations, a Farmer Field School approach facilitated by a local veterinarian is recommended, whereas in the higher potential of the lowlands, a private-sector-based extension service would be effective. Demonstration of livestock productivity enhancing practices can expose farmers and their womenfolk to practices that improve livestock productivity including value-adding procedures for dairy products but also handicrafts such as spinning and weaving (Fig . 10.7). They are part of the FFS activities in which a group of farmers each host a demonstration of one particular practice. The uptake of the practices should be monitored and new training programs mounted to meet the need.

Fig. 10.7 Handicrafts have always been part of Tajik tradition. There is scope for value-adding to wool and cashmere that at present is underutilized

Timely selling of surplus animals: Instead of using precious feed to keep their own heifers over winter, loans at favorable terms are needed to encourage farmers to buy in-calf heifers, to buy supplementary feed for them, and to finish 6-month-old small stock and 18-month-old bull calves before winter.

Encourage fattening and finishing in the irrigated lowlands: To reduce pressure on upland pastures, dehkan farmers in irrigated areas need encouragement to fatten and finish livestock purchased from upland farmers. Short-term loans could persuade more of these farmers to buy and finish animals coming off summer pastures.

Match parturition season to feed supplies and market demand for milk: Enter rams and bucks up to 1 month later in autumn so that peak nutritional needs of lactating ewes and does are met from pastures in early summer. Making autumn the peak calving period in the lowlands would help meet the winter demand for milk in urban areas.

Better access to effective veterinary health services: Further support to the veterinary sector is essential. Despite the 400 or so veterinary field units (VFUs) spread across the country, the coverage is still incomplete. Veterinarians would also give professional feedback on disease outbreaks at a particular location. Private veterinarians need training in extension methods to enable them to play a more effective role as advisors on better livestock husbandry.

Disciplined pasture management: Equitable pasture rights and timely selling of surplus animals lower pressures on overgrazed pastures. Community herds composed of animals from landless households and small *dehkan* farms enable more disciplined pasture management practices and gradual restoration of pasture productivity.

Breeding improvement: Introducing improved breeds is justified in situations where nutritional and disease environments allow the higher performing animals to express their genetic potential. In such cases, a community-based approach should be applied to ensure that animals of all owners benefit, not just owners to large herds.

Growing fodder crops: Agronomic practices that substantially increase yields of irrigated fodder crops today grown on a small area would make a substantial contribution to removing the present feed deficit. The recently established Seed Association of Tajikistan will play a central role in encouraging the private sector to produce certified fodder seed.

Feeding balanced diets: Livestock are currently fed to survive rather than for production, but small improvements in the protein/energy ratio could result in higher offtake. Even feeding small supplements of protein-rich cottonseed cake and lucerne/alfalfa hay would help, but many farmers cannot afford to purchase them. Moreover, supplies of both these feeds are low partly because areas are limited and yields are far below what is possible. Balancing livestock diets is therefore closely related to higher yields of the main protein-rich.

By 2025, the livestock sector of Tajikistan could become a larger contributor to the nation's meat and milk supply while putting less pressure on its rangeland resources. This goal requires decentralized governance of village rangeland resources, a more plentiful supply of quality fodder from the irrigated areas, and support from effective veterinary and extension services. Rangeland productivity will gradually recover as pasture user groups begin to take decisions about pasture allocation and grazing pressures. One major hurdle to achieving this goal is the growth of the rural population and the associated livestock population.

Policies that encourage rural households to become less dependent on livestock for their livelihoods will also require direct or indirect support mechanisms that encourage them to sell their livestock. For the older people, this could be achieved through more generous pensions and for younger people, creating work in the nonagricultural sector.

Acknowledgements The work reported here was generated during the ADB/GEF Rural Development project and the authors were engaged as consultants by a consortium of GITEC Consult GMBH/Donaev Management Consulting under the direction and guidance of the PMU in Dushanbe. We are indebted to the efforts of our erstwhile colleagues for their inputs into aspects of this work.

References and Further readings

FAO (2008) Country profiles- Tajikistan

Halimova N (2012) Land tenure reform in Tajikistan: implications for land stewardship and social sustainability: case study. In: Squires VR (ed) Rangeland stewardship in Central Asia. Springer, Dordrecht, pp 305–332 (Chapter 13, this volume)

IMF National Development Strategy IMF/GoT

IMF National Poverty Reduction Strategy IMF/GoT

Lerman Z (2012) Rural livelihoods in Tajikistan: what factors and policies influence the income and well-being of rural families? In: VR Squires (ed) Rangeland stewardship in Central Asia. Springer, Dordrecht, pp 165–188 (Chapter 8, this volume)

LSMS (2007) Living standard measurement study, Tajikistan. World Bank, Washington, DC

Squires VR (2012) Better land stewardship: an economic and environmental imperative if there is to be sustainable development. In: VR Squires (ed) Rangeland stewardship in Central Asia. Springer, Dordrecht, pp 275–304 (Chapter 2, this volume)
TAJSTAT 2011 Agriculture in Tajikistan 2010. Tajikistan State Committee of Statistics, Dushanbe
UNCCD 2018 National Action Plan to Combat Desertification in Tajikistan
UNICEF (United Nations Children's Fund) (2007) State of the World's Children, UN, Geneva
World Bank (2008) World development indicators 2008. World Bank, Washington, DC

Part IV
Creating the Enabling Environment

Nothing is more important than having an "enabling" policy, economic, and legislative environment. Barriers to adoption of more sustainable land management can be swept aside if suitable policies and legislative frameworks and governance rules are adopted and adhered to.

The three chapters in this part deal with issues of land tenure – the most important issue confronting land users in the Central Asian region.

Chapter 11 compares pastoral tenure arrangements in the five former Soviet republics and analyzes the thinking behind attempts to distribute land use rights in the 20 years following independence.

Chapter 12 – The importance of good governance and the role of institutions in enabling and promoting sustainable development are discussed in this chapter.

Chapter 13 is a detailed examination of the land tenure reform in Tajikistan and considers the implications of land stewardship and social sustainability.

Chapter 11
Pastoral Tenure in Central Asia: Theme and Variation in the Five Former Soviet Republics

Sarah Robinson, Christoph Wiedemann, Stefan Michel, Yerlan Zhumabayev, and Navinder Singh

Abstract Since 1991, both *de facto* and *de jure* pastoral tenure regimes diverged significantly in the five former soviet Central Asian republics (Kazakhstan, Kyrgyzstan, Tajikistan, Turkmenistan, and Uzbekistan). Four of the five republics are currently considering the introduction of pasture codes with both individual and common forms of tenure under discussion. In the light of these debates this chapter examines the evolution of pastoral land tenure and user rights in each of the five republics over the 20 years since independence. Different choices were made by policy makers that have affected two key outcomes: firstly, livestock mobility and secondly, pasture access. The situation in each of the republics is reviewed and some case studies are presented.

Keywords Landlessness • Poverty • User rights • Livestock ownership • Access • Water • Infrastructure • Transhumance • Migration • Lease agreements • Speculation

Key Points

- In the immediate aftermath of independence, a number of common patterns emerged across the region: the proportion of private animals increased, costs of inputs such as winter feed and transport rose sharply, and livestock mobility

S. Robinson (✉)
Division of Ecology and Evolution, Imperial College , London, UK
e-mail: s.robinson@imperial.ac.uk

S. Michel
Freelance consultant, 34-7 Bukhara str, 734024 Dushanbe,Tajikistan

Y. Zhumabayev
Central Asian Countries Initiative for land management (CACLIM)

N. Singh
Department of Wildlife, Fish, and Environmental Studies, Faculty of Forest Sciences, Swedish University of Agricultural Sciences, Uppsala, Sweden

V. Squires (ed.), *Rangeland Stewardship in Central Asia: Balancing Improved Livelihoods, Biodiversity Conservation and Land Protection*,
DOI 10.1007/978-94-007-5367-9_11,

decreased. Abandonment of remote pastures was accompanied by a concentration of livestock around settlements.

- Following the crisis, traditional collective herding systems quickly became re-established amongst non-state livestock owners; many pastures were used informally at first, and access was determined by former state farm boundaries, customary memory, and purchase or construction of infrastructure such as barns. But as livestock inventories recovered, attention turned to formal tenure of pasture systems.
- In Central Asia, the importance of formal legislation to pasture management was understood late in the land reform process. Attention initially focused on increasing the productivity of arable land and land codes did not distinguish between this type of farmland and pastures – thus forms of individual tenure were initially pursued in every case examined here except that of Kyrgyzstan. This republic formed separate provisions for pasture early on and recently passed a pasture code designating pasture as common property.
- In the other republics, pasture is either still allocated to large state structures or is subject to general land codes which emphasize individual forms of land use or ownership. Thus annexation of pasture areas by individuals is now commonplace in Kazakhstan and Tajikistan. In Turkmenistan and Uzbekistan legislation also favours individual management of land, but has not yet been applied to pastures, the use of which by non-state livestock often occurs in a legal gray zone.
- Barriers to livestock mobility and to use of remote pastures by smaller livestock owners are both administrative and economic, factors which are not easily disentangled. However, there is evidence from a number of the republics considered here, and from global literature on pastoral tenure, that individualized modes of pasture use may reduce livestock mobility and pasture access, in particular where average household stock ownership is low and collective herding concerns a large proportion of rural households.
- Today, Uzbekistan, Tajikistan Kazakhstan and Turkmenistan are all considering the introduction of pasture codes, with provisions for common pasture management under debate. This is then a key moment for policy makers in those countries to reflect on the lessons of the first 20 years of pastoral tenure reform in Central Asia.

1 Global Trends in Pastoral Tenure

Property rights, the ways in which pastoralists access land, are a key factor determining the impact of pastoralism on the environment. Such rights may be determined by formal legislation, traditional rules, or more usually, a mixture of both. In some cases, as in Soviet-era Central Asia, grazing patterns may be planned directly by state institutions, down to the individual migratory movements of each herd.

The recent history of pastoral tenure in developing countries has been influenced by the evolution of property rights theory in the developed world. A key point in this debate was the concern about environment degradation raised in Hardin's "tragedy of the commons." This scenario suggested that, where pasture is open to all, users will strive to increase their herd sizes as each receives immediate benefit from the resource while bearing only a share of the (delayed) costs of overgrazing as animal numbers increase (Hardin 1968). Associated with this was the idea of carrying capacity: that the maximum number of livestock which could be kept on discrete spatial area over a given time could be calculated with precision and that the resulting figures could be used to improve pasture management. Until the end of the1980s, these ideas were very influential, leading governments and agencies such as USAID and the World Bank to promote and finance pasture privatization,[1] with an emphasis on commercial ranching and the allocation of discrete parcels to individuals or households (Behnke 2008; Rohde et al. 2006; de Hann 1993).

Hardin's predictions assumed conditions of open access (where the resource has no defined boundaries or user groups). Instead, many of the pastures subject to privatization were managed as common property by user groups; they were therefore to some extent spatially bounded and subject to various internal controls on grazing (Bromley and Cernea 1989; Ostrom 1990). On privatization, such common resources often become fragmented into discrete parcels, restricting access by livestock to different seasonal pasture and water resources and affecting both animal productivity and pasture condition (Coughenour 2008). Sneath (1998) demonstrated that privatization and fencing of pasture in former Soviet Inner Asia and Chinese Inner Mongolia led to severe pasture degradation compared with neighboring Mongolia, where livestock husbandry had remained commonly managed and mobile (see also Li et al. 2007). In eastern Africa, the much-studied conversion of the Masai Mara to group ranches and then to individual sections led to a reduction in livestock mobility and a reduced range of vegetation types available to grazing households over the years (BurnSilver et al. 2008). In several countries in southern Africa, creation of fenced ranches led to unsustainable stocking rates on the remaining common land (Rohde et al. 2006).

Such examples are now common across the developing world because, although organizations such as the World Bank no longer actively promote privatization of commonly used arid rangelands (de Hann 1993), governments continue to pass tenure reforms favoring individual land title regardless of environmental conditions (Reid et al. 2008; International Land Coalition 2007). Population growth, the need to access markets and infrastructure, and desires of wealthier herders to commercialize all reinforce this trend (Behnke 2008). Yet, as the environmental impacts of these policies become apparent, there is now increasing support among the scientific community for the idea that, in poorly productive and heterogeneous environments, livestock mobility may be a more useful metric for assessing the sustainability of pasture management than carrying capacity and pasture fragmentation as much of a threat to the environment as overstocking (see discussion in Galvin et al. 2008; Li et al. 2007).

It is also important to consider the social costs and benefits of privatization of erratically productive rangelands. The privatization model is well adapted to a commercial economy, whereby individual ranchers raise large numbers of livestock primarily for sale to urban populations.[2] In regions of relatively high rainfall, equitable

[1] Here we use the term "privatization" loosely to mean formal individualized and exclusive user rights.

[2] Note that in many developed economies, extensive livestock operations frequently depend on state-owned grazing land for at least part of the year (Huntsinger et al. 2010); ranchers with individual tenure may benefit from state protection as a buffer against drought and other extreme climatic events.

privatization programs may lead to agricultural development and rising living standards as livestock raising can be intensified and pasture turned over to cropping (Behnke 2008), but in poorly developed arid parts of the world, the markets, services, and infrastructures required for intensification are poor, and investments are risky due to fluctuating environmental conditions (BurnSilver et al. 2008). For many households in these environments, the primary reason for holding livestock is for subsistence purposes and as a form of savings. During privatization, tenure rights are typically secured by households with larger herds, who can cover costs of herding individually. This has led to a loss of areas for common grazing by smaller stock owners and an increase in wealth inequalities in many areas of the world (Behnke 2008; Rohde et al. 2006; Yan et al. 2005; Wu and Du 2009). Key resources such as water sources, hay fields, and winter or dry season grazing are often privatized first or unevenly allocated, leaving some households vulnerable at certain times of the year.

However, a black and white distinction between individualized and common property-based pasture rights cannot always be made: within common pasture management regimes, individual households may hold *de facto* claims over specific wells or pasture areas for generations. Forms of individualized tenure may vary widely in the extent to which they enable users to access pastures in multiple ecological zones. The actual implementation of official privatization policies may also display a high level of variability on the ground.[3] These factors must be considered when evaluating current systems of pastoral tenure in Central Asia.

2 Pastoral Systems in Central Asia: Common Themes

Today, Uzbekistan, Tajikistan Kazakhstan and Turkmenistan are all considering the introduction of pasture codes, with provisions for common pasture management under debate. This is then a key moment for policy makers in those countries to reflect on the lessons of the first 20 years of pastoral tenure reform in Central Asia. Since 1991, both *de facto* and *de jure* pastoral tenure regimes diverged significantly in the five Central Asian republics; and it is interesting to compare how choices made by policy makers have affected the two key outcomes which interest us here: firstly, livestock mobility – *the extent to which livestock are able to move between seasonal pastures* – and secondly, pasture access – *the ability of rural households across all wealth categories to graze their animals on seasonal pastures.* In this section, we present some of the physical and economic characteristics common to the Central Asian republics which both underlie the ways in which pasture has historically been managed and help us to assess the consequences of current tenure policies.

A major feature of Central Asia is its aridity: rainfed agriculture is marginal at best and thus, outside northern Kazakhstan, high and consistent yields may be

[3] In response to pasture privatization and fencing policies in Tibet, some pastoralists fenced only key areas around dwellings while continuing to manage other pastures in groups; many minor livestock owners were marginalized by the reforms, but others benefited by leasing pasture to those with more animals (Yan et al. 2005; Goldstein 2012; Richard et al. 2006).

achieved only through irrigation; development of new lands requires large investments in canal infrastructure. For this reason, conversion of pasturelands to agricultural land is unlikely to be an economic driver of privatization as it has been in some regions of the world (Behnke 2008); intensification and commercialization of the livestock sector are likely to be more important.[4]

In contrast to non-temperate arid zones, interannual variation in rainfall in Central Asian rangelands is not extreme; rather than experiencing density-dependent crashes in drought years, numbers were historically limited by severe winters and extreme snowfall events (Robinson and Milner-Gulland 2003b).

In contrast, seasonal and spatial variation in productivity is very high and explains why historically pastoralists in Central Asia tended to follow the same broad transhumant or migratory patterns from year to year.[5] The Soviet system built on and formalized some of these migratory systems. Each state or collective farm (*sovkhoz* or *kolkhoz*)[6] had a central village plus satellite settlements or isolated barns in seasonal pastures (Vanselow, Chap. 4). Each was allocated grazing land in various parcels, sometimes located in another *raion*, *oblast*, or even republic and linked by designated migration routes. In all republics, there was a distinction between land permanently allocated to state farms and land in the "state reserve" or "land fund," which was allocated to farms on a "long-term use" or other type of temporary arrangement. Lastly, large areas of pasture were also under the jurisdiction of the state forestry department (*leskhoz*). These distinctions persisted in the land codes formulated after independence.

In mountainous republics such as Tajikistan and Kyrgyzstan, summer pastures were located in the high mountain valleys, and winter pastures were found in areas with low precipitation or on south facing slopes. In steppe- and desert-dominated republics such as Turkmenistan and Kazakhstan, within each pastoral system, summer pastures tended to be those areas with highest rainfall or access to water sources, while winter pastures were typically located on sandy areas dominated by shrubs, which could be browsed by animals in the winter despite heavy snowfall (Alimaev 2003; Kanchaev et al. 2003). Decisions about where animals should go were made by government bodies and the farm administration; salaried shepherds moved the mostly state-owned livestock (together with some private animals) to each location and were provided with transport, provisions, and logistics to this end. While some movements were based on traditional grazing patterns existing in pre-Soviet times, the provision of large quantities of winter feed, traditionally the major factor limiting livestock populations, meant that migrations could be less extensive than before and allowed livestock numbers to reach historical highs (Robinson and Milner-Gulland 2003b).

---

[4] In some regions, such as lower mountain areas of Tajikistan, large areas of former rainfed pastureland have in fact been planted since the end of the Soviet Union, often for subsistence purposes. The land may be officially converted to arable land, and incentives to privatize are likely to be higher than on land remaining as pasture. However, it remains to be seen whether such conversion has been widespread enough to seriously impact pasture availability for the livestock sector.

[5] Although within broad seasonal grazing zones, stock movements could be variable from year to year, driven by snowfall and disease (Alimaev and Behnke 2007).

[6] Henceforth, we will refer to both entities as "state farms" for brevity.

In the immediate aftermath of independence, a number of common patterns emerged across the region, albeit to widely differing extents: the proportion of private animals increased, costs of inputs such as winter feed and transport rose sharply, and livestock mobility decreased (Kerven et al. 2004). Deprived of grain from the USSR, food security quickly became an issue. Scarce arable land was turned over to crops for human consumption,[7] and supplementary feed has once again become the single most important factor limiting development of the livestock sector (e.g., see Sedik 2010, and Sedik, Chap. 9). This also means that winter grazing areas and associated shelters now constitute key resources for herders (Milner-Gulland et al. 2006). The collapse of the state and collective farm system led to a sharp drop in livestock numbers in Kazakhstan, Tajikistan, and Kyrgyzstan, while the state continued to play an important role in livestock production in Turkmenistan and Uzbekistan, where livestock numbers were less affected.

Private livestock ownership distributions became quickly characterized by a small number of households with very large herds and a large number of households with a small number of animals (see Table 11.1). These latter cannot afford the labor to watch over their tiny herds and thus rely on collective herding systems. The pooling of animals into herds totaling hundreds of head creates the economy of scale required to cover the costs of using remote seasonal pastures such as transport and shepherding. As we will see in the sections below, a fundamental issue with some of the recent reforms is that they do not take into consideration these systems.

A second and related characteristic of livestock husbandry in rural Central Asia is that, for many, it is a subsistence activity or complement to household revenue. For smaller livestock owners in Turkmenistan, "*the subsistence value of meat is greater than the cash value of small stock sales,*" (Soyunova 2003) while in Kyrgyzstan, "*animals are sold when cash is needed – to cover school fees, social obligations, food purchases, or health care needs*"(World Bank 2007). This is a factor to be considered when assessing transaction costs associated with pasture privatization.

Today, many rural areas of Central Asia are overpopulated in relation to cultivable land and water resources, and, outside Kazakhstan, urban growth has done little to relieve this pressure.[8] Livestock numbers are growing, despite adequate feed availability for winter, and animal productivity is extremely low. In such a context, pressure on pastures can only increase and animal mobility will become still more important for sustainable pasture management.

---

[7] This was often a response of individual land holders to cover subsistence needs, or in the case of Turkmenistan and Uzbekistan, a government policy aimed at reaching self-sufficiency in key staples (Hodjakov and Wright 2003; Lerman 2008).

[8] Four of the five republics are designated as low-income food-deficit nations (Babu and Tashmatov 2000); see Cariou (2002) for discussion of rural overpopulation in Uzbekistan.

Table 11.1 Data on rural livestock ownership distributions in the five republics

Country	Region and sample	Measure	Reference
Tajikistan	GBAO 2008; 1,000 hh	Average livestock ownership was nine head of small stock and two head of cattle	Mountain Societies Development Support Program (2009)
	Khatlon *oblast* 2004; 1,000 hh	10% of the sample had over ten small stock; 2% had over 20. Not one family had over 50 small stock while only three households had over ten head of cattle	Mountain Societies Development Support Program. (2004b)
	Rasht Valley 2004; 1,000 hh	76% of households had cattle, but only 3% had more than five head; 11% owned more than ten small stock	Mountain Societies Development Support Program (2004a)
Kyrgyzstan	Nationwide 2001; 3,000 hh	90% of rural households had fewer than 60 head of small stock; 22% had more than one cow.	World Bank (2003)
	Nationwide 2003; Census	Average sheep herd size estimated at 15 over all holdings declaring ownership; equivalent figure for goats was seven head	National Statistical Committee of the Kyrgyz Republic (2004)
	Chong Alai and Alai *raion*, Osh *oblast* 2005; 1,000 hh	Taking owners of small stock alone, the average herd size was 23 in the most pastoral *raion*. 10% of the sample households had over 25 head of small stock	Mountain Societies Development Support Program (2005)
Turkmenistan	Rukhabat *raion*, Ahal *oblast* 1999; 44 hh	Sample included stock owners only, of which over 80% had fewer than 100 private animals (although in Turkmenistan income is earned from herding state animals also)	Soyunova (2003)
	Rukhabat *raion*, Ahal *oblast* 2009; 13 hh	Sample of interest comprises individuals without access to state animals. In this group, small stock numbers ranged from five to 40, only one household had no stock	Jumardurdyev (2010)

(continued)

Table 11.1 (continued)

Country	Region and sample	Measure	Reference
Kazakhstan	Dzhambul *raion*, Almaty *oblast* and Moiynkum *raion*, Dzhambul *oblast*; 2003; 46 hh	Sample included stock owners only. Owners of non-mobile livestock had on average 46 small stock but made up 40% of the sample. Mobile herders owning over 300 head of small stock made up 10%	Kerven et al. (2004)
Uzbekistan	Twenty *raion* in eight *oblast* 2007; 1,600 hh	Sample stratified by *dehkan* households and peasant farms[a]. 71% of the former owned cattle (average four head); 30% owned sheep (average nine head). Among those peasant farmers specializing in livestock raising, almost all held cattle (55 head on average) and half held an average of 70 head of sheep	Yusupov et al. (2010) and Lerman (2008)

Notes: All households in the cited studies were randomly sampled, although in some cases sample frames were stratified by specific socioeconomic subgroups (e.g., stock owners only). Where samples come from very limited geographical areas, these tend to be located in predominantly pastoral areas

GBAO Gorno-Badakhshan Autonomous *Oblast, hh* households

[a]See section on Uzbekistan below. Private farmers also have household animals, which were counted in the *dehkan* statistics

3 Variation: Tenure Reform in the Five Republics

There are a number of differences between the republics which may explain divergence in land tenure policy. In the two mountainous republics, most of the rural population are engaged in both agriculture and livestock raising to varying extents, while in Kazakhstan around five million people live in areas where agricultural activities other than livestock raising are marginal or impossible (Thornton et al. 2002). The pace of economic development has also influenced the size and orientation of emerging livestock raising entities – oil wealth and associated purchasing power in Kazakhstan have perhaps favored development of larger commercial operations. Politically, the republics range from an unstable form of democracy in Kyrgyzstan to an autocratic regime in Turkmenistan, differences which underlie the diverging tenure regimes examined here.

In the following sections, for each republic in turn, we will provide an outline of the formal land tenure legislation as it affects pastures and discuss the impacts of reform on the ground in terms of (i) livestock mobility and (ii) access to pastures by various category of livestock owner. We also assess current proposals for change, as the introduction of new pasture codes is now under discussion in four of the five republics. These sections are based on a review of current legislation on land, published research, and in the cases of Kyrgyzstan, Tajikistan, Kazakhstan, and Uzbekistan, drawn on field research by the authors. It should be noted that, while numerous studies have been undertaken across many regions of Kyrgyzstan (e.g. Steimann 2011, Undeland 2005, Farrington 2005, Dörre 2012), in-depth field-based studies on other republics are available only for a small number of geographic locations. This means that many of the patterns described here may be only regionally representative. In the case of Uzbekistan, we were unable to find published studies focusing on our two main themes, so we present a thus far unpublished case study in some detail.

3.1 Kyrgyzstan: Common Property Made Law

Kyrgyzstan has undergone two phases of pastoral tenure reform since independence. Here we describe the first phase of reform and its impacts, followed by a discussion of the 2009 pasture law.[9]

Agrarian reform in Kyrgyzstan was highly successful in the sense that the overwhelming majority of those eligible actually received a share of arable land (Asian Development Bank 2008).[10] Pasture remained the property of the state and subject

[9] Much of the information presented here is taken from unpublished consultancy reports written by the lead author and based on fieldwork in three *oblast* (Robinson et al. 2001; Robinson 2007). All these patterns are also described in detail in Steimann (2011) and (Undeland 2005) cited here in the text.

[10] Parcels were initially allocated to users for 49 years, which was extended to 99 years and then finally transferred to them as private property in 1999.

to separate legislation. Up until 2009, the major law governing pasture management was Government Resolution 360 (4 June 2003), which built on a previous decree of 1999.[11] According to this legislation, pasture was split into three types, all of which were to be accessed by users through leasing arrangements:

- Village (*prisel'nye*) pastures leased from the lowest level of government, the *aiylokmotu*. These pastures comprise those directly adjacent to the village and are usually used in winter or for milking or sick animals in summer.
- "Intensive" (*intensivnye*) pastures leased from the *raion* administration. These pastures are usually located between those of the village and remote summer pastures and are envisaged for spring-autumn use, but are often used in summer (and even in winter) depending on stocking pressure and availability of remote pastures.
- Remote (*otgonnye*) pastures leased from the *oblast* administration; these are usually high-altitude summer pastures, but are not always fully used if they are very distant from settlements and where intensive pastures are deemed sufficient by users.

This system implied that herders attempting to formalize their rights to different seasonal pastures had to deal with three administrations and make three separate contracts. Leases were provided for a period of 5 years, which could be extended for another 10 and then 49 years. They were supposed to be awarded by public auction, but the costs of preparation of the requisite documents including a cadastral plan and estimates of fodder availability and optimal stocking rates usually fell upon the applicant, who was not always sure to win the bid. Thus, in reality, the formal bidding process was often replaced by a first-come-first-served approach. Once the parcel had been assigned to a user, the documentation process required a lease agreement, transfer act, and confirming certificate, all involving significant bureaucratic hurdles. Administrative capacity was poor and many local authorities lacked maps indicating boundaries between different pasture areas. These processes favored the better-off herders and rich absentee herders based in towns over local users with historical claims to the pasture (Undeland 2005).

Implicit in Resolution 360 was the assumption that pasture would be used by households in the same way as parcels of arable land distributed during the land reform process. But in Kyrgyzstan, most livestock owners form groups, each of which hires a shepherd to take livestock to seasonal pastures, usually in the summer. Most of these paid shepherds or *badachi*[12] have few animals of their own. In contrast, through the *kezu* system, group members[13] take it in turns to herd animals according to a daily rota. This system is usually observed on village pastures in winter but may also occur in more distant pastures, with shepherds rotating every 3–4 days.

---

[11] Decree No. 640, November 29, 1999.

[12] *Badachi* is used for those herding collective cattle, *koichi* is used for shepherds (looking after flocks of sheep and goats), and *jilgachi* for herders of horses; however, the term *badachi* is often used generically to refer to paid herders of all types of livestock.

[13] Often the basis for *kezu* group formation is residence in the same street (Steimann 2011).

Those owning several hundred animals usually herd their own (or pay a shepherd to herd them), and these animals are often referred to as "private livestock."[14]

These *kezu* or *bada* herds probably comprise the animals of the majority of rural inhabitants, but livestock owners participating in these systems did not usually take out leasing contracts for two major reasons: firstly, both group membership and the identity of the *badachi* could change from year to year and secondly, the groups themselves were not legal entities. Most *badachi* were not well off and unlikely to make the investment in time and money required for a pasture bid themselves. Thus, the majority of leases were taken out by those herders who had large numbers of animals and did not participate in common herding systems.

Although the first law on leasing pastures came into force in 1999, by 2007, only about 14% of Kyrgyz pastures were formally leased, although many herders obtained rental contracts with the *aiyl okmotu* but no transfer act or confirmation document from the land registration department, and were thus not listed in the statistics. For some lessees, contracts were sought for fear of losing grazing rights to others, while in other cases, leasing was part of a longer term household strategy to increase livestock numbers. Everywhere the process was fraught with irregularities: some rented a small parcel of pasture around a barn or water point but actually used much more, while others registered leases but stopped paying rent as they knew that sanctions were unlikely to be forthcoming (Steimann 2011). State revenues from pastures were thus much lower than expected.

As leasing became more common, conflicts emerged. These arose mostly between lessees, usually large livestock owners in their own right, and non-lessees and tended to be concentrated on village and intensive pasture where grazing pressure was highest. As the number of lessees increased, so the amount of pasture for common grazing decreased. Pasture allocated by *raion* or *oblast* was sometimes provided to applicants from outside the region or even from Bishkek, with no regard for previous local users of these pastures. Some even rented the land and then subleased it to villagers, although this was expressly prohibited in Resolution 360 (Undeland 2005). One household survey suggested that, of all conflicts over pasture, those linked directly to the new leasing legislation (unfair allocation, borders between plots, not allowing others to move livestock across a plot and allocation of land to those outside the community) made up about 45% (Undeland 2005).

In response to conflicts, some *aiyl okmotu* froze all new rents on village pasture or stopped providing the documents necessary for applications for intensive and remote pasture (Steimann 2011; Robinson 2007). Some even repealed existing contracts. A second reason given by some *aiyl okmotu* was that renting leads to poor pasture management, with renters staying too long on their leased parcels rather than changing area. Some local authorities, who still had to raise rent from the pasture, then simply divided the total sum due by the head of animals in the village so that each household paid its share according to livestock ownership, while others simply levied a "poll tax" regardless of head owned (Steimann 2011).

[14] These systems are all quite fluid; for example, some *badachi* are large owners themselves but will also take the animals of members of their extended families to distant pastures.

3.1.1 The New Law "On Pastures" No. 30 of January 2009

The pasture leasing system was highly criticized due to its negative impact on pasture access by the poor, the high administrative burdens which it imposed on lessees and on an underfunded administration, and for its inability to raise revenue. It also manifestly did not improve livestock mobility, but instead represented a barrier to flexible grazing patterns, imposing additional costs to movement in the form of a pasture rent charged on a hectare basis. Pressure for change came both from certain parliamentarians and international organizations such as the World Bank, which led to the development of the new law.

According to this law, land is still the property of the state, but all three types of pasture listed above have now been placed under the administration of the *aiyl okmotu*, thus removing the previous three-tier system of pasture administration. Pastures will no longer be rented out on a long-term basis and instead will be allocated to Pasture Users Associations (PUA), formed at the level of each *aiyl okmotu*. All pasture users should join the relevant PUA and obtain access rights to pasture by purchase of tickets, sold on an annual basis. The executive body of the PUA, the Pasture Committee, should develop and enforce pasture management plans,[15] monitor pasture condition, set and collect fees, issue grazing tickets, and manage revenue from pastures to invest in pasture improvement such as bridges and access roads. It should also resolve disputes involving users of the pastures under its management authority. The Pasture Committee should include representatives of pasture and both elected and executive members of local government. Existing lessees will be permitted to exchange their lease for a pasture ticket covering the maximum number of animals within the carrying capacity of the pasture area covered by their lease.

Concerns about the new law include the administrative capacity of pasture committees to make the requisite plans and to manage funds, pricing of pasture tickets, conflicts with former disgruntled lessees, and the status and borders of forestland in relation to pastureland (Bussler 2011). While allocation of pastures among groups is certainly anchored in Kyrgyz custom and practice, the building and maintenance of commonly managed infrastructure are not. Some livestock owners have already consolidated their claim to pasture through the construction of private barns; they will want to maintain pasture rights adjacent to these considerable investments, which may cause conflict if they are located on key pasture resources (Steimann 2011).

Crewett (2011) has reported some initial observations concerning the implementation of the new law: the formation of PUAs and pasture committees was supposed to be facilitated by a government-mandated agency along participatory lines, but many PUAs were formed spontaneously by pasture users or by local municipalities

[15] Pasture management plans should include maps, carrying capacity estimations, development plans, and detailed annual grazing plans to be updated every year. Annual plans should include the list of all pasture users holding a pasture ticket for that year, an inventory of all livestock for which pasture tickets have been issued, and information on livestock movement and seasonal pasture allocation.

and were usually dominated by older and wealthier herders. The facilitating organization had difficulties imposing its own rules – the process of PUA establishment proceeded to very tight deadlines and initial observations suggested that many stakeholders were away or not informed during key meetings, while local authorities themselves found the law lacked the detail required for implementation. Some amendments to the law have already been passed since 2009 in an effort to clarify outstanding issues and ensure compatibility with the constitution.[16] The law will also not solve the economic factors affecting livestock mobility: vast areas of pasture in the south and east of the country are still abandoned and the infrastructure needed to use it in disrepair (Farrington 2005; Robinson 2007).[17]

On the positive side, the 2009 law is the first in Central Asia to enshrine the principle of common property for pastures; it introduces a system by which payment for pastures is based on the number of head of livestock per user rather than on a hectare basis, thus removing penalties for using larger or more areas of pasture for grazing; it encourages mobility by designating a single management body to oversee grazing systems which include many geographically separated areas. It will be an important test case for other countries to observe.

3.2 *Tajikistan: The New Pastoral Landlords?*

In Tajikistan there is no distinction between pasture and other types of farmland; all are subject to the same legislation. For this reason, we present the broader legislative context which was designed for arable land reform and is now being applied to pastures.

According to the Land Code of Tajikistan, all land is owned by the state, but three tenure arrangements are available to users: permanent heritable land use, fixed-term use, and leasing. The law "On *Dehkan* Farms"[18] sets out institutional frameworks for permanent land use rights. According to this law, membership in a state farm confers the right to a land share equal in size to the total farm area divided by the number of eligible shareholders. However, shareholders must initiate a complex and expensive certification process to obtain full permanent title to their land in the form of a registered *dehkan* farm.[19] These reforms resulted in two main

[16] "On the introduction of changes and additions to the Law of the Kyrgyz Republic, 'On Pastures'," 2011.

[17] Livestock numbers may also be an issue here: official statistics suggested that, in 2007 total livestock numbers (in livestock units) were 34% lower than their 1990 level, although veterinary statistics indicated much high numbers (Robinson 2007). More recent official statistics suggest that livestock numbers are now close to 20% of 1990 levels (National Statistical Committee of the Kyrgyz Republic 2011).

[18] The law of May 2009 "On *Dehkan* Farms" (State Land Committee of the Government of Tajikistan 2009) is the latest incarnation of a set of laws first published in 1992.

[19] This may be literally translated as "peasant farm," although the real meaning would be closer to "private farm."

types of entity: "individual *dehkan* farms" mostly consisted of single households and their members who had successfully negotiated this certification process and "collective *dehkan* farms" in which farm managers held the full certificate and inhabitants were shareholders (Robinson et al. 2008; Lerman and Sedik 2008). In some cases, these shareholders had access to physical parcels for individual use; in others, they remained salaried workers on their own land. On some farms, during the initial chaotic years of reform, all arable land was fully converted to individual *dehkan* farms by a small number of enterprising farmers on a first-come-first-served basis, while the majority of households missed out (Robinson et al. 2008).

Until recently, the scenarios described above concerned only arable land; pasture within the boundaries of collective *dehkan* farms should also be distributed in equal shares to members, but in some regions, such as GBAO, it remained under collective management for many years due to the practice of common herding that makes splitting of pasture into shares an impractical proposition. More recently, with the intention of improving security of tenure to land, selected collective *dehkan* farms have been dissolved and all member households issued with full certification. These collective *dehkan* farms will thus cease to exist and pasture within their boundaries must be distributed in equal shares to members in the same way as arable land parcels.[20] This process of allocating equal pasture shares to households has proved highly problematic as herding systems are collective; households have widely varying numbers of animals; and pasture and water availability is spatially heterogeneous. While collective *dehkan* farms sometimes deprived members of full rights to arable land, they did play a role in pasture allocation and management; in GBAO discussions and disputes surrounding pasture distribution have delayed the certification process for arable land, the original aim of the reform (Robinson et al. 2010). In other parts of the country, as occurred with arable land, some farm members obtained pasture shares far larger than their theoretical shares under conditions of equal distribution, leaving little left for other members. Today about half of all pastures are found in *dehkan* farms,[21] but it is not clear what proportion of this area comprises collective *dehkan* farms having pastures under *de facto* collective management and what proportion is held in individual *dehkan* farms (Halimova, Chap. 13).

Large areas of pastureland are not found within the boundaries of former state and collective farms and thus not subject to equal distribution to members; indeed, some farms had no permanently allocated pastureland at all. Instead, they used State Reserve lands.[22] During the Soviet period, this pasture was allocated to each state or collective farm in large blocks for "long-term use," and in the first years of reform,

[20] It should be noted that individual households commonly secede from collectives and form their own dehkan farms; in some cases, they certify only arable land, continuing to use pasture in common with collective dehkan farm members. Such an option is not possible when collectives are *completely dissolved*, because these entities organize tax payments for pastures from all users. On closure of the collective, local officials must ensure that tax payments continue, so pasture *must* be recorded on the certificates of individual dehkan farms emerging from the dissolution.

[21] Land statistics for 01.01.2012 provided by Z. Lerman.

[22] About 11% of pastures are in State Reserve and another 11% are managed by the Forestry Department (official land statistics for 01.01.12 provided by Z. Lerman).

these areas continued to be used as common pasture. Today they are subject to fixed-term use agreements, annual lease, or permanent use arrangements. In parts of Khatlon *oblast* and the *Raion* of Republican Subordination (RRS),[23] individuals took out fixed-term use agreements on large areas of pasture on a speculative basis and then rented them back to communities and smaller *dehkan* farms (Halimova 2012), but national level statistics concerning the frequency of this type of arrangement are not available.

The law "On *Dehkan* Farms" allows applicants (theoretically any Tajik citizen) to obtain pasture on State Reserve land for permanent use on a first-come-first-served basis.[24] As with fixed-term agreements, some areas of pasture may be annexed and rented back to users; in other cases, former users may simply be excluded; Box 11. 1 provides some examples of patterns observed in GBAO.

Box 11.1 Privatization of State Reserve Lands: Examples from GBAO (Robinson et al. 2010)

In 2009 in GBAO, about 55% of pastures were located inside the boundaries of former state or collective farms and were thus eligible for division to former members, although such a distribution had not yet occurred at that time. A further 30% of pasture was allocated to collective *dehkan* farms for fixed-term use and 14% remained as unallocated State Reserve land. Of the latter two categories, some areas of pasture had already been incorporated into individual *dehkan* farms, while many applications were pending. Some applicants were shepherds of village livestock, who owned few animals themselves and continued to take to common herds to the pasture for which they had applied; others owned large herds themselves and excluded other users. In a few cases, groups of users formed multi-household *dehkan* farms using pasture in common. Some areas of State Reserve land were rented annually by collective *dehkan* farms and their shepherds, but these areas may shrink as livestock numbers grow and land is annexed by individuals into *dehkan* farms. Fear of losing pastures for common grazing prompted some *raion* such as Murghab to impose a moratorium on pasture privatization,[25] but the current transformation of collective *dehkan* farms to individual entities begs the question of how common pasture currently under collective fixed-term use agreements will be accessed, if the organizations to which these pastures were allocated disappear.

---

[23] A group of seven *raion* in the Rasht valley subordinate directly to Dushanbe rather than to an *oblast* administration.

[24] Although Article 17 seems to suggest that the area received should be proportional to the number of livestock owned by the *dehkan* farm members.

[25] This *raion* has rather different pastoral tenure arrangements than those recorded in other parts of the country (Kraudzun 2012), again underlining a high geographical variability in land reform outcomes.

The issue of livestock mobility is of course intimately bound up with the above discussion about access; the system of migration across administrative boundaries in Tajikistan more or less collapsed in the 1990s but has seen some recovery in recent years. Land statistics show that some pastures are formally allocated to livestock from other *raion* and *oblast*. These are often assigned to state livestock breeding enterprises which maintain large and highly mobile flocks.[26] As in Turkmenistan, such arrangements may also facilitate the migrations of private animals belonging to state enterprise workers. Livestock of individual or collective *dehkan* farms face greater barriers to access of pastures outside their *raion* of residence as they lack the negotiating power and organization of the state entities (Robinson et al. 2008). However, such movements may be increasing and have been observed both within Khatlon *oblast* and between Khatlon and RRS; in GBAO long-distance migrations by both common village herds and individual *dehkan* farms are widespread *within* individual *raion*. Such movements rely on access to remote State Reserve and fixed-term use pasture which, in the case of commonly owned herds, cannot be guaranteed under current laws. Where long-distance migration does not occur, forms of local transhumance may usually be observed, but according to Halimova (2011), in some *raion* close to Dushanbe, access to all pasture outside the immediate village environs has been lost, and household animals graze all year round on the same pastures.

To summarize, in some areas of the country, Tajik pastoral tenure reform can be described as a "*transfer of pastures to a few large-scale individual dehkan farms and state enterprises, which have few shareholders, whilst those dehkan farms that have large number of shareholders have access to smaller areas of pasture*" (Halimova 2012). While current legislation is in force, there is a risk that this pattern may become generalized throughout the country. Recognition of these issues is now widespread in Tajikistan and as a result, no fewer than four new pasture laws have now been drafted.[27] The present draft under consideration has clearly taken some inspiration from the Kyrgyz 2009 law "On Pastures" and includes establishment of Pasture User Associations.

However, conversion of pasture within former state farm boundaries to common land would be in direct conflict with both the current Land Code and law On *Dehkan* Farms, which specify that this land may be split into shares. Arable and pastureland are both listed on *dehkan* farm certificates, implying a theoretical reregistration of all *dehkan* farms if pasture tenure rules were to change. These problems could perhaps be circumvented by obligatory leasing of dehkan farm pastureland to PUA and

[26] In 2008, those sheep and goats belonging to state-owned enterprises comprised 10% of the national total (State Statistical Committee of Tajikistan 2009); herd sizes were the same as those in Soviet times, at 700–800 head of small stock.

[27] Two of the versions have been sponsored by international organizations; one was drafted by the Ministry of Agriculture and one by the "working group" on pasture reform, which includes various stakeholders.

nonrenewal of expiring lease agreements. But the current insistence by members of parliament that pasture already under individual tenure should not be converted in this way, and that both individuals and PUAs may apply for pasture, will reduce the likelihood that a new pasture law would guarantee seasonal pasture access to the rural population.

3.3 *Kazakhstan: Agribusiness and the Family Ranch*

3.3.1 The Process of Reform: National Trends

As in Tajikistan, Kazakhstan has not developed separate legislation for pastures and arable land. Reform started early: individuals were able to establish family farms[28] in 1990, although initially, senior staff of state farms were the main beneficiaries of this reform (USAID 2005b). Paper land shares convertible to physical plots (known as conditional land shares or CLS) were first made available to all farm employees and their families for "permanent use" in the 1995 Presidential Decree "On Land."[29] This law also introduced true private land ownership for household plots.[30] In addition to the receipt of shares through farm restructuring, land could also be incorporated into family farms through applications for State Reserve land or through purchase from other land users. The law "On Peasant Farms" of 1998, the main legal framework for family farms, allowed both temporary (leasehold) and permanent use arrangements.[31]

Family farms were not the only structure envisaged in the reform process; a number of non-state collective farm structures also emerged, including joint-stock companies, production cooperatives, and small enterprises, all of which closely resembled the structure of a state farm and which we will refer to here as "agricultural corporations." In theory, the farm management was supposed to establish a consensus on the structure of the new type of farming entity, but in reality, farm members were not presented with much of a choice; CLS were rarely redeemed for physical shares but pooled so that the former state entity could be reincorporated into an agricultural corporation (Behnke 2003). Workers wishing to leave the agricultural corporation and form family farms had the legal right to redeem their CLS certificates for demarcated land plots, but this process was bureaucratically difficult and often thwarted by corporation managers. Many lacked the machinery, inputs, and labor to farm individually. Thus, the reform process initially resulted in "destatization" rather than the creation of large numbers of family farms (Behnke 2003). In some cases,

[28] These independent units are designated as *kristianski khozyaistvo* (peasant farms), but they will be referred to here as family farms to distinguish them from the large "agricultural corporations" described in this section.

[29] Article 79.

[30] Articles 33 and 34.

[31] Originally the length of leasehold contracts was 99 years, which was later reduced to 49; the permanent use category was later converted to private ownership to reflect the 2003 Land Code (see the Law on Peasant Farms with 2003 amendments, Article 6).

farm managers persuaded the new shareholders to sell their shares, resulting in mass transfers of land to a small elite (Lerman et al. 2002). Others leased their CLS back to corporate farms or larger family farms which had managed to obtain machinery and other assets needed to work the land early on. By 2002, 18% of CLS had been transferred as the base capital for agricultural corporations, 29% into physical plots for family farms, 7% were sold, 28% were leased out, and 18% were unclaimed (Dudwick et al. 2005). In terms of area, just under 40% of agricultural land was held in family farms and the rest held by agricultural corporations. Yet the total area of agricultural land registered by farms of any type at that time was about 80 million ha – a huge drop from the 180 million ha registered as agricultural land held by state farms in 1990 (Dudwick et al. 2005).

Most of this decrease in registered land was attributable to an increase in unallocated State Reserve land, which made up 7% of land in the Soviet period and 44% in 2001 (Alimaev and Behnke 2007); many of the newly formed agricultural corporations returned land which they could not use to the state in order to avoid paying tax on it, and the majority of this land was pasture. As in other Central Asian republics, the 1995 decree "On Land" did not distinguish between pastureland and arable land; thus, those redeeming shares for physical plots were likely to receive both types of land automatically. However, in most cases, agricultural land was the premium resource sought by potential farmers. One of the major reasons for lack of interest in pasture, at least initially, was a crash in livestock numbers. During the 1990s, Kazakhstan suffered the loss of Soviet markets for livestock products, cessation of agricultural subsidies, hyperinflation in 1993–1994, and a liquidity crises as state farms were unable to pay wages (Behnke 2003). These factors led to the emergence of a barter economy in which farm wages, and indeed, most other transactions, were paid in sheep. As these were dumped on the market for cash, their value declined still further, leading to ever-increasing rates of loss. As a result, numbers of small stock in Kazakhstan dropped by 70% between 1993 and 1997 (Robinson and Milner-Gulland 2003a). As in other Central Asian countries, this was accompanied by a decline in the condition of infrastructure such as wells, barns, and roads, essential for pasture use. This reduction in stock numbers occurred at the same time as the reform process described above was progressing. While much arable land remained in large agricultural cooperatives, livestock ownership became more concentrated in households and family farms; by 2002 about 90% of animals were owned by these actors (Dudwick et al. 2005). Low stock numbers, small herd sizes, and the collapse of state support meant that livestock migration virtually ceased; vast areas of Central Kazakhstan, once used by hundreds of thousands of livestock in summer, fell out of use almost entirely (Robinson and Milner-Gulland 2003a).

The Land Code of 2003 ushered in a new phase of land reform, including true private ownership of agricultural land.[32] However, it penalized those who had been

[32] Articles 20–24 concern definitions and granting of private land, which must be purchased at set rates. Land already incorporated into family farms on a *permanent* use basis could be converted to private land at no extra charge (Article 170(3)).

leasing their CLS or subleasing physical plots held in leasehold family farms to others by forcing them to cancel these arrangements and either buy the land into private ownership; work directly on their own land share, registered as a family farm; or transfer their shares to the capital of agricultural corporations by 1 January 2005 (Dudwick et al. 2005). Non-compliance led to confiscation of shares by the state. The rationale for doing this was that these shareholders were not paying tax on this land, yet earning income on it through leasing or subleasing. As we have seen, these arrangements concerned at least 28% of all such shares nationwide. This process reduced the amount of land available to those family farms which had been leasing land from pensioners and others unable to use their shares themselves; some lost their rights through inaction or because of lack of information about the new law (Toleubayev et al. 2010; USAID 2005a). Of the formerly leased and subleased land, over half was transferred to agricultural corporations or returned to the state.[33] Few took up the opportunity to upgrade temporary use titles to full ownership because leasehold conditions were so attractive; thus, many family farms continue to hold leasehold contracts (USAID 2005a).

3.3.2 Land Reform and Pastoralism

During the 1990s, rangelands had become abundant relative to livestock inventories, and so few livestock owners actively leased pasture from the State Reserve or from the agricultural enterprises of which they were members. Rather, many used the land informally or registered leases on small areas of land for dwellings, wells, or barns and used the adjoining state or corporation rangeland for free (Alimaev and Behnke 2007). Infrastructure was variously given away as part of shares or had to be purchased depending on economic circumstances of the parent farm, but it was not always working and required large herd sizes to make its use viable, especially given high registration costs for titling (Behnke 2003). These were all reasons why members of poorer households tended to remain shareholders of agricultural corporations. In remote areas with poor access to water and winter feed, state farms had required particularly high levels of subsidy. Here, both agricultural corporations and many newly formed family farms quickly folded, resulting in out-migration and the virtual abandonment of some settlements (Robinson 2000; Behnke and Temirbekov 2003). From the beginning, three emerging types of livestock owner and mobility pattern were noted in field studies (Kerven et al. 2004; Robinson 2000):

1. Small numbers of households owning large numbers of animals who had obtained title to key infrastructure in multiple seasonal pastures and whose animals moved several times over the year.

[33] Of the leased land shares (representing 14.36 million ha) that remained after January 1, 2004, up to the deadline, about 5.8 million ha was transferred to family farms or similar structures, while 5.6 million ha was contributed to the capital of agricultural corporations, and 1.7 million was returned to the state. Thus, today CLS no longer exist. Of the subleased land plots, 65% was contributed to the share capital of agricultural corporations, and 24% went to smaller farms (USAID 2005b).

2. Those owning intermediate numbers of animals whose livestock were based outside the village at a barn and adjoining house, but did not move during the year.
3. Those owning small numbers of animals based in the village; usually livestock belonging to this category of owner were grazed around the village all year round, often herded in common herds on a rota basis known as *kyzyk*, similar to the *kezu* system in Kyrgyzstan (Behnke, personal communication).

The lack of mobility of village-based livestock has led to localized pasture degradation and low livestock productivity among this group, manifested as high levels of weight loss over the winter (Kerven et al. 2004).

Since the 2003 law, the registration of pasture areas into family farms accelerated, usually under 49-year leasehold arrangements. Most herds grazing outside village pastures are thus owned by single households holding individual title to pastureland. Collectively, these households may be significant employers in their local communities (Kerven, personal communication). However, as we have seen, many rural households returned their land shares back to the state or to agricultural corporations as they were unable to use them or were not successful in converting CLS to physical land titles by the deadline of 2005. Others may have registered only arable land or not been eligible to receive CLS in the first place. Livestock belonging to these groups still graze on land around settlements[34] (although some make arrangements with owners of family farms to have their stock kept with those herds outside the village for at least part of the year).

Otherwise, the main alternative option available today is registration of a family farm on remaining State Reserve land with all the uncertainties and transaction costs that this entails. Where such land is plentiful, herders have continued to use it without registration, but large herd sizes are required to cover transport and infrastructure in such areas (Kerven et al. 2006). Lastly, *leskhoz* land is also available in some areas and may be leased on a short-term basis.[35] For those who had contributed their CLS to the capital of agricultural corporations, although the 2003 Code reiterates that these can still be redeemed for physical plots, the withdrawal of land shares continues to be problematic, and many applications are simply not processed (USAID 2005a).

Statistics from the Kazakhstan Statistical Agency (2011) seem to confirm that family farms are becoming the most important players in the livestock sector: by 2011, 57% of all registered pastureland was allocated to this type of holding, a 25% increase in area on 2007.[36] There was no corresponding change in the proportion of arable land in family farms. There was also a transfer of livestock from households

[34] According to the Land Code of 2003 (Article 26.1), this "village pasture" cannot be purchased into private ownership; it is unclear to what extent it may be subject to leasehold title.

[35] Kerven (2012) reports annual leases in one region of the country (personal communication).

[36] Some of this increase is accounted for by increases in the total pastureland registered for use by any agricultural entity, suggesting that some land from the State Reserve is being brought back within formal tenure arrangements. The total area of pasture registered to users increased by 12% between 2006 and 2011.

to registered family farms: 68% of small stock and 81% of cattle were counted in family farms by 2011, proportions close to those attributed to households just 5 years earlier.

Kazakhstan is today considering the design of a law on pastures. One initial concept[37] recognizes overgrazing around settlements and underuse of remote pasture resources as an issue of national importance.[38] Among the policies recommended is the repair and construction of water points to open up new areas and in some cases to irrigate pasture for winter feed production. It is also suggested that pasture could be assigned to *raion* authorities, which would be responsible for allocation of short-term grazing rights through issuance of pasture tickets in some form. Local government would also be responsible for the development and maintenance of infrastructure enabling local users to exploit the remoter pastures. However, in cases where a large proportion of pastures previously within state farms are now held under individual tenure, such a system is likely to apply mainly to forest and State Reserve land.

In highly pastoral areas, the issue of commonly herded livestock concentrated around villages may become less of a problem over time if out-migration leaves mainly larger herders behind. But in areas with more diverse economic opportunities, then the question of common grazing land may remain a contentious issue: over four million sheep and goats are still held in households. Among those who hold pasture land within family farms, data on mobility are available only from case studies – there are no national statistics on the number of separate pasture areas leased by each family farm, but many may be sedentary or engage only in very limited migrations. A second issue for this group is the temporary nature of leasehold contracts, the duration of which has already been reduced once from 99 years to 49,[39] leading to worries about tenure security. These issues must be understood in more detail while considering the new law.

3.4 *Turkmenistan: Private Enterprise Under State Control*

There is little literature available on reform of the livestock sector in Turkmenistan, perhaps due to the difficulties of doing research in the country. Studies with wide geographical coverage focus on arable agriculture while those field studies reviewed here which focused on the livestock sector cover three sites, two of which are in the same province. However, it is clear that reform in Turkmenistan followed a very

[37] This concept was drafted by UNDP and is now under consideration at the Ministry of Agriculture.

[38] A speech by President Nazarbaev in December 2011 urged the development of migratory systems of livestock production (Kazakhstanskaya Pravda, 28 January 2012).

[39] The 2003 Land Law (Articles 26.1 and 101) states that "*otgonnye*" or remote seasonal pastures can only be used by family farms under the leasehold tenure arrangement and not bought into private ownership. It is not clear what proportion of the pastures currently in the State Reserve fall into this category.

different pattern to that of other republics examined so far. Although statistical data are unreliable, at the end of the 1990s, half of small stock and one third of cattle remained in the state sector; steep falls in inventories were not reported despite decreases in feed and concentrate availability (Hodjakov and Wright 2003; World Bank 2001).

3.4.1 The Reform Process

In 1990–1992, Turkmenistan increased household plot sizes through distribution of land to households and even allocated land to individuals under conditions of "private ownership" between 1993 and 1996.[40] These two measures concerned only irrigated land and in the latter case affected a relatively small number of farmers as the land provided was usually poor (Lerman and Stanchin 2003). Pasture itself is under the stewardship of government and not subject to private ownership; it may theoretically be granted under the other two existing tenure categories: use and rent,[41] but cases of pasture being accessed in this way were not reported in the studies reviewed here.

The reform which most transformed the agricultural sector was the Presidential Decree of June 1995 (and subsequent related legislation) which transformed the state and collective farms into associations of leaseholders known as *dayhan birlishik* or peasant associations, referred to henceforth as associations (World Bank 2001; Lunch 2003). Water and land remained owned by the state, but other assets were transferred to the associations. These then distributed arable land or livestock to individual members on a leasehold basis. Leaseholders provided a proportion of their output to the association or directly to state marketing organizations. In the case of livestock operations, leases concerned only flocks and herds, not the pastureland. Leaseholders could not make decisions on herd composition or marketing of state-owned animals, but the access to pastures and a proportion of young animals which accompanied leases allowed them to keep and accumulate their own flocks. The reforms thus turned the majority of the agricultural sector over to individual management, but independent decision making among these new farmers was still limited (World Bank 2001).

Leasehold contracts for livestock herding were at first characterized by considerable variability, with some shepherds being paid in cash and others in livestock offspring (Lunch 2003). Payment in live animals was more popular due to arbitrary pricing arrangements by associations and because it gave lease holders the opportunity to build up private herds. Although government targets to increase livestock numbers put pressure on associations to keep their stock and pay shepherds in cash, it appears that leasehold terms set in favor of payment in animals have persisted (Jumardurdyev 2010; Behnke et al. 2005). The terms of these agreements set expected lambing

[40] Land under the category known as "private ownership" cannot be bought and sold and may be confiscated by the state under a wide range of conditions (Lerman and Stanchin 2003).

[41] Articles 59 and 66 of the 2004 Land Code.

rates and fixed percentages of offspring accruing to the leaseholder and association, respectively. One study reports lambing rates set at an expected level of 95% for Karakul sheep and 85% for Sarajin sheep, with lambs then divided between the leaseholder and association on a 50:50 basis. However, shepherds are entitled only to small numbers of female lambs thus constraining the rate of accumulation of private animals (Jumardurdyev 2010). Behnke et al. (2005) noted that if a shepherd exceeded his quota, he could claim all surplus animals, while if he failed to reach it, he had to substitute missing animals for his own private ones – thus, incentives for increasing productivity are considerable.

Marketing of state livestock and support for livestock producers has been split over time to varying extents between the associations themselves and Turkmen Mallary, the state agency responsible for livestock.[42] Turkmenistan's system of state orders has been highly criticized as binding leaseholders into unfavorable credit, sales, and input supply contracts with the state and for distorting agricultural markets (Lerman and Stanchin 2003). However, these orders apply to a relatively small range of products, which do not include livestock, meat, or milk. Associations are able to market their own livestock on the open market,[43] but must submit a proportion of their production to Turkmen Mallary, which is also engaged in marketing (Kerven 2003). The state may also interfere in other ways; for example, presidential decrees to increase livestock numbers prevented associations from selling their stock at an economically optimal age, while a ban on slaughter of Karakul lambs caused the virtual collapse of the Karakul pelt industry (Kerven et al. 2002). Leaseholders sell their private animals and livestock produce exclusively on the open market.

According to a 2001 World Bank report, associations continued to supply leaseholders with a range of inputs and services from seeds to machinery, usually obtained from state sources (World Bank 2001).[44] However, studies in predominantly livestock raising areas report very low levels of support (Lunch 2003). The bulk of inputs such as fodder, vet services, and water are thus purchased by lease-holding shepherds on the open market. The fact that the associations receive 50% of offspring every year and yet provide little support to leaseholders in return raises the question of their function. Lerman and Stanchin (2003) have summarized them as "organizational shells" engaged mainly in administration of state-owned land, maintenance of rural infrastructure, and transmission and enforcement of state orders. They are also of marginal financial viability – income is plowed into salaries, while many are owed debts for livestock products by government marketing organizations (Kerven 2003; World Bank 2001).

[42] The ownership of a large proportion of state livestock was transferred from associations to this agency by government directives (World Bank 2001).

[43] In the case of items subject to state orders, leaseholders submit their produce directly to government marketing organizations rather than through the intermediary of the association; thus, in such cases, associations play no part in marketing of agricultural produce (Lerman and Stanchin 2003).

[44] In the case of state orders, inputs are provided by state marketing organizations, not the associations (Lerman and Stanchin 2003); see also the Law on Peasant Farms of 2007 (Article 8.2).

3.4.2 Pasture Access and Mobility: State Livestock

Access to pastures depends to a large extent on having state animals to herd. Comprising around 700–800 heads of small stock, these flocks create both the need and the means to move (Behnke et al. 2005). Yet even in the state sector, mobility is still lower than in Soviet times. Some state herds have begun to stay at one well all year round, although where seasonal and spatial variation in pasture quality and water quality is high, livestock mobility has persisted (Behnke et al. 2005). A mixture of sedentary and migratory behavior among association shepherds has also been reported by Jumardurdyev (2010), but in no cases have state animals been reported to graze around villages.

While in some cases leaseholders may be assigned wells by association directors, in others access to various seasonal grazing sites are arranged by shepherds: Behnke et al. (2005) describes how, in one district, while some shepherds used only those wells assigned officially to their own association, others negotiated access rights in other areas according to historical precedent or linked to claims over wells and houses. At a higher level, Jumardurdyev (2010) also states that associations negotiate pasture access in other districts to supplement their own pasture resources. A significant factor limiting migrations is water availability: delivery of water for dilution of saline wells by truck was reported as of the most burdensome costs borne by lease-holding shepherds (Jumardurdyev 2010).[45]

3.4.3 Mobility of Private Animals

Relatively few rural households hold leases over state flocks and herds: on farms having access to irrigated land, many association members lease land rather than livestock or are engaged in salaried employment, having a few private stock as a secondary activity. In remoter and more pastoral areas only those involved in herding state animals remain; others moved away, leaving their private stock behind (Lunch 2003). The grazing of these private animals occurs through a system known as *chekene*, a little like the Kyrgyz *bada* system described above, in which a professional shepherd is paid to herd the animals of others. The following types of arrangement were observed in the case studies reviewed here:

- Leaseholder shepherds taking *chekene* animals into their own herds, for example, as a way of covering the costs of water haulage.
- Shepherds herding exclusively *chekene* animals – thus making a living as a shepherd without herding association animals – such arrangements are common in areas where many non-herders work outside the livestock sector but own their own animals.

[45] Despite high costs and lack of support from associations, Jumardurdyev (2010) found that livestock leasehold arrangements are relatively advantageous – in one area of mixed farming, those engaged in shepherding with access to state herds had average incomes around 50% higher than leaseholders of arable land or salaried workers.

- Shepherds herding the animals of those who have moved away from the area entirely.

Mobile herds composed entirely of private animals are rare; most *chekene* animals not taken on by leaseholders are thus found close to settlements. Part of the reason for this is economic and linked to the costs of movement (Soyunova 2003). In other cases, access may be a problem: Jumardurdyev (2010) found that the presence of *chekene* herds located in remote deserts sometimes caused tensions with leaseholder shepherds. Where association flocks occupy all available wells, private animals based in villages are "locked in" and graze round village all year round, incurring high supplementary fodder costs (Lunch 2003; Behnke et al. 2005).

Overall, the Turkmen system seems to have enabled state and, at least a proportion of private animals, to use pastures away from settlements all year round with some flexibility, albeit with differing levels of mobility. New land legislation specific to pastures is currently being drafted, but is not yet in the public domain.

3.5 *Uzbekistan: A Growing but "Landless" Livestock Sector*

In Uzbekistan, three major forms of farming entity have emerged from the reform process (Veldwisch and Spoor 2008; Lerman 2008). Firstly, *dehkan farms* (whose holders are known as "*dehkan*") concern households with access only to household plots for permanent heritable use.[46] *Peasant farms* (whose holders are known as "*fermer*") are larger household farms established by state farm employees on a leasehold basis from state farmlands. There was no process of allocation of equal shares to all state farm members as occurred (or was meant to occur) in some of the other republics reviewed here. Land allocation processes openly favored rural elites so that today *fermer* make up a minority of the rural population and *dehkan* the poorer majority (Veldwisch and Spoor 2008). Lastly, "*shirkat*" (agricultural enterprises) are the successor organizations to state farms and remained structured along collective lines.[47]

Since 2003, *shirkat*, which were poorly productive and close to bankruptcy, have been progressively dismantled in favor of peasant farms, which now have replaced them as official suppliers of those products subject to state orders. Peasant farmers must produce set quantities of wheat and cotton and obtain special permission to grow other crops, which they can then sell on the open market. By 2006, the number

[46] These plots were enlarged by state decree, in the interests of food security and today have an average area of 0.2 ha, 100 times smaller than the average peasant farm (Veldwisch and Spoor 2008). *Dehkan* farmers are not subject to state restrictions on crop types to be grown and marketed; however, this type of farm may not exceed 0.35 ha of irrigated land by law, and thus, *dehkan* farmers are unable to grow much above what is needed for subsistence. They thus often work for peasant farms on a wage or cash cropping basis (Lerman 2008; Veldwisch and Spoor 2008).

[47] These three structures were first recognized in the 1998 Land Code; the legal frameworks governing each one (the Law on Dehkan Farms, the Law on Peasant Farms, and the Law on Shirkat) were passed with the Land Code in April 1998 (Lerman 2008).

of *shirkat* had fallen from 2,000 to 314 and peasant farms controlled the bulk of arable land (Lerman 2008). In contrast, most of the nation's livestock is owned by *dehkan* farmers: according to Lerman (2008), 96% of cattle and 80% of small stock are held in the individual sector, the vast majority in *dehkan* farms. The only livestock still held by the *shirkat* are Karakul sheep in desert and semidesert areas and a small number of cattle. Livestock inventories were not characterized by serious falls during the 1990s and (if state statistics are to be believed) have seen significant increases since that period,[48] mostly in the *dehkan* sector and despite reductions in feed availability (Yusupov et al. 2010).

Concerning pastures, peasant farms may be allocated land from their respective *shirkat* according to the number of livestock which they declare – they are required to have at least 30 cattle units (corresponding to 300 sheep) and may lease a minimum of 2 ha per cattle unit from the state for a period of 30–50 years. There is little information from the field about modes of pasture access among this group. In the case study areas discussed here, officially registering a farm for extensive livestock breeding is not an economically attractive option and such operations play only a minor role. Statistics also imply that the massive transfer of land from *shirkat* to peasant farms which has occurred over the past 10 years seems to have concerned mostly arable land – the share of hayfields and pastures used by these farms officially accounts for only 6% of the total (Yusupov et al. 2010).

In 1991, about 20 million ha of pasture was recorded as belonging to agricultural enterprises; today about 40% of this pasture has reverted to the state reserve (Yusupov et al. 2010). The rest (apart from the small area in peasant farms mentioned above) remains in those *shirkat* which have not been dissolved, including 103 Karakul sheep breeding operations (see next section). *Dehkan* farms, which own most of the livestock, do not have any formal entitlement to pastures and must use *shirkat* pastures on an informal basis. Where *shirkat* have fallen apart, one case study from a mountainous area has reported that the resulting "tenure vacuum" has seen the emergence of customary patterns of pasture allocation which existed in pre-Soviet times (Cariou 2002). Lastly, some pastureland is under the administration of the State Forest Agency (*leskhoz)*, which is better funded and staffed than the *shirkat* and has a much tighter control over its territory. Permission from the local *leskhoz* and the payment of a fee per head of livestock is required to obtain access to these pastures.

3.5.1 Transformation in the Karakul Sector: Pasture Held Under the Shirkat System

In Soviet times, the vast desert and semidesert areas of Uzbekistan were used as pastures mainly for the grazing of Karakul sheep and the production of Karakul lamb pelts, which were highly appreciated for luxury clothing at that time. After the

[48] Official figures for 2007 suggested a 40% increase in cattle numbers and 20% increase in small stock numbers since 1991 (Yusupov et al. 2010).

breakdown of the Soviet Union, the former *sovkhoz* specialized in Karakul sheep breeding were transformed into *shirkat* under the semiprivate company Uzbek Karakul, which is required to fulfill governmental plans for production. However, the quality of the Uzbek Karakul sheep breeding has deteriorated; demand for Karakul lamb pelts and world market prices have both decreased significantly. As a result, the number of Karakul sheep owned by the *shirkat* has been shrinking. At the same time, the number of livestock owned by peasant farms and households (*dehkan* farms) has increased dramatically.

In some regions, the number of household livestock far exceeds the number of *shirkat* livestock. Private households usually graze their livestock on the pastureland leased by their *shirkat* free of charge and without legal entitlements. For the moment the *shirkat* have had to accept this situation – they lack both the capacity and a legal framework to control pasture access effectively, but there is a growing awareness that the current system is not viable in the long term.

However, the discussion about land tenure reform and the future role of the *shirkat* and households is just only beginning. Different models are under consideration (private vs. collective leasehold of pastureland, remuneration of the *shirkat* for pasture use vs. leasehold directly from the state). In Box 11.2, we present case studies from two regions, which differ in pasture productivity, population density, and the relationship between the *shirkat* and local households.

4 Summary and Conclusions

4.1 Land Reform

During the 1990s, a reduction in livestock numbers combined with general economic breakdown led to the collapse of migratory systems, abandonment of remote pastures, and high stocking rates around settlements across Central Asia. These trends were most marked in Kazakhstan and Kyrgyzstan where state animals virtually disappeared along with associated state structures in the rural areas, leaving private livestock with no formal support. In Uzbekistan and Turkmenistan, livestock numbers did not collapse. But even state herds have become less mobile than in the past, and formal systems for seasonal pasture access by private animals are lacking. Availability of working wells also limits use of pastures in many desert and steppe areas.

Traditional collective herding systems quickly became reestablished in all five republics following the crisis; in some cases, shepherding of pooled animals is undertaken by participating households on a rota basis, and in others, a professional shepherd is employed. Those owning relatively large numbers of animals either herd these themselves or pay someone to do so. Common herding systems are particularly developed among mobile flocks in Tajikistan, Kyrgyzstan, and mountainous parts of Uzbekistan but also exist around settlements in desert-steppe areas. Many pastures were used informally at first, and grazing access was variously determined by former state farm boundaries, customary memory, locations associated

Box 11.2 Case Studies from Uzbekistan

i. *Foothills and semidesert zones:* Farish district is located in Jizzakh region, some 250 km southwest of Tashkent; Bogdon *shirkat* is located in the eastern part of the district covering an area of about 75,000 ha with around 20,000 inhabitants, mainly concentrated in the foothills of the Nuratau mountain range in the south. The foothills and plain between these mountains and the Aidar and Tuscan lakes in the north are covered with semidesert vegetation. In 2011, household livestock resident on this territory outnumbered the *shirkat* livestock – some 67,000 sheep and 10,000 cows belonging to *dehkan* households graze on *shirkat* land around settlements, whereas approximately 25,000 *shirkat* sheep graze in the plain in the north. The 3,000 sheep officially registered as belonging to peasant farms (*fermer*) are almost negligible.

The *shirkat* herds are based at barns located at wells in the plains; these animals are mostly sedentary and only in dry years, when the pastures become totally unproductive in summer, are livestock transferred to other areas (mostly to irrigated lands), where the shepherds have to pay peasant farmers for the use of land. Occasionally the state (in the form of Uzbek Karakul) organizes livestock migration, maintaining elements of the Soviet system. During a harsh drought in 2011, livestock in southern Uzbekistan was close to starving, and about 3,000 *shirkat* sheep were brought to the pastures of the Bogdon *shirkat* by train.

Most *shirkat* shepherds are able to graze their private livestock with the *shirkat* sheep. Existing systems for paying the shepherds include both fixed and variable in-kind monthly payments and an arrangement comprising an obligation to fulfill plans combined with the right to keep all the newborn sheep exceeding the plan. Here, nonfulfillment of the plans entails replacement of the difference with private livestock. The *shirkat* is theoretically obliged to provide its shepherds with veterinary services, winter fodder, and equipment, but in reality, few services are provided. As in Turkmenistan, many *shirkat* herdsmen are *de facto* independent farmers whose main incentive to graze *shirkat* sheep is the free access to *shirkat* pastures for their own livestock.

In contrast to *shirkat* animals and those of their herders, livestock belonging to *dehkan* households are almost completely village based during all the year, grazing around each settlement and returning there every evening. The three schemes for this type of grazing are familiar from other republics reviewed here and include individual grazing (where a member of the family grazes the livestock of a single household every day) and herding of collective flocks on a rota basis or using a paid shepherd. In addition, some livestock are put out to graze without a shepherd and return to the village on their own.

(continued)

Box 11.2 (continued)

Most minor livestock owners in most villages on the territory of Bogdon *shirkat* practice collective grazing with a paid shepherd, whereas large livestock owners usually graze their flocks individually.[49] As none of these owners have official land use rights, there are no formal rules for grazing, but most (collective or individual) pasture users have a notion of which territory belongs to them by customary law. However, there are conflicts about pasture use between:

- Pasture users from different villages who claim the same territory for themselves.
- *Shirkat* herdsmen and private pasture users from the villages close to the plain.
- Collective pasture users and big livestock owners (who do not follow the rules which have been developed by those who have joined the collective grazing).

Such conflicts usually remain unresolved or are solved through unofficial intermediation by the village authorities. The *shirkat* has lost control of one third of its territory to the private pasture users from the villages, and yet has to pay tax on this land. The management has been unable to enforce payments for the use of this pasture, firstly because there is no legal basis for the use of *shirkat* pasture by private households and secondly no backing from the local authorities (which recognize the importance of livestock to villagers' livelihoods) for such payments.

The pastureland in the higher ranges of the Nuratau Mountains is under *leskhoz* management and is more productive than the semidesert pastureland in the foothills and plains. Villages in the mountains are close enough to this territory for their livestock graze there in the day during the summer, returning to the village in the evening. Even if they have to pay a fee to the *leskhoz*, they consider this worthwhile due to the higher quality of *leskhoz* pastures in comparison to the free-of-charge *shirkat* pastures. Only in exceptional cases are livestock taken up to night camps in *leskhoz* territory. The *leskhoz* tends to give priority access to its pastures to big herds from other districts or cities, which follow a well-organized and transhumant grazing scheme and pass through the Nuratau mountain range from east to west during the spring months. The source of these migratory livestock is unclear; but they are not *shirkat* livestock, implying that private households and farmers are still able to undertake long migrations to access seasonal pastures in certain cases.

(continued)

[49]The situation differs slightly in the western part of Farish district on the territory of the Farish and Kyzylkum *shirkat*: here, collective village-based grazing is rather uncommon. Most households either graze their livestock individually, have it graze without a shepherd, or give it to a *shirkat* shepherd who has a barn in the steppe.

Box 11.2 (continued)

ii. *Desert areas:* Our second case study, from Romitan district, in Bukhara region, is rather different. The study site concerns the former territory of Kyzyl Rovat *sovkhoz* which covered over 220,000 ha of sandy desert. There are only three villages close to the Amu Darya river with a population of 960 inhabitants, whereas the desert is almost unpopulated. Today about 60% of this land is managed by the Kyzyl Rovat *shirkat*, 35% by the *leskhoz*, and 5% is part of peasant farms. In contrast to the above example, most livestock belong to the *shirkat* (7,200 sheep), while the local population owns about 2,000 sheep and 500 cows. Of the 52 wells constructed within the overall Kyzyl Rovat land use area during the Soviet period, only three were operational on *shirkat* territory in 2011, leaving most pasture abandoned. *Leskhoz* territory is also not used for the same reason. Livestock mobility is severely constrained due to this lack of wells. The quota system for Karakul pelts and debt servicing both create disincentives to increase livestock production and thus to repair wells. Those with registered peasant farms in contrast have invested in wells on their land and are able to keep higher stocking densities on their holdings. Small stock belonging to households are grazed with *shirkat* animals, while cattle graze, unsupervised, close to settlements by informal agreement with the *shirkat*.

with particular households during the Soviet time, and purchase or construction of infrastructure such as barns. The initial focus of formal land reform was on arable land, but as this progressed and private stock numbers rose, attention turned to formal tenure of pasture systems.

Of the cases examined here, only in Kyrgyzstan have pastures been designated as common property. In the other republics, pasture is either still allocated to large state structures or is subject to general land codes which emphasize individual forms of land use or ownership.[50] As livestock numbers grow, these land codes and associated legislation are increasingly being applied to pastures, but even today, and particularly in remoter areas, much pasture use by non-state livestock still occurs in a legal gray zone. A real understanding of pasture use patterns can be obtained only through detailed field studies, which are few and far between. High regional variability in the outcomes of land reform makes it very difficult to make general statements about republics from these case studies, so the results presented here represent only a rough outline of current trends.

[50] It should be noted that the establishment of "private farms" by groups is possible in both Kazakhstan and Tajikistan, but has not generally been taken up for the purposes of pasture management. This may be due to the fact that arable land comes under the same registration process and also to the unstable membership of common herding groups.

4.2 Livestock Mobility

Following the worst years of the crisis, in the mountainous republics common herding systems allowed livestock mobility to increase quite quickly in the form of transhumance to high-altitude summer pastures, but migration to more remote pastures across administrative boundaries remains limited. In Kyrgyzstan, the new law on pastures may remove some of the administrative barriers to such movement, but economic constraints continue to restrict the use of some remote pastures. In Tajikistan, only the few remaining state herds and relatively small numbers of private animals continue such long-distance movements. In that republic, mobility may also become affected by the annexation of common pasture by individuals for exclusive use as pasture reform accelerates (Kurbanova, Chap. 7)

In predominantly pastoral areas of Kazakhstan and Turkmenistan, many left regions with poor water and winter feed supplies – leaving behind mainly households with large numbers of private animals (or access to state ones). In Turkmenistan, such households, who lease large state flocks, have fewer economic or administrative restrictions on pasture access than Kazakh pastoralists. The latter use a combination of leasing and informal use of state land to access grazing resources, but the number and scale of movements undertaken is directly proportional to herd size, which is highly variable. A shortage of viable wells in both these two republics, and in Uzbekistan, restricts the area of pasture which is available for use.

4.3 Pasture Access

In Kyrgyzstan, it is unlikely that the new pasture law will eradicate tensions between those participating in common herding systems and owners of large individual herds; domination of the new Pasture Users Associations by wealthier herders may be inevitable (Crewett 2011; Kraudzun 2012). But the existence of a legal basis for common property means that some access to seasonal pastures by all categories of livestock owner has a greater chance of persisting in this republic than in Kazakhstan or Tajikistan.

In more highly populated or agro-pastoral areas of the desert-steppe-dominated republics, where significant numbers of smaller livestock holders exist, they generally have poor access to pastures – in Turkmenistan and some areas of Uzbekistan because state herds occupy most of the wells, and in Kazakhstan, because much of the pasture and the infrastructure which makes it usable has already been purchased or registered into a family farm. Those with small numbers of animals do not generally register grazing land and so must either arrange to place stock within larger herds or graze their animals around the village all year round. In Tajikistan rural households have very small numbers of animals and collective herding systems are thus well developed, making individual permanent use a particularly poor choice of tenure type. Pastures are now being annexed by individuals and, in some areas of the country, access has become a critical issue[51] (Kurbanova, Chaps. 7 and Halimova, Chaps. 13), although in other regions, *de facto* collective access persists in spite of the legislation.

5 Conclusions

From the above discussion, we may tentatively conclude that common herding systems (and associated tenure arrangements) may be most appropriate where large numbers of households are engaged in livestock raising as a secondary activity. However, even where this is not the case, individualized forms of property right may be poor at facilitating access to multiple noncontiguous parcels of land, especially if each must be registered separately. In Kazakhstan, households having exclusive access to only one pasture area are likely to have higher supplementary feed costs and to experience greater problems of livestock disease and productivity than those who are mobile (Kerven et al. 2004). Transaction costs associated with land registration are high in all republics and a disincentive to livestock mobility.

In Central Asia, the importance of land reform legislation for pasture management was understood late in the reform process. Attention was focused on increasing the productivity of arable land, which is why forms of individual tenure were initially pursued in every case examined here. Kyrgyzstan formed separate provisions for pasture early on, which was an enabling factor in the passing of a pasture code. Today, the other four republics are considering the introduction of pasture codes, with provisions for common pasture management under debate. Yet, because there is no formal distinction between pasture and arable land in the underlying land codes and laws governing the establishment of household farms, policy makers will find it very difficult to introduce pasture-specific common property regimes without modifying other existing legislation. In Kazakhstan, where much pasture is already under individual forms of tenure, and in Tajikistan, where that process is now beginning, it may now be possible to introduce common property rules for pasture only on the state reserve, which is often very remote from settlements. In Uzbekistan and Turkmenistan, it seems that further reform of state farms and remaining *shirkat,* which currently hold legal tenure over vast areas of pasture, is inevitable, and it is in these republics that the largest transformations in pasture access and allocation may be still to come.

Acknowledgment We would like to thank Ilka Starrost and Heino Hertel for putting the authors in contact and Carol Kerven and Zvi Lerman for their advice and data.

[51] A recent (unpublished) study by the World Bank, which presents results from 1,800 surveyed farmers in 18 *raion*, found that pasture access and rotation was ranked among the top five problems most commonly cited by farmers (World Bank 2012).

References and Further Reading

Alimaev II (2003) Transhumant ecosystems: fluctuations in seasonal pasture productivity. In: Kerven C (ed) Prospects for pastoralism in Kazakhstan and Turkmenistan: from state farms to private flocks. Routledge Curzon, London/New York

Alimaev II, Behnke R (2007) Ideology, land tenure and livestock mobility in Kazakhstan. In: Galvin K, Reid R, Behnke R, Hobbs N (eds) Fragmentation in semi-arid and arid landscapes. Springer, Heidelberg

Asian Development Bank (2008) The impact of land reform on agriculture, poverty and the environment in Kyrgyzstan. Asian Development Bank http://www2.adb.org/Documents/Reports/Consultant/38079-KGZ/38079-KGZ-TACR.pdf

Babu S, Tashmatov A (eds) (2000) Food policy reforms in Central Asia: setting the research priorities. International Food Policy Research Institute, Washington, DC

Behnke R (2003) Reconfiguring property rights in livestock production systems of western Almaty Oblast, Kazakstan. In: Kerven CK (ed) Prospects for pastoralism in Kazakstan and Turkmenistan: from state farms to private flocks. Routledge Curzon, London

Behnke R (2008) The drivers of fragmentation in arid and semi-arid landscapes. In: Galvin K, Reid R, Behnke R, Hobbs N (eds) Fragmentation in semi-arid and arid landscapes. Springer, Dordrecht

Behnke R, Temirbekov S (2003) A combined year three annual report: work packages 1 and 5. In: *DARCA. Unpublished report*

Behnke R, Jabbar A, Budanov A, Davidson G (2005) The administration and practice of leasehold pastoralism in Turkmenistan. Nomadic Peoples 9(1&2):147–169

Bromley W, Cernea M (1989) The management of common property resources: some conceptual and operational fallacies. In: *World Bank Discussion Papers No. 57.*: World Bank

BurnSilver S, Worden J, Boone R (2008) Process of fragmentation in the Amboseli ecosystem, Southern Kajiado district, Kenya. In: Galvin K, Reid R, Behnke R, Hobbs N (eds) Fragmentation in semi-arid and arid landscapes. Springer, Heidelberg

Bussler S (2011) Community based pasture management in Kyrgyzstan: a pilot project in Naryn region. GIZ, CAMP alatoo, Bishkek

Cariou A (2002) L'évolution géographique récente des zones rurales de piémont et de montagnes en Ouzbékistan. *Cahiers d'Asie centrale* (Numéro 10):271–291

Coughenour M (2008) Causes and consequences of herbivore movement in landscape ecosystems. In: Galvin K, Reid R, Behnke R, Hobbs N (eds) Fragmentation in semi-arid and arid landscapes. Springer, Dordrecht

Crewett W (2011) Decentralized pasture governance in Kyrgyzstan, challenges for implementation. Paper presented at the conference: *Pastoralism in Central Asia: status, challenges and opportunities in mountain areas*. 13–18 June, 2011 Bishkek

Dörre A (2012) Legal arrangements and pasture-related socio-ecological challenges in Kyrgyzstan. In: Kreutzmann H (ed) Pastoral Practises in High Asia. Springer, Dordrecht/Heidelberg/New York/London

Dudwick N, Fock K, Sedik D (2005) A stocktaking of land reform and farm restructuring in Bulgaria, Moldova, Azerbaijan and Kazakhstan. FAO, Rome

Farrington JD (2005) De-development in Eastern Kyrgyzstan and persistence of semi-nomadic livestock herding. Nomadic Peoples 9(1&2):171–197

Galvin KA, Reid RS, Behnke RH Jr, Hobbs NT (eds) (2008) Fragmentation in semi-arid and arid landscapes, consequences for human and natural systems. Springer, Dordrecht

Goldstein MC (2012) Change and continuity in a nomadic pastoralism community in the Tibet Autonomous Region. In: Kreutzmann H (ed) Pastoral practices in High Asia. Springer, Dordrecht/Heidelberg/New York/London

Halimova N (2011) Pastureland Tenure in Tajikistan: assessment and recommendations. In: *Report for the sustainable pasture, arable and forest land management, rural development project in Tajikistan* ADB and GEF

Hann C de (1993) An overview of the World Bank's involvement in pastoral development. Paper read at Donor Consultation Meeting on pastoral natural resource management and pastoral policies for Africa organised by UNSO (United Nations Sudano- Sahelian Office) December 1993, at Paris

Hardin G (1968) The tragedy of the commons. Science 162:243–1248

Halimova N (2012) Land tenure reform in Tajikistan: implications for land stewardship and social sustainability: a case study. In: Squires VR (ed) Rangeland stewardship in Central Asia. Springer, Dordrecht, pp 305–332 (Chapter 13, this volume)

Hodjakov O, Wright IA (2003) New patterns of livestock production. In: Kerven C (ed) Prospects for pastoralism in Kazakhstan and Turkmenistan: from state farms to private flocks. Routledge Curzon, London/New York

Huntsinger L, Forero LC, Sulak A (2010) Transhumance and pastoralist resilience in the Western United States. Pastoralism 1(1):9–36.

International Land Coalition (2007) *Mobile livelihoods, patchy resources & shifting rights: approaching pastoral territories*, www.drylands-group.org/Articles/1317.html

Jumardurdyev D (2010) Socio-economic survey of the Karakum Pilot Region. Unpublished report for GIZ

Kanchaev K, Kerven C, Wright IA (2003) The limits of the land: pasture and water conditions. In: Kerven C (ed) Prospects for pastoralism in Kazakhstan and Turkmenistan: from state farms to private flocks. Routledge Curzon, London/New York

Kazakhstan Statistical Agency (2011) Kazakhstan in 2010. Astana

Kerven C (2003) The privatisation of livestock marketing in Turkmenistan. In: Kerven C (ed) Prospects for pastoralism in Kazakhstan and Turkmenistan: from state farms to private flocks. Routledge Curzon, London/New York

Kerven C, Russel A, Laker J (2002) Potential for increasing producer's income from wood, fibre and pelts in Central Asia. *Socio-economics and policy research working paper 45*: ILRI

Kerven C, Alimaev II, Behnke R, Davidson G, Franchois L, Malmakov N, Mathijs E, Smailov A, Temirbekov S, Wright IA (2004) Retraction and expansion of flock mobility in Central Asia: costs and consequences. Afr J Range Forage Sci 21(3):91–102

Kerven C, Alimaev II, Behnke R, Davidson G, Malmakov N, Smailov A, Wright IA (2006) Fragmenting pastoral mobility: changing grazing patterns in post-Soviet Kazakhstan. USDA Forest Service Proc RMRS-P-39:99–110

Kraudzun T (2012) Livelihoods of the new livestock breeders in the eastern Pamirs of Tajikistan. In: Kreutzmann H (ed) Pastoral practices in High Asia. Springer, Dordrecht/Heidelberg/New York/London

Kurbanova B (2012) Constraints and barriers to better land stewardship: analysis of PRAs in Tajikistan. In: Squires V (ed) Rangeland stewardship in Central Asia. Springer, Dordrecht, pp 129–164 (Chapter 7, this volume)

Lerman Z (2008) Agricultural development in Uzbekistan: the effect of ongoing reforms. *Discussion Paper* No. 7.08, The Hebrew University of Jerusalem. http://ageconsearch.umn.edu/bitstream/37945/2/lerman-uzbek.pdf

Lerman Z, Sedik D (2008) The economic effects of land reform in Tajikistan. Policy studies on rural transition no. 2008-1, FAO Regional Office for Europe and Central Asia. http://www.fao.org/fileadmin/user_upload/Europe/documents/Publications/Policy_Stdies/Tajikistan_en.pdf

Lerman Z, Stanchin I (2003) New contract arrangements in Turkmen agriculture: impacts on productivity and rural incomes. *Discussion Paper* (11.03)

Lerman Z, Csaki C, Feder G (2002) Land policies and evolving farm structures in transition countries. World Bank, Washington, DC

Li WJ, Ali SH, Zhang Q (2007) Property rights and grassland degradation: a study of the Xilingol Pasture, Inner Mongolia, China. J Environ Manage 85:461–470

Lunch C (2003) Shepherds and the state. In: Kerven C (ed) Prospects for pastoralism in Kazakhstan and Turkmenistan: from state farms to private flocks. Routledge Curzon, London/New York

Milner-Gulland EJ, Kerven C, Behnke RI, Wright IA, Smailov A (2006) A multi-agent system model of pastoralist behaviour in Kazakhstan. Ecol Complex 3:23–36

Mountain Societies Development Support Programme (2004a) 2004 Final Survey of Rasht Valley, Tajikistan for the Tajikistan Rural Poverty Reduction Project

Mountain Societies Development Support Programme (2004b) 2004 Baseline household income survey of Khatlon: community based agriculture sector development for Tajikistan

Mountain Societies Development Support Programme (2005) 2005 Baseline household survey of Alai and Chong Alai

Mountain Societies Development Support Programme (2009) Gorno-Badakhshan household income survey 2008

National Statistical Committee of the Kyrgyz Republic (2004) Results of the first agricultural census of Kyrgyz Republic: volume 1. Livestock in 2003, Bishkek

National Statistical Committee of the Kyrgyz Republic (2011) Sel'skoe Khozyaistvo Kyrgyzskoi Respubliki [Agriculture of the Kyrgyz Republic]. Bishkek

Ostrom E (1990) Governing the commons: the evolution of institutions for collective action. Cambridge University Press, Cambridge

Reid RS, Galvin KA, Kruska RS (2008) Global significance of extensive grazing lands and pastoral societies: an introduction. In: Galvin K, Reid R, Behnke R, Hobbs N (eds) Fragmentation in semi-arid and arid landscapes. Springer, Dordrecht

Richard C, Yan Z, Du G (2006) The paradox of the individual household responsibility system in the grassland of the Tibetan plateau, China. USDA Forest Service Proc RMRS-P-39:83–91

Robinson S (2000) Pastoralism and land degradation in Kazakhstan. Department of Biological Sciences, University of Warwick

Robinson S (2007) Report of the pasturelands ecologist for ADB Agricultural land improvement project, Kyrgyzstan. Asian Development Bank

Robinson S, Milner-Gulland EJ (2003a) Contraction in livestock mobility resulting from state farm re-organisation. In: Kerven C (ed) From state farm to private flocks: prospects for pastoralism in Kazakhstan and Turkmenistan. Routledge Curzon Press, London

Robinson S, Milner-Gulland EJ (2003b) Political change and factors limiting numbers of wild and domestic ungulates in Kazakhstan. Hum Ecol 31(1):87–110

Robinson S, Karasartov S, Bobukeeva M, Gorborukova LP, Bush G, Fitzherbert A (2001) Pasture and land tenure in Kyrgyzstan. In: *Report for DFID project 'Sustainable Livelihoods for Livestock Producing Communities'*

Robinson S, Higginbotham I, Guenther T, Germain A (2008) Land Reform in Tajikistan: consequences for tenure security, agricultural productivity and land management practices. In: Behnke R (ed) The socio-economic causes and consequences of desertification in Central Asia. Springer/ Dordrecht

Robinson S, Whitton M, Biber-Klemm S, Muzofirshoev N (2010) The impact of land reform legislation on pasture tenure in Gorno-Badakhshan: from common resource to private property? Mt Res Dev 30(1):4–13

Rohde RF, Moleele MM, Mphale M, Allsopp NB, Chanda NR, Hoffman MT, Magole L, Young E (2006) Dynamics of grazing policy and practice: environmental and social impacts in three communal areas of southern Africa. Environ Sci Policy 9:302–316

Sedik D (2010) The feed-livestock Nexus in Tajikistan: livestock development policy in transition. *Policy studies on rural transition* No. 2010-1. FAO Regional Office for Europe and Central Asia. http://www.fao.org/fileadmin/user_upload/Europe/documents/Publications/Policy_Stdies/Livestock2010_en.pdf

Sedik D (2012) Livestock management problems and policies in Tajikistan: implications for land stewardship. In: Squires V (ed) Rangeland stewardship in Central Asia. Springer, Dordrecht, pp 189–212 (Chapter 9, this volume)

Sneath D (1998) State policy and pasture degradation in Inner Asia. Science 281:1147–1148

Soyunova O (2003) The costs and returns of change. In: Kerven C (ed) Prospects for pastoralism in Kazakhstan and Turkmenistan: from state farms to private flocks. Routledge Curzon, London

State Land Committee of the Government of Tajikistan (2009) Dar Borai Khojagi Dehkoni [On Dekhan Farms]. Dushanbe, Tajikistan: Sarparast

State Statistical Committee of Tajikistan (2009) Sel'skoe Khozyaistvo Respubliki Tadjikistan [Agriculture in the Republic of Tajikistan]. Dushanbe

Steimann B (2011) Making a living in uncertainty. Agro-pastoral livelihoods and institutional transformations in Post-Soviet Rural Kyrgyzstan. In: Müller-Böker U (ed) Human geography series 26. University of Zurich, Bishkek/Zurich

Thornton PK, Kruska RL, Henninger L, Kristjanson PM, Reid RS, Atieno F, Odero AN, Ndegwa T (2002) Mapping poverty and livestock in the developing world. ILRI (International Livestock Research Institute), Nairobi

Toleubayev K, Jansen K, van Huis A (2010) Knowledge and agrarian de-collectivisation in Kazakhstan. J Peasant Studies 37(2):353–377

Undeland A (2005) Kyrgyz livestock study: pasture management and use. http://landportal.info/sites/default/files/kyrgyz_livestock_pasture_management_and_use.pdf

USAID (2005a) Kazakhstan land administration report http://landportal.info/sites/default/files/kazakhstanlandadministration.pdf

USAID (2005b) Assessment of the implementation of the interim provisions of the land code http://pdf.usaid.gov/pdf_docs/PNADF080.pdf

Vanselow KA, Kraudzun T, Samimi C (2012) Land stewardship in practice – An example from the Eastern Pamirs of Tajikistan. In: Squires V (ed) Rangeland stewardship in Central Asia. Springer, Dordrecht, pp 71–90 (Chapter 4, this volume)

Veldwisch GJA, Spoor M (2008) Contesting rural resources: emerging 'forms' of agrarian production in Uzbekistan. J Peasant Stud 35(3):424–451

World Bank (2001) Turkmenistan: an assessment of leasehold-based farm restructuring. *Europe and Central Asia environmentally and socially sustainable development series* No. 500. http://www-wds.worldbank.org/external/default/WDSContentServer/IW3P/IB/2001/06/29/000094946_01062004023753/Rendered/PDF/multi0page.pdf

World Bank (2003) Kyrgyz Republic: enhancing pro-poor growth. Poverty Reduction and Economic Management Unit Europe and Central Asia Region http://web.worldbank.org/WBSITE/EXTERNAL/TOPICS/EXTPOVERTY/EXTPA/0,,contentMDK:20268977~menuPK:435735~pagePK:148956~piPK:216618~theSitePK:430367~isCURL:Y~isCURL:Y,00.html

World Bank (2007) Kyrgyz Republic livestock sector review: embracing the new challenges http://www-wds.worldbank.org/external/default/WDSContentServer/WDSP/IB/2007/03/14/000090341_20070314160221/Rendered/PDF/390260KG0Lives1iew0P09028701PUBLIC1.pdf

World Bank (2012) Perceptions of farmers and farm workers on land reform and sustainable agriculture in Tajikistan: summary overview report (draft)

Wu Z, Du W (2009) Pastoral nomad rights in Inner Mongolia. Nomadic Peoples 12(2):13–33

Yan Z, Wu N, Yeshi D, Ru J (2005) A review of rangeland privatisation and its implication in the Tibetan plateau, China. Nomadic Peoples 9(1&2):31–51

Yusupov YB, Lerman Z, Chertovitskiy AS, Akbarov OM (2010) Livestock production in Uzbekistan: current state, issues and prospects. Nasaf, Tashkent

Chapter 12
Governance and the Role of Institutions in Sustainable Development in the Central Asian Region

Victor Squires

Abstract The central purpose of this chapter is to demonstrate that adaptable, flexible, and appropriate institutions are central to sustainable development. The paradigm of agent-based social systems underlines the importance of better frameworks to support the role of institutions in sustainable development. The identified sustainable development strategies may be adapted to local conditions by countries and communities. Each of these strategies relies on judicious modification of preexisting institutions. A brief case study of the legal and institutional framework in Tajikistan is presented. This chapter shows *inter alia* why and how institutions—the glue of society that defines community—are critical to sustainability. Without them, sustainable development would remain in the domain of rhetoric.

Keywords Agent-based social systems • Development • Quality of life • Property rights • Human capital • Social self-governance • Environmental change • Trade regimes • Biogeophysical perspective • Interplay • Uncertainty • Mass communications • Ecosystem capacity • NGO • CBO • Tax • Enabling environment • Ethics • Social mores • *Khashar* • Donors • Foreign aid

Key Points

- Today in the western world, governance ideals and many practices support facilitation of collective actions through diverse institutional frameworks. This is in contrast to the ideology in the Central Asian countries which favors big, controlling

V. Squires (✉)
College of Grassland Science, Gansu Agricultural University, Lanzhou, China
e-mail: dryland1812@internode.on.net

GITEC Consulting GMBH, Cologne, Germany
e-mail: dryland1812@internode.on.net

V. Squires (ed.), *Rangeland Stewardship in Central Asia: Balancing Improved Livelihoods, Biodiversity Conservation and Land Protection*,
DOI 10.1007/978-94-007-5367-9_12, © Springer Science+Business Media Dordrecht 2012

government. Most governments determined, performed, and/or controlled many (if not most) collective actions, routinely dominating people, communities, organizations, and markets.

- Governance is not just a theoretical consideration. Coordination between systems at different levels of aggregation is aided by having fewer levels of governance and by clearly defining the responsibilities of each. Community sustainability can be choked off by inappropriate regional, national, or international institutions, or by a deficiency in technical and financial resources.
- Development is about social change with a strong implication of improvement. Thus, development may be understood as the process of making collective life more human. Development is the process of creating effective institutions to organize and regularize interactions so that risk is reduced and daily life made more predictable.
- Ultimately, development is a matter of changing minds, how people think and what they expect, changing legal and political systems, and developing new social practices.
- Social adaptation is the changes in instruments, ideas, and institutions that reduce community dependence on the affected part of the environment or to cope with external environmental stress.
- Unsustainability and sustainability are both a matter of how individuals and communities think about and their perception of the value of the world around them.
- At the most general level, institutions are constellations of rules, decision-making procedures, and programs that define social practices, assign roles to the participants in such practices, and govern the interactions among the occupants of those roles. Defined in this way, institutions constitute an important feature of the landscape in all areas of human endeavor.
- Although institutions are by their nature conservative, resisting change and supporting consistent practices, they should not be ungainly but should be ready to adapt for change. Institutional development is the conscious change wrought by policies following strategies, usually through the formal institutions of government, administrative regulations, and laws that assign the exercise of decision-making power in economics and politics.
- When institutions deal explicitly with human/environment relations, it is normal to refer to them as environmental or resource regimes. In thinking about large-scale environmental changes that have significant anthropogenic components, it is natural to focus first and foremost on the roles that these environmental and resource regimes play both in causing environmental problems and in constituting the principal components of solutions to such problems.
- Those seeking to understand the effectiveness of institutions are unlikely to make progress if they approach the problem like chefs, assuming that relatively simple recipes will work well, in the sense of producing similar results under a wide range of circumstances. Rather, they will need to adopt the perspective of physicians who know that a wide range of conditions may affect the health of specific individuals and that diagnosing particular cases necessarily requires skill in identifying the particular combination of conditions at work in each case.

- In solving environmental problems, we face a double problem: to date unknown dimensions of uncertainty and at the same time lost possibilities of solutions. The first aspect focuses on specific patterns of risk perception and the problem of an increasing lack of knowledge in terms of what we should know in order to make decisions. The second aspect focuses on what follows from basic features of modern society.
- Sustainable development demands at least a partial return to the traditional adaptation of social institutions to ecosystem needs. Every community, however, faces a distinct task in developing adaptive institutions, due to the uniqueness of each community's natural and socioeconomic environment. Natural environments and their importance to the community vary with ecosystem, the community's dependence on its environment, its resources, and so on.
- There is no development without change; sustainable development is continuous, long-term change. The only appropriate assumption is that development is more sustainable if environmental stresses are reduced and the ability to cope with them continually increased. Fewer stresses permit a more purposeful development strategy, and increased coping ability allows the community to pursue its strategy without distraction.

1 The Development Process

The term "development" is used to mean variously the following: economic growth, increasing human rights, more education, better quality of life, and much more (Chap. 12). As suggested by one critic of a narrow economic interpretation, development is about making life "more human" for as many as possible. Development is about social change with a strong implication of improvement. Economic wealth may help, as will political participation. Technology plays a part, and the rule of law has it uses. But if we have learned anything from an examination of more than 50 years' practice of development, it is that development is a complex process that we poorly understand, that occurs for myriad reasons and fails for as many reasons; it is fragile and unpredictable (World Bank: World Development Reports). We also know that it involves change in social institutions from religion to education to economy.

Ultimately, development is a matter of changing minds, how people think and what they expect, changing legal and political systems, and replacing cultural common beliefs with new social practices. We now realize that land tenure systems, property rights, ideas about individualism and collective behavior, enforcement of contractual obligations, among many other factors, influence whether development occurs and what effect it has.

In addition to the complexity of dynamic interaction between dynamic ecological and social systems, the subjective collective understanding of "development" and "environment" continues to change. The concept of development reflects the evolution of collective ideas. An acceptable level of economic inequality at one time may be unacceptable a few years later. Economic growth has become less important, and human rights have become more important aspects of development in many western societies. As scientific knowledge about the sensitivity of ecosystems to human incursions accumulates, individual and collective ideas about their usage will change.

1.1 Adaptation of Institutions

Development may be conceived of as the adaptation of institutions in response to such changes as (1) concepts of what is possible and desirable in human existence and (2) in ecological and socioeconomic environments that are changing in often unpredictable ways. Institutional changes produce physical, financial, and human capital resulting in greater social adaptation (see below).

Individuals experience development through their community. This is where life chances are formed from strategies, policies, and projects, and where human rights become real. The community forms the person and determines their livelihood and lifestyles. Thus, it is at the community level that human rights can be most effectively made a part of sustainable development. It is at the local level that abstract concepts become facts of life, where policies have direct effect on personal existence, and where the "humanness" of development becomes a part of personal life.

Sustainable development comes from a process of *social adaptation* designed to permit the greatest possible local adaptability and flexibility within flexible national and international institutions. Social adaptation focuses on the internal processes by which a community can adapt to environmental stress without changing its basic structure. It extends the concept of vulnerability by assuming that communities are dynamic and can use multiple strategies, including internal changes, to respond to environmental stresses.

1.2 Social Adaptation

Social adaptation extends the concept of vulnerability by using ideas from agent-based models.[1] It assumes, *inter alia,* that social systems are dynamic; that their behaviors and their responses to external stress, internal strategies, and policies are unpredictable; and that the appropriate unit of analysis is the community. In social adaptation, communities display the following attributes:

- Are unpredictable.
- Are, in their ideas and institutions, products of their history.

[1] Agent-based models also called individual-based models. Agent-based models could help to evaluate policies. The search for general principles underlying the internal organization of such systems often uses bottom-up simulation models such as agent-based models. Agent-based models could help to evaluate policies.

- Are limited in their responses by organizational inflexibility.
- May use physical capital assets to protect themselves from the effects of environmental stress. These are instruments.
- May change its institutions, formal or informal, to allow them to adapt to external stress.
- May not change the essential character or structure of those institutions that constitutes them as a community.
- Are affected by changes in individual and collective ideas through their effect on institutions.
- May use strategies to improve the process of sustainable development and policies to set measurable short-term goals. The effects of strategies and policies are unpredictable.

The response strategies of social adaptation groups fall into three categories: instruments, ideas, and institutions:

1. *Instruments* are the technical and physical means that a community uses to choose its way of life in the face of environmental change.
2. *Ideas* are the values and beliefs that drive the community and define collective long-term strategies and the related short-term policies, the local meaning of development, and how the community should relate to ecological and socio-economic environments.
3. I*nstitutions* embody ideas and define the practical daily relationships between individuals that constitute the community and implement strategies and policies (see below).

Social adaptation refers to the changes in instruments, ideas, and institutions that reduce community dependence on the affected part of the environment or to cope with external environmental stress.

Instruments. These include the technical and practical means of resource extraction and conversion into products and community defense against environmental stress. They include physical capital like farming equipment, factories, roads, flood levees, food storage, and water treatment, which may contribute to community sustainability. They also include financial capital that can acquire the necessary physical capital and human capital in the form of technological and technical ability. Many Central Asian countries, especially after the collapse of the Soviet Union, have long suffered from an inability to operate, maintain, and repair imported plant and equipment. Modern physical equipment is of no use without the infrastructure (parts and supplies depositories) and the capability to keep the equipment operating (requiring technology transfer) and to adapt it to local needs (technological mastery).

Technological innovation is commonly, and mistakenly, presumed to be a preserve of the developed countries. Economists have long recognized that the advanced technology of developed countries is produced in response to the production and market conditions of those countries. Aimed at product differentiation in a crowded market in a production environment with high labor and low capital costs, they usually are incremental, labor saving, and capital intensive. Instruments offer some measure of protection against environmental stresses, but they are no panacea.

Ideas. Beyond mere survival instincts, ideas are the basis of individual human action. Ideas are the beliefs and values that make us social animals; the concepts of right and wrong, good and bad; the beliefs in abstract symbols that hold meaning for us and become our goals; and the source of judgment and selection between alternative strategies and actions. Ideas govern how we perceive and understand the world and choose our behavior.

1.3 Different Views of Sustainability

Unsustainability and sustainability are both a matter of how individuals and communities think about and value the world around them (Emadi, Chap. 5). Traditional communities in developing countries have usually developed in harmony with local ecosystems. The idea of a separation between people and their environment makes little sense for them. In those communities, the idea of people as "part of" a larger whole that contains all of life and the systems that support it, is lived and expressed in their cultures. Before the Soviet era, the culture of Central Asian farmers and herders accepted that people were embedded in "nature" (Rahimon, Chap. 3).

There is an extensive literature that argues that the only way to reduce consumption of natural goods over the long term is to change the ways that people value nonhuman life. There is a belief that only a change in the ideas that people hold will change their behaviors sufficiently to make sustainable development possible. A new consciousness could assign greater importance to abstract concepts like "environment" and "equity," and less to material accumulation. It would place humanity within the system of natural life not outside it or above it. This is the basic concept embodied in land stewardship (Squires, Chap. 2). In developing countries, the self-actualization that is fundamental to many of these ideas is an impossible dream for all but the very few, buried as they are under the realities of a daily struggle to survive.

Ideas about human rights, equity, and justice also are critical to sustainable development. For development to be worth sustaining, it must be founded on human rights, but not to the exclusion of value in other forms of life. Protection and enhancement of rights for everyone is the essence of development, and preserving ecosystems cannot be at the cost of humanity for humans.

Changes in ideas are more problematic. Cultural practices that may be central to community cohesion and even to its material health can clash with the ideas embodied in products, services, and work practices imported into the community. Traditional culture may oppose many aspects of development as defined in the west. For example, in many countries, women are not allowed to participate fully in society or to participate in decision-making, and traditional culture may privilege an elite that opposes transparency and full participation in decision-making (Kurbanova, Chap. 7). Where traditional practices inhibit the sustainability of community, the community must make the ultimate decision to abandon or modify traditional ideas and practices. In some cases, this may threaten ideas that are fundamental to the community. In part, this is the problem in Islamic countries such as those in Central Asia, especially Afghanistan where traditional interpretations of the Koran reject modern institutions (Rahimon, Chap. 3).

The common weakness of current concepts is the belief that a community is a fixed entity and that response to external stress should ideally avoid internal change. Such static thinking has no place in achieving sustainable development. There is no development without change; sustainable development is continuous, long-term change. The only appropriate assumption is that *development is more sustainable if environmental stresses are reduced and the ability to cope with them continually increased.* Fewer stresses permit a more purposeful development strategy, and increased coping ability allows the community to pursue its strategy without distraction.

Instruments, institutions, and ideas interact. As ideas evolve, for example, as a result of the environmental movement, institutions change. Institutions that embody and reinforce beliefs and values direct individual beliefs and behaviors. As institutions come to reflect a growing environmental ethic, technology is progressively directed to solving environmental problems. But as institutions change, reflecting ideas that may still be only marginally accepted, they become more embedded in society, guiding individual and public behavior. Institutions influence the quantity and quality of technological innovation, and ideas embodied in technology may influence the dissemination of other ideas and the construction of institutions.

As the prime tool-wielding animal, humans have always preferred the technological fix to more social solutions. In part, this has driven development and encouraged technological innovation. Advances in agriculture countered food shortages by managing and working in greater harmony with the environment. From that grew settlements, then villages, towns, and cities in which capital could be accumulated and from which blossomed "civilization" in all its art and industry. Technology has continued to be the tool of choice in mastering the environment and adapting it to human needs and wants. Technology is best used to fix technological problems; as a solution to social problems, it may have enormous unintended consequences. To fix social problems directly requires massive coercion or enormous persuasion to change individual motivations, neither of which is practical or effective.

Although social problems are rarely clearly defined, technology may reduce or change a problem to manageable proportions. For example, hunger could be eliminated by more equitably distributing the available global production. Because this requires subversion of food markets and massive aid inputs, this social solution has not been attempted. Instead technology has been used to increase global production. Despite continued maldistribution, the expansion of production has reduced hunger but often with serious unintended consequences. The green revolution and biotechnology have neatly avoided the social fix but have created ecological and social problems out of the technical solution. The consequences of this technological fix for a social problem include exhausted soil, pollution of water resources, increased crop vulnerability to disease, deforestation, industrialization of food production, increased demand for fossil-fuel fertilizers, and loss of rural livelihoods.

This technology-driven "fix" was a special feature of Soviet-style economies, and the legacy of that approach persists in many Central Asian countries.

The agent-based paradigm[2] focuses on the role of rules in ordering social interactions and creating the thing that we call "community," the sense of belonging to a collective entity that benefits all. There is, however, a strong correlation between the strength of social institutions and the level and rate of development (as measured through the human quality of life index and gross domestic product). Development is the process of creating effective institutions to organize and regularize interactions so that risk is reduced and daily life made more predictable.

2 Nature and Role of Institutions

At the most general level, institutions are constellations of rules, decision-making procedures, and programs that define social practices; assign roles to the participants in such practices; and govern the interactions among the occupants of those roles. Defined in this way, institutions constitute an important feature of the landscape in all areas of human endeavor. Thus, marriage is a social institution governing relations among members of family units; markets are economic institutions dealing with interactions between buyers and sellers of goods and services; electoral systems are political institutions guiding the interactions of voters and elected officials. As these examples suggest, institutions can and do vary greatly along numerous dimensions, including the nature and number of their members or subjects, the character and scope of the social practices they initiate, the degree to which they are formalized in legally binding or other official formulations, their location on a spectrum running from newly formed to long-established arrangements, the extent of the organizational apparatus established to administer them, and the degree to which they are embedded in larger systems involving both other institutions and culturally determined behavior.

Institutions guide economic processes and scientific and technological developments. They also embody and reflect, and then form and develop, ideas about what is socially and individually beneficial and appropriate. Institutions are the key to social adaptation and sustainable development. By governing social interactions, institutions largely define the direction of development of instruments and ideas. This section describes the critical central role of institutions in sustainable development. Social institutions are central to sustainability: they reflect and influence ideas and direct the production of instruments. They embody ideas and frame social interaction—scientific research and economic competition—from which new technology and new ideas originate. Institutions are the rules that mediate agent interaction in the agent-based social paradigm. These rules may be informal or formal.

[2] Agents in this context are the various elements in society who are capable of making individual and independent decisions that can impact on systems. See also earlier footnote.

But in either case, they should be dynamic, changing as governments prescribe or in response to social and environmental pressures. A more detailed case study of legal and institutional aspects in Tajikistan is provided below. It focuses on the role of institutional and legislative reforms in facilitating land stewardship.

2.1 The Role of Institutions in Sustainable Development

Institutions loom large both in causing and confronting large-scale environmental changes. Much of the interest in this regard focuses on environmental/resource regimes or institutions that deal explicitly with human/environment relations. But the interaction of these regimes with other institutional arrangements must be considered as well. Major challenges in this field involve (a) evaluating the proportion of the variance in ecological conditions attributable to institutions, (b) pinpointing the determinants of the effectiveness of institutions, and (c) framing guidelines for the design of institutions to deal with specific problems.

This chapter shows *inter alia* why and how institutions—the glue of society that defines community—are critical to sustainability. Without them, sustainable development would remain the stuff of rhetoric. It is only in myriad communities that the abstractions of sustainable development take on a meaning related to the practicalities of life. But human and natural systems also must be taken into account. Institutions that guide local decision-making operate within local and national contexts. Communities of all sizes can increase their adaptive capacity by increasing technological innovation and ecological efficiency, by changing the commonly accepted beliefs about the value and uses of natural goods, and by increasing social equity and political participation. Sustainable development is a matter of changing many processes simultaneously, which requires continuous institutional development.

For two reasons, this chapter does not analyze in detail the state of knowledge about national and regional institutions in every Central Asian country. First, the variety of institutional forms and social and cultural contexts would make any formal analysis impractical and, as far as the author is aware, none has been successfully attempted. Second, it is more important to think about the nature of the concept of sustainable development and how institutions are central to its achievement. As sustainable development is an abstraction that describes no current community form or process, it is important to realize that sustainable development will be a process of trial and error, continuous reevaluation, and adjustment that is quite unlike present society. Sustainable development is a conscious process that involves individuals, communities, and local, regional, and national government organizations. This process, however, will be driven by uncertainty and evolving ideas and beliefs about the nature of development and the appropriate use of and relations to environment.

It is essential to recognize at the outset the existence of limitations on the roles institutions play in this realm and of complexities that make it dangerous to generalize from one setting to another, regarding the design of institutions intended to govern human/environment interactions. Institutions constitute a crosscutting force in this realm. They determine a portion, sometimes a large portion, of the course that human/environment relations take in a wide range of settings. But in every case, institutions operate in conjunction with other driving forces (e.g., demographic, economic, and technological forces) that affect large-scale environmental processes independently or interact with institutions to create a complex web of drivers.

Moreover, institutions themselves operate at many levels of social organization and vary greatly in terms of the consequences they produce (see case study below). What works perfectly well in one social setting (e.g., local common-property systems) may be inoperable or lead to unsustainable uses of ecosystems in other settings (e.g., global arrangements dealing with climate change). Institutions that yield acceptable results during some stages of their existence may contribute to the occurrence of significant environmental problems during other stages. The challenge is to develop procedures that will allow us, at one and the same time, to separate out the effects of institutions from the impacts of other driving forces and to enhance our understanding of the ways in which institutions interact with other drivers to cause large-scale environmental changes in some instances and to contribute to preventing or ameliorating such changes in other instances.

Ideally, sustainable development should recognize the uniqueness of local ecosystems and human societies, make use of local abilities, and avoid what is not available locally: in developing countries, the most advanced technology and financial capital. In traditional societies, community institutions have developed in response to the immediate environment. Institutions governing the access to grazing land, the sharing of water resources for agriculture, and so on are adaptations of social systems to ecosystems that permit sustainability (Jacobs, Chap. 6). Traditional communities have developed institutional arrangements appropriate to their survival.

3 Institutions and Environmental Change

Although institutions are by their nature conservative, resisting change, and supporting consistent practices, they should not be immune to change. Institutional development is the conscious change wrought by policies following strategies, usually through the formal institutions of government, administrative regulations, and laws that assign the exercise of decision-making power in economics and politics. Unexpected effects are common, even for commonly visited issues, such as poverty and income taxes. Policies are more likely to change problems than solve them and are frequently revisited. Implementation is a further variable, the effects of which are unpredictable.

If competition and extremes of both wealth and poverty encourage excessive consumption, institutional modification to reduce inequality and social competition would be beneficial both socially and ecologically. Such formal institutions common in developed countries as progressive income taxes are intended, in part, to reduce economic inequality. Institutions also evolve with no conscious social effort from changes in technology or ideas.

In the language of economics, sustainability has to recognize and reflect the relative scarcity of human and physical capital in the locality. At the same time, nations are abstract constructs of diplomacy, war, and commerce that usually cover multiple ecosystems and often encompass multiple cultures. Thus, the institutional structures that organize ideas and activities in local regions within countries become critical to sustainability. As adaptability and flexibility are the essence of sustainability, those institutions must support flexible decision-making that is responsive to learning. Institutional change may be effective in directly responding to environmental stress, though this is rare. Governments may only indirectly influence the informal institutions that are the warp and weft of daily life. Informal institutions evolve in response to environmental changes, including changes in formal institutions.

Most accounts of the causes of large-scale environmental changes invoke a major role for the nation's institutions. Emissions of ozone-depleting substances such as chlorofluorocarbons, or CFCs, and greenhouse gases such as carbon dioxide, for instance, are commonly regarded as unintended by-products or, in the language of economics, externalities of the operation of structures of property rights that do not compel owners/users to take these environmental side effects into account in their private calculations of benefits and costs. Much the same is true of the clear-cutting of forests on the part of harvesters who operate under systems of land tenure that do not force them to pay attention to collateral damages inflicted on local people and on ecosystems or long-term costs (e.g., the consequences of releasing carbon stored in trees) arising from consumptive uses of forest products. For their part, depletion of the rangeland forage base and associated disruptions of ecosystems are regularly interpreted as consequences of rules governing the harvesting of living resources (e.g., open access rules) that do not give individual harvesters effective incentives to limit their activities in the interests of conserving stocks for the future (Squires, Chap. 2).

Yet institutions also figure prominently in most accounts of strategies for preventing large-scale environmental changes or coming to terms with them once they have occurred. Proposals for the protection of forest ecosystems frequently highlight adjustments in prevailing systems of land tenure designed to strengthen the rights of nonconsumptive users of forest products in relation to the rights of timber harvesters. Many recent efforts to break the vicious circle leading to accelerated land degradation—often described in terms of the metaphor of the tragedy of the commons—center on changes in the rules of the game that are being enshrined in law (see Halimova, Chap. 13) that are designed to affect outcomes by allowing individual users to reap the benefits of actions aimed at ensuring sustainable use now and in the future (Squires, Chap. 2).

When institutions deal explicitly with human/environment relations, it is normal to refer to them as environmental or resource regimes. The traditional arrangements dealing with the management of irrigation systems in small-scale societies, the more elaborate arrangements governing the uses of public lands at the national level, and the international regimes designed to protect the ozone layer and the Earth's climate system are all examples of environmental or resource regimes. In thinking about large-scale environmental changes that have significant anthropogenic components, it is natural to focus first and foremost on the roles that these environmental and resource regimes play both in causing environmental problems and in constituting the principal components of solutions to such problems. Yet it is essential to recognize from the outset that institutions dealing with other human activities can and often do produce significant environmental consequences. But any number of other arrangements may have far-reaching environmental consequences as well. It follows that considerations of the institutional dimensions of environmental change cannot deal *exclusively* with studies of environmental or resource regimes.

3.1 Effectiveness of Institutions

There is great variance in the effectiveness of institutions or, in other words, the extent to which they determine the course of human/environment relations. Some institutions are largely ignored by all those nominally subject to their rules and decision-making procedures. Others prove far more effective during some stages of their existence than other stages. Still others appear to yield decisive solutions to the problems that give rise to their creation. As a result, those interested in large-scale environmental changes have a strong interest both in explaining apparent successes, such as the ozone regime, and in determining whether these successes offer lessons of interest to those concerned with other large-scale environmental issues, such as climate change or the loss of biological diversity.

In every case, however, there are major analytical and methodological problems facing those seeking to prove conclusions about the effectiveness or ineffectiveness of specific institutional arrangements. Central to this challenge is the danger of arriving at conclusions that are based on spurious correlations. To illustrate, suppose a problem like oil pollution at sea arises, an explicit regime is created to solve the problem, and the problem subsequently subsides. Can we conclude with confidence from this evidence that the regime has proved effective? Not necessarily. Despite the correlation between regime creation and problem-solving, the forces responsible for alleviating the problem may lie elsewhere. Even more likely is the prospect that institutional responses will constitute just one of a suite of interacting forces, including technological advances, demographic processes, economic incentives, and political pressures, that together determine the behavior of relevant actors with regard to particular issues. It may make sense in such cases to single out institutional forces for special attention, especially when there are good reasons to believe that institutional reform constitutes a necessary condition for solving the problems at hand. But the more basic challenge is to improve our understanding of systems of interacting forces and the roles institutions play as elements in these systems.

Where there is consensus on the proposition that an institution makes a difference, we come next to the issue of formulating criteria to be used in evaluating the performance of the relevant institutional arrangement. Those interested in large-scale environmental systems will find it natural to approach this issue initially from the perspective of sustainable development or ecosystems management. Do regimes governing access to grazing lands or arrangements dealing with international trade in endangered species, for instance, contribute not only to the maintenance of sustainable harvests of the resources in question but also to the avoidance of nonlinear or chaotic changes in the broader ecosystems to which these resources belong? This biogeophysical perspective on effectiveness is obviously essential. But, at the same time, it is important to ask questions about the degree to which institutional arrangements produce results that are efficient and that conform to various standards of equity (Jackson, Chap. 15). In fact, sustainability, efficiency, and equity are likely to be closely linked under real-world conditions.

Given the costs of dealing with large-scale environmental problems, success in the pursuit of sustainability will be determined, in considerable measure, by the extent to which we succeed in finding ways to achieve the desired results as inexpensively as possible. Given the difficulty of coercing key actors, into adjusting their behavior to avoid or minimize environmental problems, more attention needs to be paid to the structure and functions of institutions. Moreover, the search for solutions that all concerned can accept as fair or just and therefore deserving of respect, looms large as a condition governing success in the pursuit of sustainability.

3.2 *Institutional Dimensions of Widespread Environmental Change*

Among those interested in the institutional dimensions of widespread environmental change, three analytic themes have surfaced recently as matters deserving particular attention. These themes are often described as the problems of (i) fit, (ii) interplay, and (iii) scale. The problem of *fit* revolves around the proposition that the performance of institutions in environmental terms is determined, in large measure, by the congruence or compatibility between the attributes of the relevant institutions on the one hand and the principal properties of the ecosystems in question on the other. Sensitive monitoring mechanisms and a capacity to adapt institutional arrangements quickly to ecological changes, for instance, are important in dealing with ecosystems prone to sudden, nonlinear changes. Similarly, the priority attached to the operation of compliance mechanisms should be a function of the capacity of the ecosystems in question to tolerate violations of the rules governing human uses of the relevant goods and services (Squires 2012, Chaps. 2 and Leake, 18). The problem of *interplay*, by contrast, centers on the proposition that institutional arrangements regularly interact with one another, even though it may seem convenient to treat them as self-contained entities for purposes of analysis. Understanding institutional interplay, therefore, is clearly a challenge that looms large for those concerned with the institutional dimensions of widespread environmental change.

The problem of *scale* arises from the fact that institutions affecting large environmental systems operate at a number of levels of social organization ranging from traditional practices governing the harvesting of local stocks of forage and woody species for fuelwood through national arrangements dealing with human uses of natural resources located on public lands to international regimes addressing global problems such as climate change and the loss of biological diversity. Under the circumstances, it is natural to ask whether we can scale up and scale down in the dimensions of space and time in our efforts to understand the operation of institutions. Can we apply lessons drawn from the study of small-scale, local systems to the analysis of environmental regimes operating at the national level? Or do differences in the character of the actors involved or the nature of the relationships among them make it doubtful whether propositions developed at one level can be applied with suitable adjustments at other levels? Because anthropogenic forces affecting ecosystems occur at all levels of social organization, any comprehensive account of the institutional dimensions of environmental change must deal with processes at work at each of these levels. But this does not validate the conclusion that knowledge of the role of institutions developed at one level can be applied in any straightforward manner to processes at work on other levels.

The fact that institutions generally constitute one, among a suite of driving forces, which interact with one another in complex ways gives rise to a second major complication. The impacts of institutions may be substantial under a variety of circumstances. Yet the same institutional arrangements (e.g., common-property regimes) may generate different consequences depending on the character of the constellations of driving forces within which they are embedded. Under the circumstances, the facts that the universe of cases of distinct institutions is virtually always limited and that there is considerable heterogeneity among the members of this universe produce a third complication for those seeking to demonstrate causal links in this realm. Crafting appropriate financial mechanisms and systems of implementation review is obviously important in many, perhaps most, instances. Institutionalizing the rules of the game in the sense of embedding them in social practices that actors engage in as a matter of course is an important means of maximizing the behavioral effects of institutional arrangements. It would be hard for individual actors to escape from many long-standing local regimes dealing with common-pool resources, even in the unlikely event that they thought to make a conscious effort to do so. The interplay of regimes with one another is unquestionably a factor of considerable importance in determining the effectiveness of environmental regimes. It is always a mistake, therefore, to concentrate exclusively on the creation of specific arrangements without paying attention to the surrounding institutions with which these arrangements are likely to interact. The larger ecological and socioeconomic settings within which institutions operate are major determinants of effectiveness. The incidence of mismatches between regimes and their settings makes it clear that those responsible for creating institutional arrangements need to pay much more attention to this factor than they have in the past. It also highlights the importance of avoiding any tendencies to assume that "one size fits all."

We already know a number of useful things about the institutional dimensions of environmental change. It is clear, for instance, that systems of private property, public property, and common property all lead to sustainable human/environment relations under some conditions but to large-scale environmental problems under other conditions. The task ahead, therefore, involves refining our understanding of the conditions under which one structure of property rights or another can be used to alleviate specific problems rather than engaging in sterile debates about the overall superiority of one system of property rights or land tenure over the others (Robinson, Chap. 11). Similarly, there is little doubt that the importance of issues relating to compliance varies dramatically from one institutional arrangement to another and that there are a number of paths leading to compliant or noncompliant behavior on the part of various groups of subjects. While the arguments of those who stress the importance of enforcement are valid under some conditions, enforcement in the ordinary sense of imposing sanctions on violators is not the key to achieving high levels of compliance in every situation. These insights do not lend themselves to packaging in the form of simple and invariant design principles. But taken together, they constitute a significant body of knowledge that is relevant to the efforts of those seeking to avoid or ameliorate large-scale environmental changes.

At the same time, it is apparent that we need to learn a lot more about the roles institutions play in causing and confronting environmental changes. In part, this is a matter of enhancing our understanding of institutions as such. Why are some environmental and resource regimes more successful than others? How can we pin down the causal roles of specific environmental arrangements? Partly, it is a matter of illuminating the interactions among institutions and other driving forces that together determine the course of human/environment relations. How do rules and decision-making procedures interact with demographic, economic, and technological forces? Do institutional arrangements that work perfectly well in some settings lead to outcomes that are unsustainable in other settings? In some respects, the challenge of understanding the interactions among driving forces that determine the course of human-dominated systems is daunting. But progress in meeting this challenge may yield particularly large payoffs for those seeking to illuminate the institutional dimensions of environmental change.

3.3 *Problems of Uncertainty*

In solving environmental problems, we face a double problem: so far unknown dimensions of uncertainty and at the same time lost possibilities of solution. The first aspect focuses on specific patterns of risk perception and the problem of an increasing lack of knowledge in terms of what we should know in order to make decisions. The second aspect focuses on what follows from basic features of modern society. The crucial point is that these features constitute systematic limits to individual problem-solving by means of strengthening individual morals and responsibilities (land stewardship). As a result, there are systematic limits in the prevailing efforts to integrate ecological and sustainability criteria into decision-making by means of information and moralizing. One answer to that is a change of the institutional framework and resulting incentive structures, so that it becomes easier to respond to changing ecological scarcities. But the crucial question remains: how to narrow the gap between prevailing short-run rationalities of the economic and the political decisions, on the one hand, and the requirements of long-run sustainability, on the other. Institutionalizing new platforms of communication and participation may be a suitable option to tackle this problem.

Whenever we face problems in modern societies, we observe a tendency in mass communication to explain them as a result of inadequate behavior due to inadequate values or lack of awareness. Consequently, strategies of problem-solving under discussion are, in public opinion, usually based on information, enlightenment, and appeals to morals in order to generate the desired behavior voluntarily, for example, by instilling land stewardship (Squires, Chap. 2). And if this strategy fails, political regulations are usually expected to be an appropriate substitute.

We can also observe these patterns as far as environmental and sustainability problems are concerned. Strengthening environmental awareness and environmental regulations are the preferred strategies. These strategies are basically familiar to western political culture. But we have to go beyond them if we really want to approximate sustainability in Central Asian societies. In western societies, behavior and overall results are more and more determined by the (moral) quality of institutions and resulting incentive structures, and less by the (moral) quality of motives and individual morals. In contrast to the size of the task, and compared to the knowledge needed in order to solve these problems, our available knowledge is rather poor. Moreover, problem-solving mechanisms, which were effective in the past, like influencing behavior by norms commonly shared, fail more and more. The crucial point is that, nowadays, appealing to morals in order to strengthen those norms will be successful only under certain conditions and rather by exception than as a rule. Because of certain characteristics of modern societies, we tend to focus on the moral quality of institutions and their resulting incentive structures. But implementing an ecologically favorable institutional change into the market and political system demands new platforms of communication and participation to bring about and also to make use of an intensified environmental awareness, which is more sensitive to questions of suitable institutions.

Sustainable development demands at least a partial return to the traditional adaptation of social institutions to ecosystem needs. But every community faces a distinct task in developing adaptive institutions, for each community's natural and socioeconomic environment is unique. Natural environments and their importance to the community vary with ecosystem, the community's dependence on its environment, its resources, and so on. Each community also occupies a unique location in socioeconomic space. Communities at different levels of development face unique challenges in constructing adaptive institutions. Communities are able to partially protect social institutions through technological innovation, financial power, and centralized policies. But in the context of the substantial uncertainty about ecosystem capacity and social needs, adaptability and flexibility remain the essence of sustainability. Community institutions must support flexible decision-making that is responsive to learning, and those communities that are most successful in building such institutions will be the least vulnerable to the natural and socioeconomic change around them. For every community, the overarching strategy is to increase its adaptive capacity. What is possible and appropriate for each community can best be determined within the community, where necessary with appropriate technical assistance. The basic objective should be to increase the stock of instruments, reconsider ideas, and develop flexible institutions that are more responsive to changes in the socioeconomic and ecological environments.

Less developed communities, poorer in human and physical resources and less resilient to environmental change, may develop institutions to increase social capital,[3] whereas more developed communities are able to partially protect social institutions through technological innovation, financial power, and centralized policies. For example, Tajikistan and several other neighbors were assessed as the least adapted to impending climate change impacts (Fay et al. 2009; Oxfam 2009).

But in the context of pandemic change in globalization and the substantial uncertainty about ecosystem capacity and social needs, adaptability and flexibility remain the essence of sustainability. Community institutions must support flexible decision-making that is responsive to learning and those communities that are most successful in building such institutions will be the least vulnerable to the natural and socioeconomic change around them.

Many traditional communities have collapsed because their institutions—developed in response to their environment—have failed to respond to changes in that environment. Ideally, communities should increase physical, financial, and human capital while not depleting social capital. The emphasis on community-based strategies and political participation responds to the need to maintain and build social capital in the face of institutional changes that arise from changed markets and production technologies. Each community needs to increase its social adaptive capacity[4] within its ecological and socioeconomic environment.

The first to recognize damage may be those who live by and depend on the ecosystem, though the "tragedy of the commons" warns otherwise. The commons is a social arrangement that permits public or open use of a limited resource. The tragedy is that the absence of governance (regulation) of resource use leads to exploitation by rational individuals that will destroy the resource for all. It is individually rational to maximize resource use even if that leads to a total collective loss. Diminishing production from a resource may actually encourage more rapid exploitation as each producer seeks to maximize its gains before the resource is eliminated. Too often, as seen in fisheries, ecological damage is only apparent when collapse is imminent.

Effective sustainable development policies will require scientific support. Orthodox science may not be able to define sustainable development or predict the behavior of ecosystems or social systems, but it still has a large role to play in identifying alternate policies to implement chosen strategies and to measure the effectiveness of policies.

[3] Features of social organization such as networks, norms, and social trust that facilitate coordination and cooperation for mutual benefit.

[4] The capacity of a system to adapt if the environment where the system exists is changing.

4 Facilitating Land Stewardship: The Role of Institutional and Legislative Reforms: A Case Study from Tajikistan

4.1 Legal Situation

The Republic of Tajikistan has a relatively well-formed body of law that establishes governance, management, and rights of access and use frameworks directly related to pasture, arable, and forest lands in the country. The existing framework includes the Constitution, various legal codes (Land Code, Forestry Code, Civil Code, etc.), laws (Law on Dehkan Farms,[5] Land Use Planning Law, Law on Public Associations, etc.), presidential decrees, government resolutions, and various subsidiary regulations or guidelines issued by ministries or other government bodies within their realm of authority. Within this framework, the most common areas of concern are that:

(a) The provisions are often vague.
(b) They fail to clearly map lines of authority of the different government institutions responsible for implementation that do not overlap (roles and responsibilities of good governance frameworks).
(c) They lack clear procedures that if followed would assist in implementation.

Naturally such issues are not conducive to economic development of any country. Fortunately, the GoT has shown that it is committed to reforms aimed at strengthening and improving the legal framework related to the agriculture sector in Tajikistan.

While there are many relevant legal instruments, the most important existing within the current legal framework that relate to pastureland management, and the agriculture sector more generally, include the following:

Land Code (2008). This is probably the most important piece of legislation, as it maps out the current system of land use rights that an individual or legal entity may possibly acquire[6] and the framework for how they might be acquired. It covers all the categories of agricultural and nonagricultural as well as the State Reserve and Water and Forest Funds. This important law has gone through a number of amendments in order to strengthen it over recent years, with more changes already approved by the GoT that will be considered for adoption by the legislature in 2012.

Law on Dehkan Farms (2009). Another important law for the sector, the Law on Dehkan Farm, establishes the rights and procedures of Dehkan farmers that have

[5] Dehkan (literally peasant) farms arose after the collapse of the Soviet Union and land previously in collective or State farms was allocated to former employees (see fuller explanation in Halimova, Chap. 13, this volume).

[6] Both the Constitution and Land Code clearly indicate that the State owns all land in Tajikistan, but that individuals and legal entities may acquire, use, manage, and transfer the land use rights listed in the Land Code, which are rights "*in rem*" to the land under the law.

acquired land use rights existing within a joint Dehkan farm,[7] a family Dehkan farm, or an individual Dehkan farm. It includes many important procedures which establish the relationship between farm shareholders and a farm chairman acting as the manager, the farm's dissolution, and a provision that clearly states the government cannot interfere in the operation of Dehkan farms except as provided in existing legislation.

Forestry Code (1993). The Forestry Code provides the governance framework for all forest resources in Tajikistan, particularly those located in areas that are classified under the Land Code as State Forest Reserve land. This legislation is particularly important because many areas that are considered by the GoT as pastureland exist within areas categorized as State Forest Reserve. A new Forestry Code, drafted with technical assistance provided by GIZ, was adopted by parliament in 2011. Many provisions of the now previous code were simply unworkable and did not support mechanisms for the sustainable management of the remaining limited forest resources in the country.

Civil Code (1998): The Civil Code provides the framework for a number of legal norms in the Republic of Tajikistan that relate to commercial law transactions, such as the ownership and transfer of property, entering into contracts, and other matters that can have either a direct or indirect impact on agricultural land management issues.

Land Use Planning Law (2008). Provides a framework for local government authorities to conduct land use planning within their areas of geographical authority. This is important in any future role they may play in any system of local pastureland management planning.

Law on Self-Governance of the Local Mahalla Councils[8] *(2008)*. This legislation lays out discretionary powers of the local communities and provides them with mandates relating to their roles in running social and economic activity. This includes the formation of sustainable natural resources management groups and committees, establishment of commercial structures, maintaining cooperation with international organizations, and setting up coordination bodies incorporating members of the mahalla councils located on other territories.

Law on Self-Governance of Jamoats[9] *(2009)*. This important legislation lays out the powers and authority of the lowest level of government currently recognized in Tajikistan and provides these authorities with important mandates relating to their roles in the land use rights allocation and management process, collection of

[7] Joint Dehkan farms are commonly referred to as collective Dehkan farms. This terminology was changed during the most recent amendments to the law in 2009 in order to reflect better the actual land use rights of the shareholders in such a farm, which are joint and severable (a member of a joint Dehkan farm may have his interest in the land separated in order to establish a family Dehkan or individual Dehkan farm).

[8] Mahalla is the local governing group see Glossary

[9] Commonly referred to as the Law on Self-Governance of Town and Township.

taxes, and other matters that impact on agricultural land management. There is a requirement to draft new detailed terms of reference for individual jamoats based on local realities.

GoT Resolution 111 (2007). This resolution reiterates what is already stated in the Law on Dehkan Farms, meaning that farmers have the right to grow what they want and to market their products where they wish.

GoT Resolution 481 (2008). This resolution formalizes the program of improvement and rational utilization of the rangelands of Tajikistan in 2009–2015.

GoT Resolution 406 (2009). This resolution formalized the procedure for addressing needs and challenges in the agriculture sector of Tajikistan through the establishment of six high-level working groups in the areas of agriculture, water, land, alternative financing, local governance, and social protection.

4.2 Institutional Structures and Support Services

This section describes the institutions involved in sustainable land management in Tajikistan. These may be generally grouped into the following categories: (i) public, (ii) private, and (iii) donor. The section also identifies the problems of these institutions, while measures to reform them are highlighted below.

4.2.1 Government Institutions

The main GoT institutions serving agriculture include the Ministry of Agriculture (MoA), Ministry of Water Resources and Land Reclamation (MWRLR), Committee for Land Management, Geodesy and Cartography (CLMGC), and Committee for Environmental Protection (CEP). These are supported by local government counterparts at oblast and raion level plus several research institutions of the National Academy of Agricultural Sciences. Policy and strategy are formulated by the president's office and the parliament of the country, while GoT policy is regulated and implemented by the line ministries.

The key ministries/committees and agencies dealing with rangelands are briefly described below.

The MoA is the primary institution responsible for planning and management of agricultural development with oblast and raion departments of agriculture implementing policies and program and delivering other functions in rural areas. The ministry has several subdivisions and agencies such as Crop and Livestock Management Department, Department of Planning and Finance, Mechanization Department, Construction Department, State Veterinary Inspection, and Pasture Trust. They provide farmers with technical and extension services such as pasture-condition monitoring and rational pasture utilization and rehabilitation (under the Pasture Trust).

The Ministry of Land Reclamation and Water Resources (MLRWR) is a key state executive body of power in the sphere of water resources and land reclamation for policy management and implementation. MLRWR has broad powers in the sphere of pasturelands sector as follows:

- Makes decisions for the proper utilization and protection of water resources, construction of waterworks, rural water supply, and pasturelands irrigation
- Develops and implements short- and long-term state programs
- Streamlines a proper combination of state-centralized capital investments and business funds into the sector
- Supports and maintains the state-run irrigation systems
- Coordinates activities centralized and decentralized rural water supply systems and pasturelands irrigation

The Committee for Land Management Geodesy and Cartography (CLMGC) has many detailed functions. For instance, it (i) issues and keeps a register of land use right certificates, (ii) demarcates farm boundaries, (iii) provides maps, (iv) monitors pasture use, and (v) evaluates land ensuring the articles of the Land Code are upheld.

The Committee for Environmental Protection (CEP) is responsible for environmental protection of forestry resources and there are two departments:

i. Forestry and specially protected territories consisting of the state enterprise on (a) forestry and hunting and (b) specially protected territories (national parks and reserves)
ii. Hydrometeorology

The CEP is headed by a chairman who is appointed by the president. The chairman has two deputies each responsible for one department. The main functions of the CEP include:

1. Monitor use and condition of forestry and wildlife resources including forestry enterprises and nurseries, forestry-hunting enterprises, and forestry reserves.
2. Protect and preserve special territories such as national parks and reserves including wildlife and biodiversity.
3. Ensure that forestry and wildlife resources are used in an environmentally friendly manner and well preserved.
4. Develop policy and legislative framework and set norms and standards for environmental protection and monitoring.
5. Carry out environmental appraisal of various projects and other proposed interventions in the forestry and other sectors.
6. Carry out regular hydrometeorological observations of climatic conditions, including temperature and precipitation.

The oblast and, hukumats, and jamoats, at their respective level, have responsibility to (i) implement government policies and program and enforce laws, (ii) ensure that socioeconomic targets set by the central government are achieved, (iii) ensure timely tax collection and payment of wage and pension, and (iv) verify and approve applications for Dehkan farm establishment (see Table 9.10 Description of legal responsibilities for pasture management in Tajikistan and Kyrgyzstan in Sedik, Chap. 9).

4.2.2 The Main Institutional Problems

(a) Lack of institutional capacity to deliver effectively mandates and services with assigned functions are often nominal and not delivered.
(b) Lack of funds to implement development programs and provide technical services to farmers.
(c) Salaries since September 2011 have been raised significantly (minimum salaries now equivalent to US$500); it is now hoped that government can be more selective in staff recruitment and now encourage more opportunities for professional growth.
(d) Lack of opportunity for staff to upgrade their skills and knowledge.
(e) Lack of ownership of donor-funded projects due to lack of capacity to absorb effectively project outputs and associated capacity building/institutional strengthening activities.

The MoA, for example, is small with most of its resources taken up in regulatory affairs dealing with animal health, seed, and support for improved plant varieties/hybrids and animal breeds. The ministry therefore lacks the financial and human resources required to deliver effectively all of its intended functions. This often prevents proper implementation of long-term development policies and programs especially when they are ambitious, which is often the case.

The MWRLR also has inadequate financial resources for operation and maintenance and combined with weak enforcement of water-use fee regulation has resulted in deterioration of physical infrastructure and reduced water delivery efficiency. The MWRLR is therefore unable to maintain its infrastructure especially given that a large percentage of it is pump irrigation is often nonoperational and costly to maintain.

4.3 NGOs/Private Sector Institutions

These include village and informal community institutions, local NGOs, and private sector service providers. Formal and informal grassroots organizations as well as service providers would comprise some of the stakeholders for projects since they would be demand-driven and community-based.

The NGO sector is widely represented internationally and locally. International NGOs (INGOs) currently involved in agriculture include the Aga Khan Foundation (AKF), Mission East, ACTED, Oxfam, GAA, and Helvetas. Local NGOs engaged in agriculture include the Agriculture Training and Advisory Centre (ATAC), *Chorvodori va Baitori,* and Center of Biotechnology. These NGOs are mainly subcontracted by donors such as the World Bank, IFAD, USAID, EU, UNDP, and FAO to implement projects and program and often compete with specialized (technical) agencies such as FAO and UNDP. They act like commercial consulting companies.

Several agencies have been assisting the GoT since 1997 to undertake initiatives to encourage the sustainable management of upland resources and to assist the mobilization of upland rural communities to identify and address their problems. These include (i) AKF in Gorno-Badakhshan (GBAO), (ii) ADB support for afforestation around Lake Sarez, and (iii) World Bank, UNDP, FAO, and NGOs such as German Agro Action (GAA).

Many of the international and local NGOs who operate throughout Tajikistan are appropriate to be selected as implementing partners owing to (i) strong presence, long-term, and solid experience of working at grassroot level with communities, (ii) experienced and qualified staff, (iii) flexibility and mobility in delivering project activities, and (iv) low cost and efficiency. The disadvantages, however, include (i) low quality of technical and financial reports, (ii) delay in submission of technical and financial reports, and (iii) overexpansion of areas of specialization to cover all agriculture's subsectors.

Private sector institutions in agriculture are represented by several input[10] suppliers, processors, and service providers.[11] Private sector institutions, however, are underdeveloped due to the lack of an enabling environment, and this is especially true for those involved in agriculture. Many of the institutions[12] are treated like any other commercial enterprise and are subject to heavy taxation and government regulation and control.

The slow pace of reform, demanding licensing and inspection requirements, weak infrastructure, and high business risk continue to be barriers to greater participation by the private sector especially in rural areas. Agriculture however is a high-risk industry due, for example, to unpredictable weather conditions and the volatility of world market prices such as those for cotton (particularly 2011).

4.4 Donors/Foreign Aid Organizations

Tajikistan relies heavily on donor aid in the form of loans and grants to develop agriculture since the GoT allocates limited resources to develop agriculture including implementing its reform and maintaining its infrastructure. The main donors include the ADB, EU, FAO, GIZ, IFAD, ICARDA, UNDP, USAID, and WB.

Problems faced by donor programs include:

- Donor-funded projects are often implemented as stand-alone ones, and little if any cooperation between donor projects is observed even between those funded by the same donor. There are consequently few opportunities to (i) draw on

[10] Such as seed and fertilizer.

[11] Including private veterinarians, animal breeding, and advisory.

[12] Such as input suppliers, processors, and service providers.

lessons learned, (ii) take advantage of synergy, (iii) jointly utilize resources, and (iv) avoid duplication.

- Donor-funded projects often provide minimal ownership by local communities and government of project activities. Government implementing partners often lack capacity, staff, commitment, and resources to ensure that project activity continues well after the end of donor funding.

4.5 *Community-Based Organizations*

The GoT together with the international donor community recently contributed to development using community-based organizations. Examples of this may be seen in certain projects funded by the ADB, EU, UN, and WB and implemented by NGOs such as German Agro Action, Mercy Corps, and Save the Children. These projects helped to strengthen traditional social institutions such as mahalla committees as well as the management capacity of district and jamoat-level development committees.

Community mobilization in Tajikistan has two forms:

(a) The traditional natural informal grouping when community leaders, aware of a particular issue, help to unite the community around resolving a problem. This does not, however, lend itself beyond small self-help initiatives.[13]

(b) More institutionalized which is normally instigated by donor-inspired programs and projects where problems and solutions need to be reviewed in a more structured and democratic way. Structured community groups are required to make decisions and importantly to manage funds.

Mahalla is a form of local self-governance at the village level but is rather informal. The main law guiding formation and functioning of mahallas is the law of the Republic of Tajikistan on "Local self-initiative bodies" (last revised 2008). The main purpose of establishing mahallas is to enable local communities to address urgent social and economic issues faced by the communities such as resolution of land disputes, fixing infrastructure (irrigation canals, water supply schemes, bridges, schools, etc.), and joint grazing of livestock.

The mahalla still retains an ancient custom called the *khashar* which is the peoples' assembly that identifies problems and their solutions. Khashar is voluntary, and community members never refuse to participate in social work. It is "helping people by the people" as opposed to "helping people by the GoT." Khashar is for the practical implementation of post-emergency or development activities such as the construction of destroyed homes, post-disaster road rehabilitation, and bank and shore protection work.

---

[13] These would, for example, deal with a broken bridge or road and the current grazing/herding arrangements.

Most of the national and international NGOs[14] have been involved in capacity building through community mobilization and group formation. This helps to strengthen traditional social institutions like the mahalla committees as well as strengthening the management capacity of the district and jamoat rural development committees. The latter played an important role during the recovery phase after the civil war when the country was in a critical situation, and it was necessary to instill into people faith and hope for a peaceful future.

Some of the positive aspects resulting from recent community mobilization activities include:

1. Creation of new social institutions such as pasture-user associations, forest-user committees, and meat and milk producers associations
2. Continued development and support for associations like water-user associations, associations of Dehkan farmers, and parents and teachers associations
3. Improvement of existing public institutions
4. Capacity building through local initiatives
5. Strengthening of the capacity of traditional community institutions such as mahalla committees
6. Creation and development of different types of new community organizations including NGOs

Experience of mobilization activities in Tajikistan and elsewhere suggests the following conclusions:

(a) The effectiveness of project-inspired community institutions wanes or ceases after projects are completed.
(b) The dominance of influential members in committees tends to suppress the enthusiasm of other members.
(c) Participating communities may not realize sufficient benefits to make participation and project investment worthwhile. Communities may also not wish to wait a long time from planning to implementation given prior experience with donors of promising funding once a plan is completed and then funding only one project rather than the entire plan.
(d) Institutional and policy development activities would include the creation of working groups, capacity building of implementing partners, provision of support services/inputs, supply of rural finance, and development or strengthening of CBOs. Proposed activities include the following: creation of working groups, capacity building of implementing partners, provision of support services, and strengthening of CBOs.

For example, a functional analysis of the Ministry of Agriculture (MoA) was carried out by an international consultant recruited by the World Bank to produce recommendations for institutional reform. The report suggested the reform of key agricultural institutions to define their role as policy and strategy makers rather

[14] Such as GAA, AKF, Mercy Corps, Save the Children, CAMP Kuhiston, Azal, and Latif.

than development activity implementers. The MoA, for example, could become responsible for developing national policies and strategies and creating an enabling environment for private sector and farmers.

Existing land legislation was amended several times in the last 10 years to advance land reform but with little success. The current land reform is incomplete, and law implementation and enforcement have been slow and ineffective. The following changes to the Land Code were proposed to the GoT in December 2010:[14]

(a) Land use rights may be acquired and owned by individuals and legal entities of the Republic of Tajikistan.
(b) Land use rights are subject to buying, selling, gifting, exchanging, pledging, and other transactions.
(c) Holders of those rights can enter into land use transactions without government interference.
(d) Individuals may not be deprived of their land use rights except for cases stipulated in the land legislation.
(e) Owners of land use rights have the right to independently transfer their rights in favor of third persons.
(f) There can be no discrimination regarding access to land use rights based on gender.
(g) Only farmers and farm organizations can hold "ownership of land use rights" to agricultural land, and all currently issued land certificates remain valid.
(h) The government cannot arbitrarily terminate land use rights that have already been issued to farmers and farms except for cases stipulated in the land legislation.

Other land legislation will need to be amended in line with changes proposed to the Land Code relating to the Laws on (1) Dehkan Farm, (2) Mortgages, (3) State Registration of Immovable Property, (4) Land Valuation, (5) Land Management (sometimes called Law on Land Planning), (6) Land Reform, and (7) Civil Code.

So, in the case of Tajikistan, and doubtless other Central Asian countries as they transition to a market economy, there will be much trial and error before the institutional framework evolves to a more workable solution. In the meantime, there is attempted reform, and, as the following section shows, greater transparency and accountability are features of good governance. Having a sound legal framework is also important (Hannam, Chap. 17 and Robinson, Chap. 11).

5 Governance

Today in the western world, governance ideals and many practices support facilitation of collective actions through diverse institutional frameworks. This is in contrast to the ideology in the Central Asian countries which favors big, controlling government (Rahimon 2012, Chap. 3). Most governments determined, performed,

[15] The amendments to the Land Code were initiated by the DCC, and the process was led by USAID Land Reform Project.

and/or controlled many (if not most) collective actions, routinely dominating people, communities, organizations, and markets.

In the west, facilitation of public-values-oriented social self-governance and civil society is one principal concern of public administration today. Another is facilitation of responsible market economies. The third is governmental administration that is conducive to facilitative-state integrity, effectiveness, efficiency, and economy. Each of these three dimensions is briefly analyzed below, following a note of concepts that apply to all three. In both theory and practice, facilitation often means that public administration (and government generally) should leave people and markets and their interactions alone, but of course there is no perfect compliance with this ideal. Instead there is support of dispersed authority among governance rule systems and restructured public administration to facilitate their effectiveness.

Social self-governance and civil society involves balancing transparency and access with privacy and security, and this is a growing public administration challenge. Globally, public administration and the private sectors alike often function in contexts of crime, corruption, and terrorism that are both international and domestic. Governmental openness is increasingly one means supported in the west to combat these problems. Corruption reduces predictability, increasing the risk and uncertainty of investment. Transparency of formal governance institutions reduces transaction costs and increases the predictability of investment, thus lowering its costs. Lack of transparency is an especially effective block to inward foreign domestic investment.

Government under law and governance through law is the ideal that the western countries strive for. Administrative law standards as essential to assure legal certainty in public administration are the following: the rule of law, procedural fairness, timeliness, administrative discretion, proportionality, and professional integrity. A standard of rational bureaucracy is that authority should be commensurate with responsibility. That endures as a basic tenet.

Governance is not just a theoretical consideration. Coordination between systems at different levels of aggregation is aided by having fewer levels of governance and by clearly defining the responsibilities of each. The coordination of activities between many levels of aggregation is the most complex part of implementing the vision of nested systems of communities, regions, and nations. Community sustainability can be choked off by inappropriate regional, national, or international institutions or by a deficiency in technical and financial resources. There are also many opportunities for crosscutting conflict between overlapping jurisdictions. Regional authorities designed as the optimal unit of ecosystem management may conflict with local communities that make their livelihoods from the ecosystem.

6 Summary and Conclusions

Generally, management of rangelands has aimed at optimizing short-term benefits from the production of food, fiber, and fuel. However, we have compromised the ecological integrity of global ecosystems and caused negative impacts on our social environment by not accounting fully for environmental and social costs. For humans to live sustainably, we must manage natural resources in a way that prevents their depletion and protects their potential for self-replenishment.

Natural systems are comprised of numerous complex interactions that are difficult to anticipate, predict, or control. An adaptive management framework is needed to respond flexibly to ever-changing circumstances in a manner that uses both our scientific knowledge and local experience to best achieve sustainable goals. The ecosystem service approach should not be viewed as a replacement of, but rather as a compliment to, adaptive resource management (Squires, Chap. 2.) A sound and structured decision-making process is essential so that logical decisions are made that best achieve a sustainable goal.

Managing to retain resilience is paramount to sustain social ecosystems. Support entities and policy makers must also adopt an adaptive management *modus operandi* and work with managers toward achieving sustainable management in a participatory manner to sustain livelihoods and our natural and social capital.

Institutions loom large both in causing and confronting large-scale environmental changes. Much of the interest in this regard focuses on environmental/resource regimes or institutions that deal explicitly with human/environment relations. But the interaction of these regimes with other institutional arrangements must be considered as well. Major challenges in this field involve (a) evaluating the proportion of the variance in ecological conditions attributable to institutions, (b) pinpointing the determinants of the effectiveness of institutions, and (c) framing guidelines for the design of institutions to deal with specific problems.

Acknowledgements I am grateful to my erstwhile colleagues Robert Oberndorf and Kuvat Bapeav on the ADB/GEF Rural Development Project in Tajikistan for some of the information they developed. Specifically, Section 4 about the governance and institutions in Tajikistan is due to their work and to efforts of the national consultants who facilitated access to sources and personnel.

References and Further Reading

Emadi MH (2012) Better land stewardship to avert poverty and land degradation: a viewpoint from Afghanistan. In: Squires V (ed) Rangeland stewardship in Central Asia. Springer, Dordrecht, pp 91–108 (Chapter 5, this volume)

Fay M, Block RI, Ebinger J (2009) Adapting to climate change in Eastern Europe and Central Asia. World Bank, Washington DC

Halimova N (2012) Land tenure reform in Tajikistan: implications for land stewardship and social sustainability: a case study. In: Squires VR (ed) Rangeland stewardship in Central Asia. Springer, Dordrecht, pp 305–332 (Chapter 13, this volume)

Hannam I (2012) International perspectives on legislative and administrative reforms as an aid to better land stewardship in Central Asia. In: Squires V (ed) Rangeland stewardship in Central Asia. Springer, Dordrecht, pp 407–430 (Chapter 17, this volume)

Jackson R (2012) Attitudes toward endangered wildlife protection and barriers to implementation. In: Squires V (ed) Rangeland stewardship in Central Asia. Springer, Dordrecht, pp 357–380 (Chapter 15, this volume)

Jacobs M, Schloeder C (2012) Afghanistan's extensive livestock producers. In: Squires V (ed) Rangeland stewardship in Central Asia. Springer, Dordrecht, pp 109–128 (Chapter 6, this volume)
Kurbanova B (2012) Constraints and barriers to better land stewardship: analysis of PRAs in Tajikistan. In: Squires V (ed) Rangeland stewardship in Central Asia. Springer, Dordrecht, pp 129–164 (Chapter 7, this volume)
Oxfam (2009) Reaching tipping point: climate change and poverty in Tajikistan. Oxfam International, Dushanbe, 22 p
Rahimon RM (2012) Evolution of land use in nomadic culture in Central Asia with special reference to Kyrgyzstan and Kazakhstan. In: Squires V (ed) Rangeland stewardship in Central Asia. Springer, Dordrecht, pp 51–70 (Chapter 3, this volume)
Squires V (2012) Better land stewardship: an economic and environmental imperative if there is to be sustainable development. In: Squires V (ed) Rangeland stewardship in Central Asia. Springer, Dordrecht, pp 31–50 (Chapter 2, this volume)
World Bank (2006) World development reports. Equity and development Washinton, DC
World Bank (2007) World development report. Development and the next generation Washinton, DC
World Bank (2008) World development reports. Agriculture for development Washinton, DC

Chapter 13
Land Tenure Reform in Tajikistan: Implications for Land Stewardship and Social Sustainability: A Case Study

Nargis Halimova

Abstract This chapter is a brief overview of the implications of land reform for tenure and use of pastureland in Tajikistan. It starts by describing the land tenure system, the modes of the land use rights prevailing on agricultural land, and the appraisal of land tenure security. Further, the analysis focuses on the increasing privatization of land threatening the use of common pastureland on which villager's have long relied. The social and ecological impacts of the changed tenure system as a result of land reform after the 1990s and its consequence for transhumance in the absence of proper modes of tenure and institutional arrangements are discussed. The current trend of land legislation reform is outlined.

Land reform is likely to have adverse social and environmental impacts, unless carefully redesigned to meet the needs of small stockholders. The plight of rural landless livestock holders is highlighted, and the relevance of communal tenure arrangement is stressed. Conclusions and recommendations are made toward a sustainable land policy, which can ensure a balance between the economic benefits and socioeconomic and ecological principles of sustainable land use.

Keywords Land tenure system and modes • Land use rights • Dehkan farms • Common tenure • The social and ecological impacts • Elites • Legislation • Privatization

Key Points

- Land tenure has been singled out as the most important issue concerning the sustainability of farming on the rangelands of Tajikistan and may be throughout Central Asia. Appropriate land policies, therefore, are crucial to ensuring economic sur-

N. Halimova (✉)
GITEC Consulting GMBH, Cologne, Germany
e-mail: halimova.nargis@gmail.com

V. Squires (ed.), *Rangeland Stewardship in Central Asia: Balancing Improved Livelihoods, Biodiversity Conservation and Land Protection*,
DOI 10.1007/978-94-007-5367-9_13, © Springer Science+Business Media Dordrecht 2012

vival, for decreasing land-related conflicts and putting communities on a steady course of sustainable development. The current tenure system is inappropriate both for environment and social sustainability and development; when the tenure of the significant part of the pastureland is concentrated in few large private farms for lifetime-inheritable use right and majority small livestock holders are excluded, development of a sustainable and inclusive land policy is a challenge.

- The government of Tajikistan initiated the land reform program in the early 1990s to promote privatization and national economy. The programs were mainly targeted privatization of land, but the importance of securing tenure over communal common property was disregarded. There are many flaws and gaps in the legislation governing land, and the government is working through these problems to improving the situation. Much remains to be done to improve the present inequity in land allocation and remove the perverse incentives that contribute to accelerated land degradation and impose severe limitations on the adoption of better land stewardship.
- From the reforms of the 1990s emerged the current land tenure system in Tajikistan, which to a certain extent could be beneficial for a smaller group of people regarding the arable (cultivation) land. However, it has failed greatly to ensure the social and environmental pillars of sustainable land policy. If not carefully designed, any reform in any country adversely affects access of rural poor to the land and challenges the efforts of governments and development partners to alleviate poverty. Mortgages and "sale at despair" conventionally are the main causes of rural landlessness in many countries of the world if the reforms tend merely to favor privatization and commercialization. Creation of a land market alone cannot secure land (or user rights) for a majority of the small farmers. The poor households may face many difficulties to participate in the land market such as lack of access to information, lack of money, and unequal market power. This can easily lead to consolidation of land in the hands of a few elite groups, further land speculation, and a higher proportion of landless rural poor.
- The rapid increase in the number of landless stockholders is an emerging challenge for rural development. Securing of land tenure by the poor is conventionally counted important for rural poverty alleviation. However, in Tajikistan, the current trend is merely increasing private land tenure that greatly diminishes the area available for common grazing and tramples the rights of households who formerly used it. This has had an adverse impact on the livelihoods of many rural householders, which have smaller herds but no land shares in any farm. Though during the land reforms of the 1990s, the majority of households had at least one member who should have received an individual land share from the dissolved *kolkhozes*, many have missed out due to the socioeconomic issues and legal gaps.
- There are many social, environmental, and land tenure factors that make it difficult for land users in Tajikistan to undertake land restoration. Poverty and harsh rental rates imposed by landlords who often acquired the previously common-use pasturelands by questionable means and the shrinking resource base (pasture for their livestock) are a fact of life for many of the 75% of people who live in rural Tajikistan. In a country with major rural poor population, the land tenure

reforms should include some safeguards to protect these categories. Unless carefully written, the reforms can threaten poor householders' access and tenure to land as unintended consequences.

1 Introduction

1.1 Land Tenure: The Why and the Wherefore

Land tenure is the relationship, whether legally or customarily defined, among people, as individuals or groups and state, with respect to land, that is, it is rules invented by societies to regulate behavior and define how property rights to land are to be allocated within societies. Land tenure rules define how the access is granted; rights to use, control, and transfer land; as well as associated responsibilities and restraints. In simple terms, land tenure systems determine who can use what land for how long and under what conditions. Land tenure includes a bundle of rights, historically established and applied in different countries in variety of modes, such as private or state ownership, common tenure, private use rights, perpetual or fixed-term use right, leasehold, and primary or secondary users.

The land tenure system can be formal in a statutory system or informal in a customary system. In statutory system, the rights of the user are formally recognized by the state and supported by the title deed or formal lease. In customary systems, the tenure is enshrined by the communal and customary structures. The secured land tenure refers to clear and accessible legal, institutional, and social mechanisms ensuring the land user's continuous access to the given plot and access to clear legal procedures on transferability of land rights. The secured land tenure depends on the country-specific context. It can be secured in both statutory and customary tenure systems if the mechanisms are clear and accessible to the land users and their absence, conversely, makes the tenure insecure in either system.

Tajikistan's land tenure system belongs to statutory system (see below). The state has exclusive ownership of land, and the private individuals and entities can have use rights. Tajikistan legislation includes the pastureland in the category of "agricultural land." This is different from the situation that prevails in other Central Asian countries where pasturelands (rangelands) are a separate category (Robinson, Chap. 11). The pastureland is divided according to season of use, with more than half the area as summer pastures (Strong, Chap. 10). It is estimated that tenure to about 68% of pastureland is granted to entities in the form of private *dehkan* farms (51.4%), cooperatives, and joint stock companies. However, paradoxically, based on statistics, over 90% of livestock in the country are owned by individual households (Kurbanova, Chap. 7).

The development partners (SIDA, CIDA, USAID, IFAD, and WB) recognize that land tenure by the poor is important for rural development and poverty alleviation in developing countries. The common property is the only asset of the poor due to the limited opportunities to take part in the land market and compete with the better-off people in acquiring the land. Although the importance of common property as secured land tenure for most people in developing countries is widely recognized, it still receives relatively little attention in land policy reforms, including those undertaken in Tajikistan. For many years policy-makers were convinced by the arguments put forward in the now-classic "tragedy of the commons" (Garrett Hardin 1968) that common-use resources were doomed to overexploitation, whereas privatization of property leads to more sustainable use. However, many modern scholars argue that this is not always true. In her researches on the analysis of economic governance, Nobel winner Elinor Ostrom demonstrated how the common possession and use of resources such as forests, fisheries, oil fields, and pastures can be more successful by people who use them, than by government or privatization (Ostrom, 1990).

In some developed countries, it is believed that outright land ownership is psychologically necessary before some farmers will make long-term investments and raise their interest to conserve the soil through crop rotations. However, currently, many authors argue for common property or comanagement solutions and reject individualized tenure options (Banks 2003; Fernandez-Gimenez 2002). Particularly in Tajikistan, a country with scarce land resource, where 93% of its land territory are mountains; population of 7.73 million (2011) (of this 40% children below 14 years old), 46.7% living in poverty (2009). About 75% of total population living in rural area depend on livestock breeding for livelihoods; and other environmental factors prevail. Private ownership of pastureland in Tajikistan would fail the test of inclusive and sustainable land policy. Though over 50% of pastureland in Tajikistan is in private tenure (by primary land users), the prevalence of farmer owner–operators is rare. Rather, the pastures are mostly used by informal tenant farmers, who lack incentives for soil conservation.

This assessment is made by the analysis of existing land relations and institutions which enshrine them and their social, economic, and environmental consequences. A review of land tenure problems emerging from land reform (how they affected the poor's access and environment); national laws and government programs; studies of other projects; formal land records; field visits and consultations with the farmers on their perspectives (including constraints in use of land and interviews concerned officials. Consideration of anecdotal evidence also formed the opinions and conclusions presented here particularly the lack of proper land registration and cadastre system to make the formal land records readable.

2 Land Reform in Tajikistan: The Story So Far

2.1 First Steps and Basic Principles

The term "land reform" has a variety of meanings. It may involve the restoration of land rights to previous owners, a process known as land restitution. This occurred in countries in transition when former private rights in land were restored. Land reform may involve the redistribution of land rights from one sector to another, for example, by taking land from the state or from individual owners of large estates and giving it to people who have no land. The impact on the land may be preplanned, but it may also result from property tax reforms that alter the value of land and in consequence its use (Land Administration Guidelines: UN 1996).

Development partners assisted Tajikistan to undertake the land reform in the early 1990s in order to promote privatization and market economy. However, the reform was not well designed, and earlier versions of the law fostered a situation where there was a clear and severe imbalance between the key pillars of sustainability, in particular the social and environmental sustainability. The programs were merely biased toward privatization, but the importance of securing tenure over communal common property was disregarded. The programs of international donors (including the World Bank) pushed the government for "privatization" of land. However, as we see the consequence now, these programs had ignored the social and environmental principles of sustainable land policy.

Though the law does not admit private ownership of land as such in legal terms, de facto the tenure of most pasturelands are acquired by a few elite groups, and access of pastures by the majority of rural residents is shrinking. The new landlords (i.e., the primary land users who received the Land Use Certificate, as a document proving the legal title to the land parcel, hereinafter "Land Certificate") are entitled to arbitrarily decide whether to lease out pastures to landless villagers and at what rent, based on the principle of "freedom of contract" in a market economy. Thus, the programs of international donors funded ostensibly to alleviate poverty and increase the livelihood of the poor in fact may result to depriving them of the access to the natural assets, such as pastureland, that they had used under common property regimes.

The design of the land reform programs did not ensure the balance between the key principles of sustainability (see below). The reform was rushed with a lack of proper mechanisms to ensure the equal participation of all groups. The area of pastureland available for communities and households with smaller herds has shrunk or in many cases became unavailable for communities despite the fact that about 75% of Tajikistan population lives in rural area and depends on subsistence agriculture for livelihoods. The reform led to a loss of access to pastures historically used by local village-based peoples. It also affected adversely the mobility of livestock due to lack of access to the water source, right of way through fragmented parcels, economic inability, or reluctance of some new landlords to take the livestock to the remote areas. The significant part of pastureland has been granted to those elite groups who are either high-ranking government officials or connected to this group. The practices of short-term leases are imposed by new landlords (primary land users (see below)). By their actions in denying access to grazing land in an agrarian economy where few other sources of income are available, there is the imposition of a system that is almost feudal. Most of the natural resources are concentrated in the hands of a few elites, while other land users are overexploited or eliminated.

3 The Tajikistan Land Tenure System

All land in Tajikistan is owned by the state. The Land Code allows the lease, mortgage, and inheritance of user rights, but not their sale.

3.1 The Land Tenure Modes

From the reforms of the 1990s the following forms of land tenure emerged (land use rights) that can be provided to individuals and entities:

- "Lifetime-inheritable use rights"
- "Perpetual use rights"
- "Fixed-term use rights"
- "Lease"

Those holding the first three of the aforementioned forms of tenure are the "primary land users." They get land title upon such registration in the State Committee on Land Management, Geodesy, and Cartography (SCLMGC) and receive a Land Certificate for a given land parcel. Those who lease the land are the "secondary land users" (Land Code, 2009).

3.1.1 How the Agricultural Land Was Granted

Agricultural land was granted to individuals and entities mainly through the following processes:

Liquidation of kolkhozes – distribution of land shares to its members: The kolkhoz land was distributed to its members for lifetime-inheritable use right in equal shares. These could be from 0.10 ha to few hectares depending on population density and land scarcity in each region. The small size of individual land shares and lack of technical and economic capacities include factors that made the farm members to pool their shares and establish joint dehkan farms. The size of individual land share is defined but cannot be demarcated and all farmland is used jointly. The law provides right to any farm member to leave the farm and establish or own a family farm. In this case, the land surveyors of local land committees shall define the most suitable plot for partition and demarcate the land share of individuals leaving the joint farm.

Random allocation of pastureland from state reserve based on individual application: The decision about the land allocation is taken by the district chairman. It is based on proof of livestock population by the applicant (i.e., as many livestock population the applicant has, as many pastures may receive). The pastures in large estates were granted to private individuals/entities. The process lacked procedures to make the land distribution open and public. This led to the socially unfair and asymmetric distribution of common resource pastures. Only a small group of people with financial and political advantages could prove such imbalanced criteria and have access to the process. Often, those who received pastureland in this manner have also formed dehkan farms.

Random distribution of land for personal supplemental plots: At national level, about 44,000 ha, of which 25,815 ha was irrigable, were granted to 409,000 rural households as supplemental plots per 0.15 (arable) or 0.5 ha (rainfed) based on Presidential Decrees in 1995 and 1997. A total of 75,000 ha was meant to be distributed; hence, it is known locally as "presidential land." The aim was food security, so included is a provision that the land could be taken back if unused for the approved purpose. Due to this noncompliance or incomplete distribution, the above-estimated distributed area is much lower than provided for by the decrees. The decrees do not specify the modes of tenure, but in practice, the tenure is for lifetime-inheritable use right. The distribution process was random and lacked any defined criteria.

4 Land Tenure Security: A Critical Appraisal

Individuals or entities received pastures under the fixed-term use right, lifetime-inheritable use rights or leasehold. The issues affecting the security of these tenure modes for arable and pastureland will be discussed in the following paragraphs in the context of key conventional indicators of secured land tenure:

- *Continued access* – long-term use
- *Protection* – available and accessible clear mechanisms of tenure protection when jeopardized
- *The transferability* – available and accessible clear mechanisms, freed from authorities' interference

4.1 The Lifetime-Inheritable Land Use Right

This appears to be the most secure; nevertheless the tenure is jeopardized by the legislative gaps that disregard the socioeconomic conditions experienced by the legitimate land right holders. One of the critical gaps is the undeveloped legal procedure for inheritance of land rights, which is theoretically allowed by the Land Code. The Civil Code regulates the inheritance and requires certain legal procedures to follow in defining the legitimate inheritor and requires notarization for inheritor's right to be established for the given private property. Since the land is state owned, and as such is not provided as immovable property in the Civil Code, its procedures on inheritance and notarization are inapplicable for transfer of land use rights.

The undeveloped land lease market impedes leasing land parcel/shares or transferring it to family members by individual absentee land right holders who cannot cultivate the land personally due the wide range of socioeconomic issues such as:

- Continuous labor migration abroad
- Translocation to another village (particular for females at marriage)
- Death, illness, and disability
- Motherhood and lack of facilities in rural area for the children's daycare,
- Lack of awareness about their own land rights or land shares in newly formed farms
- Undeveloped legal procedure and limited resources/empowerment to enforce one's land rights

The further fragmentation of farms formed from the former *kolkhozes/sovkhozes* is a continuous process. Many rural residents who, due to the mentioned socioeconomic issues, were absent at the moment of the subsequent farm fragmentation lost their land shares with no possibility to regain it. The land shares that they would have otherwise been entitled to have ceased to exist and been redistributed to the existing farms or to more privileged groups. Thus, many legitimate rural residents lost land tenure due to lack of possibility to transfer it to the family or lease it out (Kurbanova Chap. 7).

The land cadastre and land title registration system is sporadic and undeveloped that also jeopardize the tenure security. The registration procedure is far from the "single window," expensive, and time-consuming, which delays issuance of Land Certificate for the farmers, reportedly sometimes for months and years. This often creates conflicts between the land users and also with the local officials since the rights to use land commence upon the state registration and issuance of Land Certificate. The farmers need to cultivate land while their application for registration is processed. Often officials responsible for processing applications, due to corrupt interests, are not interested in facilitating the preparation of land within 25 days[1] but artificially delay the process.

The cases about the discrepancies between the actual physical boundaries of the parcel and the prepared maps for it were also revealed. Registration is based on a "first-come-first-served" principle. Particularly in the light of the issues discussed above, care should be taken to avoid the situation that when registering the rights of an individual served first, it does not preclude the legitimate land rights of others (see Box 13.1).

The legal gaps, unavailability, and inaccessibility of legal procedures to enforce and protect the land rights challenge the tenure of users. The norms of the Land

[1] As provided by rules on the procedure of surveying, registration, and issuance of the certificate for the land use right and the document for the individual land share and the replacement and the repeal of the certificate for the land use right and the document for the individual land share approved by the Resolution of the Government of the Republic of Tajikistan, July 2, 2009, № 374 (author's unofficial translation).

Box 13.1 The Pitfalls of a First-Come-First-Served System: An Example from Vardaht raion

During the author's visit in Vahdat district for a meeting with the farmers and officers of the District Land Committee in June 2011, it was found that a woman's land share/plot of 0.3 ha in irrigated land close to her house and adjacent to her neighbor's land plot has been registered and mapped to the Land Certificate of the neighbor. She had received this land plot from the dissolved kolkhoz and has been using it for many years but had not yet registered. She had been raising the issue with the officers of the District Land Committee who admitted the error but had no capacity to return it, instead offered another plot that is found unattractive by the woman. Now she can do little if anything to return her legitimate right to her land plot since the neighbor has secured the state registration and received the Land Certificate covering her plot too. An approach toward a more systematic registration system and modernized cadastre technologies would minimize such risks.

Code about protection of land tenure often are underdeveloped. For instance, Article 50 envisages a role to the local land committees to restore the violated rights of legitimate land users:

Article 50. Restoration of violated land rights

A violated right should be restored in the following cases:

(a) The local land committee invalidates an act in non-compliance with the legislation and that infringes upon the land rights and interests of land users protected by the law.
(b) Unauthorized occupation of the land plot.
(c) In other cases provided for by the laws and other legal and normative acts of the Republic of Tajikistan.

The district chairman endorses acts, that is, decision on allocation to or withdrawal of land right of individuals or entities. The local land committee having a twofold subordination, both to the Republican Land Committee and to the district government, actually more to the latter, is not in position to "invalidate" any "noncompliance" decision of district chairman. There is little, if any, it can do to protect or restore the violated rights of land users. The protection provisions of the Land Code, particularly the point (*a*) of Article 50, are unenforceable and fail to reduce the vulnerability of land use rights. The point (*b*) *unauthorized occupation of the land plot,* of mentioned article, is more confusing. It is actually written as if the land committee can restore the rights even for unauthorized occupation, thereby quasi motivates unauthorized occupation. Moreover, the claims about infringement of land rights are reviewed according to civil law by court.

To reduce the farm tax burden, there are cases when rural residents by different means are excluded from the farm's register (often female and poor members).

Another critical issue affecting the legitimate land rights of rural residents is lack of awareness about own land rights and/or limited capacity to enforce them. Many shareholders have misperceptions about their position in the farm even though their names are registered in the farm's Land Certificate. They refer to farm chairperson as the landowner rather than as co-shareholder. They are charged high lease fees for the use of own land shares in the farm and are treated like lessees and workforce. "*Shahodatnomai sahmi zamin*" – appendix document to the farm's Land Certificate that verifies the farm member's share often – is not delivered to them intentionally to keep the members uninformed. The farm chairperson should be elected by the general meeting of shareholders, but normally this process is bypassed or the outcome is a foregone conclusion involving district officials.

4.2 The Fixed-Term Land Use Right

This tenure mode is split into a short term (up to 3 years) and long term (from 3 to 20 years). Based on anecdotal evidence, an insignificant portion of pastureland is granted for fixed-term use right. The farmers have concerns that about every 10 years they have to overcome the burden of "paper work" to renew their tenure though they have Land Certificate and even fear of risk that their application for renewal of tenure may be rejected. The law provisions are not explicit about the tenure period. Land Certificate indicates the tenure is given for long term, but does not provide specific number of years (period) during which the long-term tenure is valid. The Land Code defines "long term" as "from three to twenty years." Neither the Land Certificate nor the Land Code provides the specific period of tenure validity. It is believed that the legislature has provided such a provision to reserve the state's right to these lands for implementation of policies and possible redistribution. On the other hand, lack of specific period of tenure creates grounds to challenge tenure any time from the year four, corruption incentives, especially in the case of any replacements in high-rank official of district administration as experienced by farmers. Many farmers lack confidence if their tenure will be renewed or not. The legal gaps and broad powers of government authorities affect the confidence and motivation of both primary land users and tenants for sustainable land use but encourage maximizing the profits in the short term by overexploitation of land. Individuals and enterprises that grasped vast areas of pastures by shrinking the tenure of the poor villagers actually use its insignificant part, if any, for grazing own livestock but lease it out to villagers. Often, these leasing practices are negligent and do not comply with capacity of pastures.

4.3 The Lease Right

Legislation allows short-term and long-term (up to 20 years) lease of land. The rural householders and smaller-scale dehkan farms lease pastures from the large landlords, state reserve, and forest land. The leases typically are informal for 1 year with possible annual renewal, lack regulations and safeguards for protection of pastures and tenants from exploitative practices, and make tenure rights of lessees weak. The landlords (primary land users) pay annual tax in average about five somoni/ha (about US$ 1 as of September 2011) but charge the lease rate per livestock head/season. In some areas, 20–25 somoni are charged per cow/month (about US$ 6). Often number of grazing cows exceeds the capacity of pastures, and actually no one controls the compliance. The landlords even collect high income from leasing, lack interest to invest for improvement of pastures, but seek to maximize the profit in short term. The landless villagers are concerned that large holdings of pastures by private entities have only land speculation purpose – lease it out to the needy villagers who lack other sources of livelihoods.

To lease the pasture from another district for moving the livestock based on seasonal needs (transhumance), the farmers need not only negotiation with the landlord but also approval of hosting district government. The approval process lacks transparency and clear procedures, which discourages the livestock mobility. The present system simply encourages exploitation and impedes the social and environmentally sustainable development. The state granted tenure of pastures to private individuals and entities at the price of diminishing the common use but with the expectations that the new "owners" will take care of the land, but due to mentioned factors, this is not happening. The benefits expected therefore from changing common land use regime to the private use right cannot be realized.

4.4 Pastureland Area and Tenure Facts

Of approximately 4.7 million ha of agricultural land, about 82% (3.9 million ha) are pastures, 15% (0.7 million ha) arable land, 2% (109, 671 ha) of perennial forage, and 0.5% (21 264 ha) are grasslands. Of 3.9 ha of pastureland, about 51% (2.0 million ha) are summer pastures, 18% (707, 476 ha) winter pastures, 18% (691, 837 ha) spring–autumn pastures, and 11% (434, 496 ha) round-year pastures. Only 2% (70 839 ha) of total pastureland are provided with a reliable water source.

- Possession of Pastureland by Entity Forms
  At the end of 2010, in the country, more than 51% (about 2.0 million ha) of pastureland were allocated to the farms for lifetime-inheritable use right. About 11% is retained in state reserve fund, 9% in the forest fund, and about 2.5% in state agricultural enterprises and under the jamoats authority. However, data about the area of pastureland retained under the jamoats authorities, which could be offered to the villagers as common resource, is unavailable (such data is not collected or stored). Almost all the state farms are liquidated, and only about 1.7% of the land is still in the collective farms.
- Possession of Pastureland by Small and Large Farms
  In order to develop future policy reforms, we should have clear picture about the percentages of pastures granted under the different tenure modes. Regrettably,

the present statistics provides only general data. It is also impossible to analyze the existing data to identify information on interesting parameters, including determining the area of pastures granted in large tracts to private farms since the land information and administration system is primitive. The data patterns, collection, and storage require improvements and computerization that would allow data processing. The digitized cadastral system would present a clear picture about the land tenure situation. The gap with unavailability of formal data was filled up by discussions with relevant officials and farmers that concludes a significant proportion of pastures in the country are granted to private holdings typically in large estates of 500 ha and above.

- Possession of Pastures Under Different Tenure Modes

  The formal data about the percentage of pasture granted under different tenure modes, such as lease, lifetime-inheritable use right, and fixed-term use right, is also unavailable. According to informal estimations, about 68% was apparently granted for lifetime-inheritable use right to the private dehkan farms and other enterprises. By the end of 2010, based on formal data over 51% of pastureland were granted to dehkan farms. This means at least over half of the pastures are granted for lifetime-inheritable land use right since under the law, the dehkan farms receive land under this tenure mode.

5 The Current Tenure System Versus the Key Pillars of Sustainable Land Policy

In development of land policy, a balance between the following pillars of sustainable land policy should be ensured:

1. Efficiency and promotion of economic development
2. Equity and social justice
3. Protection of the environment and promotion of sustainable land use

The reforms of the 1990s have created current land tenure system, which to a certain extent could achieve the first pillar for a smaller group of people who cultivate arable land. According to statistics, some fragmented private farms increased crop production. However, it has failed greatly regarding the pastureland and two latter pillars of sustainable land policy. If not carefully designed, any reform in any country adversely affects access of rural poor to the land and challenges the efforts of the government and development partners to poverty alleviation. Mortgages and "sale at despair" conventionally are the main causes of rural landlessness in many countries of the world if the reforms tend merely to favor privatization. Creation of a land market alone cannot secure land access for a majority of small farmers. The poor households may face many difficulties to participate in the land market (lack of access to information and money, unequal market power, and so on). This can easily lead to consolidation of land in the hands of a few elite groups, contribute to land speculation, and to a higher proportion of landless rural poor.

The analysis points to the fact that land reform process so far has resulted in the random and socially asymmetric allocation of pastureland. Many rural households have no land shares either for pastureland or for arable land. The reforms focused merely on privatization leading to socially uneven distribution of important common resource such as pastures. As a result, the rights of rural households owning smaller herds and their access to their traditional grazing lands are being eroded. They can only rely on an annual and informal lease from the primary land users who now control large areas often of formerly communal pastures. Rent practices are neither environmentally sustainable nor socially justified. In addition to environmental impacts, such a system makes the rich richer, but most of the rural poor become poorer.

The transhumance system that utilizes pasturelands on a seasonal basis by altitudinal migration of flocks and herds is essential for sustainable use of pastures in Tajikistan. Historically it was applied before and during the Soviet time. The farms of one oblast were assigned pastures to use in another oblast. The state farms and *kolkhozes* had higher livestock inventories and, correspondingly, were assigned larger areas of pastures. These allocations of pastures between the oblasts were reviewed in 1993 and 2003. Yet the transhumance principle has been reserved. The farms of one oblast preserved the seasonal pastures in another oblast. For this purpose, 53,088 ha of land were allocated for herds' drove and resting area. However, based on consultations, the most part of it is inaccessible as these areas also were "privatized" as individual estates; facilities are poor and destroyed by floods and landslips.

The undertaken land reform had negative impacts to the transhumance system. First, often the tenure status of the pastureland in the hosting oblast has changed. Only few farms emerged after partitioning of large Soviet farm could preserve the *kolkhozes'* pasturelands in another oblast. Now accessing the seasonal pastures beyond their own farm and oblast depends on negotiation with hosting district authorities and heads of the respective host farms who have tenure, which in current system is accessible only to the large private stockholders who have the "influence" and the technical infrastructure to facilitate long-distance livestock movement. Such challenges discourage the transhumance particularly by the owners of smaller herds. Nevertheless, transhumance is still run by large livestock owners who have economic resources and are in a position to negotiate the access in hosting oblast. However, that constitutes a relatively minor proportion of total livestock inventories, and solutions must be found particularly to help the smaller livestock owners that jointly are the largest owner of livestock in the country.

Of the total 3.9 million ha of pastureland, about 68% are granted to private dehkan farms, cooperatives, and other enterprises, typically in large tracts. Many of these "owners" are reluctant to use their remote and seasonal pastures, so a large part of the forage biomass is underutilized. Based on statistics, it would appear that approximately 90% of the livestock in the country are owned by the householders with a low proportion owned by the dehkan and state farms. However, the statistics do not always reflect the real situation. Based on field consultations, the view was formed that in reality dehkan farms might own at least 30% of livestock. Dehkan farms may underreport the livestock population they own to avoid being levied taxes.

Two other factors are at work as follows: (i) the householders strive to increase the livestock numbers to improve their livelihoods and income and (ii) the demographic growth means that the number of households increases over time. In such a plight, the lack of proper tenure and institutional arrangements such as the herding institutions impedes access to pastures by rural households who are more in need for pastures to make livelihoods and who, collectively, are the owners of most livestock in the country.

The households who own the majority of livestock use the easily accessible pastures close to the villages. These pastures may also now belong to private holdings and dehkan farms that typically lease out these former common-use pasturelands for annual use by the communities. The livestock population grazing in the pastures close to the settlements normally exceeds greatly the capacity of resource (reportedly 10–12 sheep/goats). Moreover, grazing there occurs almost the year round, which does not let the grass recover. As a result, much land is now of low productivity and some former pastureland is infested with toxic plants or otherwise inedible plants.

6 Privatization: End of Commons and Beginnings of Major Landlessness

The rapid increase in the number of landlessness stockholders is an emerging challenge for rural development. It is conventionally accepted that the secured land tenure by the poor is important for rural poverty alleviation. However, in Tajikistan, the current trend is merely increasing private land tenure that greatly diminishes the area available for common grazing and trampling the rights of households who formerly used it. This has had an adverse impact on the livelihoods of many rural householders, particularly children's illnesses due to malnutrition (40% of total population are children below 14 years old), which have smaller herds but no land shares in any farm. Though during the land reforms of the 1990s the majority of households had at least one member who should have received an individual land share from the dissolved kolkhozes, many have missed out due to the socioeconomic issues and legal gaps discussed earlier (Kurbanova, Chap. 7).

The proportion of households with no land share also is increasing due to the demographic variables and great number of newly emerging families as children mature, marry, and start families of their own. As a result, the majority of the rural households have limited opportunity to access the pastures since the rights of the secondary users often are weak and unprotected particularly with the current widespread practice of offering informal leases, when it is agreed for 1 year, normally every year, and when rents are arbitrarily decided by the primary users who control large areas of pastureland. Moreover, the primary land users typically prefer leasing to the entities such as dehkan farms (as middleman), rather than dealing with individual households, even though most of livestock is owned by village-based householders (Box 13.2).

Box 13.2 Case Study: Emerging Landless Households and Communities

A mahalla chairperson during a group discussions raised concern about the conflictive situation his community faced recently. Reportedly, the pasture area that has been used by the community traditionally has been "privatized" formally by a private entrepreneur from Dushanbe city, who has undergone the formal state registration for the given land and received the Land Certificate. Now the villagers are denied access to the pasture – a community common grazing area, which they have been using traditionally. Moreover, there are about 100 households of this village where every household depend on livestock breeding for livelihoods, not one household has any land shares in any Dehkan Farm. There is also no alternative pasture land available that the given community could use.

Mahalla – is an informal local self initiative body at the village level

Shafti Mijgon Village, Varzob District, June 2011

7 Current Land Reform: A Critical Appraisal

The reform of the 1990s has drastically changed the land tenure system in Tajikistan. De jure the land is retained in state ownership, but the form of use right has been changed from the common to private. The balance was changed between public and common tenure in favor of the private use right. During 2008–2010, amendments were made to legislation to improve land tenure security focused on privatization and open land market. The Law "on State Registration of Immovable Property and Rights to It" (2008), the Law "on Mortgage" (2008), the Law "on Dehkan Farms" (2009) were enacted. The Land Code was amended (2009) including provisions on order of land acquisition for the state and public needs and rights for compensation based on market value and the concepts of inheritance and mortgage of land use rights. However, these provisions remain unimplemented as clear legal procedures for their enforcement remain undeveloped.

In 2011, a draft Law "on Amendments to the Land Code of the Republic of Tajikistan" was developed by the working group on land reform and submitted to the government. As the previous amendments, these (version of March 2011) merely approach the sale and mortgage of land but disregard the existing legal gaps impeding the transferability of land rights (undeveloped legal procedure for inheritance, notarizations of transfer of land rights required by legislation, and transferability of land shares of absentee land shareholders to family members).

It also lacks the social safeguards for small-scale farmers, the poverty stricken, or even farm members unaware of their own land rights. The "social safety net" that is briefly mentioned in Government Agriculture Reform Program (ARP) (April 2011) says about need for establishing a monitoring system on reform process for "corrective measures." But it fails to specify what of corrective measures and on what principles they will be based. It is critical to anchor the principles of each policy and reform from the onset and make it clear to everyone. The ARP considers "the need to conduct a study to identify the issues and recommendations on informal employment" as sufficient safeguards. This is rather a weak statement with no provision of specific safeguards to help the poor (majority of rural population), to retain the land tenure for their livelihoods, and to avoid land speculation by elites.

In a country with 75% rural poor population, 46.7% poor, 40% children below 14 years old, and lack of other income sources, the land tenure reforms should provide safeguards to protect these categories. Unless carefully written, the reforms can threaten poor householders' access and tenure to land as unintended consequences. The mortgage and "distress sale" are key conventional reasons for the rural poor landlessness if reforms are biased toward privatization. The opening up of land market alone will not provide the majority of small farmers' access to land but can easily result in land consolidation/speculation and consequently to landlessness of rural people. This can adversely affect rural households (in particular women), and the government and development partners' efforts to alleviate the rural poverty will fail. It is important to safeguard against undue influence by small groups of powerful people who stand to gain most from the reforms. The drafted amendments to the Land Code are biased to privatization but pay no attention to securing the land tenure as common (communal) property which is an only secured asset of the poor (Ostrom 1990s). The legislative environment continues the trend toward privatization of land but disregards the importance of common resource pastures to meet the needs of villagers.

The land sale market is a component of many economies. But each country's specific context, resource bases, historical patterns, and peoples' income sources differ. Therefore, merely creation of land market and privatization of the common-use natural resource such as pasture is inappropriate in Tajikistan. The current reforms in neighboring countries (Mongolia, Kyrgyzstan, and China) retain the pastureland owned by the state and perpetual tenure given to the communities (Hannam, Chap. 17). Though those countries have accepted the private ownership for other land categories (residential and arable (cultivation) land), the pastureland retained as state property and tenure is granted to each community (Robinson, Chap. 11). If the mentioned amendments to the Land Code will be adopted to allow merely the land sale with no social safeguard mechanisms, this will result in the loss of land access and tenure by the poor who are majority in the rural areas of Tajikistan and further adversely affect their livelihoods.

The mentioned amendments on principal issues refer to government resolutions not yet drafted/enacted. For instance, they mention necessity of establishing key principles for prevention of monopolistic activity in land market but fail to provide what these principles are. Article 17 (2) point 2 mentions that there should be *restrictions* in marketing of land use rights for the land of special and natural conservation areas (e.g., forest fund, the land reserved for water conservation, cultural heritages, natural reserves, national defense and security, industry, communication, and transport). But it fails to provide the meaning and "boundaries" of such restrictions to ensure its common understanding, that is, what transactions are allowed or disallowed with these land categories? The secured land tenure requires fixing key principles in law and making rules clear to everyone. The law provisions should be explicit and clear, and the text of law should be written as it is meant to avoid the ambiguity in its interpretation and implementation.

The drafted amendments to the Land Code introduce new concept – *ownership of land use right* by private individuals/entities (though the land as such, based on Constitution, is exclusively owned by state) to allow the land sale and mortgage. Many local legal scholars including the representatives from the Ministry of Justice of Tajikistan are skeptical about compatibility of this new proposed concept – "ownership of land use right" with other legal concepts in legislation of Tajikistan (and the legal system of the region). Like the previous amendments, the proposed ones try to approach the specific articles of the Land Code, which will challenge implementation. But a comprehensive approach of all provisions or drafting of new Land Code would be advisable to ensure harmony with all legislation of Tajikistan.

If land reform merely focuses on privatization, it will meet a failure. Clear safeguards should be established to protect the poor farmers from the "despair sale" and mortgage. The safeguards and principles should be fixed in the law, not resolutions, since the latter are often subject to influence of powerful private interests.

8 Two Key Aspects of Sustainable Land Management

8.1 Sustainable Land Policy

In lack of policy and special law on pasture management, during the recent reforms, the laws designed for arable (cultivation) land have been chaotically applied in granting tenure to pastureland. Such a policy and law remains undeveloped still. The *Program for Improvement and Effective Use of Pastures in the Republic of Tajikistan for 2009–2015* cannot serve as policy. It just provides schedule for cultivation of some pastures to prevent degradation using government funds. Though it fails to specify the selection principles of pastures to be improved by the government funds, for instance, is it only the state pastures or the private pasture estates that are eligible for state funding? The opinions that the large pasture estates in fact belong to some of high-rank officials and the lack of pasture selection principles/criteria may suggest conflict of interest in use of government funds for pasture improvements. The program does not discuss any policy and principles, neither the tenure nor the institutional arrangements in its management of pastures.

Undeveloped policies and regulations lead to unsustainable and impoverishing use of forest land and products as the Soviet Forest Code fails to meet the modern political and socioeconomic changes. Forest land that also can be leased as pastures by individuals and entities is retained in management of state forest agencies under the Committee on Environment and Nature Protection. Lack of an appropriate taxation system and informal short-term leases and lack of proper institutions to control the compliances resulted in impoverishment of forests. The procedures of leasing the forest land, the roles and functions of local and national forest agency in relation to the tenants, lacks clarity. GIZ assisted drafting a new Forest Code that was approved by the parliament in 2011.

8.2 Land Cadastre and Unified Registration System

The cadastre is an information system consisting of two parts: maps or plans showing the size and location of all land parcels and text records that describe the attributes of the land. It is distinguished from a land registration system in that the latter is exclusively about legal titles. A system for recording land titles, land values, land use, and other land-related data is an indispensable tool for a market economy and sustainable management of land resources. All industrialized nations with a market economy maintain some sort of land register system that fulfills the above requirements (UN Land Administration Guidelines 1996). An effective land cadastre and administration system provides security not only for landowners but also for all partners, national and international investors, banks and traders, and governments. It is also an instrument for development of national land policy and mechanisms to support national economy.

Tajikistan is still to establish such a system, while almost all republics of former Soviet Union have it. The registration functions in Tajikistan still are shared by different agencies (SCLMGC, Inter-Districts Bureau of Technical Inventory (MBTA), Ministry of Justice (MOJ), and jamoats). The different agencies hold registers without coordination and unified centralized database. This creates an indispensability to keep and support parallel registration systems and information storage. Further, the ways of keeping records need improvement since in each agency, the numbers and indexes on immovable property are run differently. The MOJ register of mortgages is based on the names of the parties, the MBTI records are based on the address of the property, and the SCLMGC's records are based on a land plot identifier. The registers are paper based and the public has no access to the system. The existing manual systems frequently limit the opportunities for implementing optimal solutions. There is no computerized network to connect all registers in the country and support running the registration in "online" mode. The laws on land lease and land valuation still are based on Soviet economy principles and obviously cannot meet today's changes.

One of priority tasks for effective land administration in the country is establishment of unified agency for registration of all forms of immovable property and rights to it. This requires combining the functions and registers of different agencies currently running registration separately into one body, which would be more efficient and streamlined.

Countries in transition are recommended to investigate the possibility of implementing integrated land information systems, where the formal registration of legal information as well as technical information is supervised, controlled, and operated by one public authority, and not split between two or more ministries and authorities.

This does not exclude distributed solutions with practical activities being undertaken in regional or local offices (UN Land Administration Guidelines 1996).

Unified registration system shall be provided with modern information technology and staff capacity building accordingly. Introducing a new land administration system, including the implementation of formal land information registers, is a huge and time-consuming process. The reforms of legal framework, organizational structure, financial mechanisms, and technical issues are closely interconnected, while the first three are more complex to solve than technical issues.

SIDA was active in establishing such a system in Tajikistan (2005–2010) and assisted in preparing the Law on Registration of Immovable Property and Rights to It (2008), provided with principles of unified registration system and "single window," base steps for setting unified registration agency, human resources development plan, and registration module for the entire country trialed in pilot districts (Tursunzade and Hissar). In 2010, SIDA's focus on Central Asian region was diverted, and the undertaken milestones on establishment of modern cadastre and land information system in Tajikistan stagnated.

9 Conclusions

Neither the private nor common tenure appears secured in the country due to absence of policy with clear forms of tenure over pastureland. *De jure,* the land is exclusively owned by the state, but the practice and use rights of the primary land users are common with the private property regime rather than the state/public. They can act as landowners and exclude communities in use of pastures and may make arbitrary decisions about leasing pastures with monopolized fees though prior to reform these pastures are traditionally being used by the communities. At the same time, the existing restrictions in sale of land use rights are uncommon for a private property regime. The Land Code was drafted in the middle of the 1990s (enacted in 1997), when the legislators were more influenced by the soviet land tenure principles and found it hard to decide in favor of an open land market. As a result, the Land Code and related legislation lack explicit provisions to assume the land has *de facto* became private property or is still public. In trying to improve the security of land tenure as a private property, Tajikistan should at the same time accommodate the needs of landless households and the growing number of landless livestock owners and ensure their continuing access to pastures.

All of this is against a background of the burgeoning rural population who are dependent on livestock as the key source of livelihoods. It is advisable to establish and develop the communal tenure of pastures throughout Tajikistan that accommodates not only the private economic individual interests but also the social and environmental variability and needs of growing small landless livestock owners–householders. These are the key factors that retained the pastures as communal property rather than as private property in the neighboring countries (Kyrgyzstan, China, and Mongolia) in recent reforms (Robinson, Chap. 11 and Hannam, Chap 17). Securing the tenure through privatization of natural common resources is not always the best solution for rural development; it benefits elite groups who can influence the process or have capacity to exploit it. In other words, securing the land tenure for *the right people* has to be paramount. However, the proposed amendments to the Land Code trended to strengthen the position of these landlords at the expense of all others.

Theoretically, Tajikistan could have a blend of tenure arrangements and coexistence of private tenure and communal common tenure (to accommodate the ecological and social variability through spread of wealth to majority, along with private economic benefits). The proposed allocation of very remote and inaccessible pastures to potential investors for long term (25 year) lease, even if based on clearly developed criteria and procedures is problematic. Who should build and pay for the required infrastructure (water points, roads, bridges)? However, in practice, ensuring a proper allocation process is unfeasible due to wide corruption and lack of transparency and clear principle of such pastureland allocations.

Still a large proportion of the pastureland in Tajikistan is under fixed-term use right that can be granted to communities for communal tenure, upon expiration of fixed-term use. The advantages include (1) every household (former or new) within the given community can access pastures subject to compliance of developed regulations, (2) avoiding high costs involved in private property regime such as land surveying, registration of individual titles, and excluding access through policies, fences, etc. Conversely, the management and stewardship of common property can be more efficient due to lower registration and exclusion costs that can be shared by a community. Unlike the private property regime, the risk to jeopardize the rights of others during land right registration based "on first-come-first-served" principle is avoided.

There are many studies and examples supporting the advantages of community tenure principle. In summer 2011 during meetings, the herders in Norin Oblast of Kyrgyzstan were asked about their perceptions of previous system (prior to Kyrgyzstan Pasture Law (2009), elite groups obtained long-term tenure for vast area of pastures and subleased to communities) and the current (state owns pastures but grants the tenure to communities as common resource to be managed by pasture users' groups). The herders and the local specialists in every community we met emphasized the advantages of the current system, and some are highlighted below:

- The pastures' ecological conditions improving due to users' group management and planning, compared to overgrazed conditions they had before.
- The social cohesion between the herders improved since now every community is granted certain pastures to use and care about. Before, they competed for lease of best pastures from the individual landlords.
- The pastureland speculation by the powerful elites is eliminated who used to charge herders/communities high rates, which increased the social tensions and insecurity.
- While full privatization of the land appeared to be successful in Western countries, it is not feasible in Tajikistan, where:
  - Livestock requires mobility in order to balance the variability of the available fodder particularly in different seasons.

– Limited land resource but minimum of about 75% of total population depends on agriculture and livestock breeding with lack of other sources of income.
– Socioeconomic variables focusing on private economic interest should not, however, obscure the crucial role of land tenure and land policy for equity and social balance as well as environmentally sustainable development.

10 Recommendations

The current tenure system is inappropriate both for environment and social sustainability. A land reform is needed that would make a balance and ensure the sustainable and inclusive land policy. Though the significant part of the pastureland is concentrated in few entities for lifetime-inheritable use right, a "soft" land reform toward a more sustainable, inclusive, and even spread of pasture tenure can be achieved through a well-designed taxation. Tax can be applied to achieve a wide range of policy objectives; it is an instrument for sustainable land policy and socially desirable land tenure in many countries. Progressive land taxes would be a good instrument to avoid hurting the vulnerable households but discourage the large landowners from keeping tenure of large pasture estates just for speculative informal leases to landless households and often makes these pastures underutilized. It would promote impetus in land use right market, increased state revenues from sale dues, and spread tenure to a wider public and the farmers–operators. The land taxes however must be carefully designed to avoid adverse impacts to the poor and vulnerable households; otherwise, this category will be forced to sell land use rights to avoid high tax payments.

Progressive taxes for pastureland would target the landlords that have lifelong inheritable tenure for large area of pastureland but sit on informal leases. Their tenure purpose is more speculative – leasing to major landless rural residents. This category of landlords would be forced to return the pastureland (or part thereof) to the state voluntarily, which can then be redistributed to each jamoat to create communal pastures for all residents of the given jamoat. The following are priority recommendations to consider in development of a strategy for the long-term sustainability of the rangelands in Tajikistan. The following recommendations are made and elaborated below:

• *Policy and Pasture Law*
The current insufficiently regulated and spontaneous tenure arrangement cannot remain unattended. The pasture tenure granted randomly and subsequent practices do not permit socioeconomic and environmental sustainability. Lack of clear policy will cause the situation to deteriorate. Policy and pasture law provides principles of sustainable and inclusive land policy. These along with economic growth, cater for social and environmental sustainability. Clear and specific regulations on tenure and stewardship provided in a specific pasture law are required soon to avoid the threats from environmental degradation and social stratification. Only within appropriate tenure and institutional settings can proper

land stewardship and sustainability be possible. The process of pasture law drafting however should be more open to public debate. Closer cooperation and coordination with the different stakeholders must be ensured to benefit from their contribution and address the concerns of every party.

- *Redistribution of Pastures for Communal Tenure*
  The government should put efforts into making pastureland available for communities to improve the livelihoods of rural population. Development organizations and many scholars and researchers argue for common property solutions and reject individualized tenure options. Common property in developing countries also bears an insurance function especially for the poor in case other income sources fail. Common-use pasture should not be confused with open-access resource. In common pasture the access is granted only to those living in the geographic boundaries of the given community, not to outsiders. The land available for the community pasture tenure can include the recovery of pasturelands currently under fixed-term use right and lifetime-inheritable use right. It should be based on principles:

  – The ownership of pastures retained by state and perpetual tenure granted to community as common resource on *kishlak* or *jamoat* level.
  – The community common pasture is not an open pasture – it cannot be used by the outsiders it is reserved for those living in the given community.
  – The community is granted perpetual tenure to pastures and it cannot be sold or challenged by any state body except in case of state needs provided for in law.
  – No individual, family, or group land share should apply in community common pastures.

These are critical principles to accommodate the social and environmental variability and avoid the continuous registration of individual rights as well as conflicts. It ensures that all people living in community have equal and continuous access to the common pasture and are safeguarded from losing the equal access right. The tenure should be granted/registered to the given community (jamoat) but not to any individual/entity names. The families newly migrated/emerged in the community should have immediate equal tenure as long as they live in the given community. The pastures formerly being used by a community should be reallocated to given community, and minimum area of community pasture shall be defined by the technical experts based on local factors, physical landscape, land accessibility and availability, social and environmental variability, and so on.

In Tajikistan context, the communal pastures are the best approach also due to extended and perpetual male migration abroad, but women, children, and elders remained back home. It is not feasible to expect every shareholder to participate in farm management decisions and review progress reports, in current approach that includes individual land shareholding. Many of these shareholders will never be able to manage their farm shares properly despite the continuous trainings or awareness-raising activities. However, they always need access to pastures and therefore need more simple regulations than those based on legal and business concepts of share and shareholding. Tajikistan may consider the Kyrgyz experience where the state exclusively retains the ownership of pastures and grants tenure to communities as a common resource. In Tajikstan. unlike in Kyrgyzstan, the past reforms have seen the greater part of the pasturelands granted as "lifetime-inheritable use right" to individuals/private farms. This approach is still feasible for Tajikistan; the state can regain the pastures through the progressive taxation from the individual tenure (of lifetime inheritable) and redistribute these lands to the communities for common perpetual use rights.

- *Progressive Taxation for Pastureland Tenure*
  The current pastureland tax is extremely low (less than US$1/ha/year) though the landlords' revenues made from lease to residents and sale of other livestock products are very high. For example, the cost of meat (with bones) is 30 somoni (about US$ 7, as of October 2011) and has rising tendency. The tax in effect should be the state's rental charge for the land. The current extremely low pastureland tax rate set in 2006 remains unchanged despite of changes of many factors, including significant raising prices of livestock products. The taxes should be increased and spent also for pasture improvement, building infrastructure to make them accessible. A progressive taxation policy will encourage greater reallocation of land as being just a "landlord" collecting rent will be less profitable. Larger landlords would thus be encouraged to:

  - Become a farmer owner–operator and use more distant pastures; the state will collect more taxes from sale of animal products produced by these farms.
  - Return the surplus pasture to the state, which can be reallocated to jamoats for community common use by the rural households through the pasture users' groups.

  Progressive tax will be important and can be done based on area of landholding. At the lower end, the smaller land area, there should be lower flat tax so that the poor are protected, and at the higher range, you can take the tax all the way up to 50%. A series of increments would be defined, and each increment would attract a higher tax level. Tax rates for some increments could be trebled. Such tax should be applicable only to land. It is recommended not to impose any taxes on buildings and assets of the farms, such as machinery to encourage investments for mechanization.
- *Institutional Reform and Decentralization for Local Management of Pastures*
  The state institutions authorized to monitor compliance in use of pastures are technically and financially weak. The alternative community institutions have never been in Tajikistan at least since the beginning of Soviet system. The communities and pasture users are not involved in pasture management. The management of natural resources by those who use it has been demonstrated to be more effective than by the state or private institutions (Ostrom 1990). The institutional reforms with decentralized authorities that would increase the pasture users' involvement in management and planning of pastures would be appropriate to develop in Tajikistan.

Option A: Pasture Planning and Management by Jamoat Officials
Its advantages include easier and quicker setup with minimum capacity-building activities since the jamoat structures are already in place. Jamoat has a well-established structure that addresses different community issues and supports implementation of governmental and nongovernmental programs in communities. However, the jamoat administration has many other tasks, and community pasture management would be a burden.

Option B: Pasture Planning and Management by Community-Based Organizations–User Groups
The advantage is that the community involvement in management of resource will be established which is proved to be the most effective measure in governing the natural resources by those who actually use it (Hua, Chap 14). All stakeholders can be part of such community groups including local officials for assistance in the preparation of participatory pasture use plans, defining pasture user fees and other regulations. The challenge is that, though in the longer term it would be a sustainable approach, at the onset it requires much institutional and capacity-building support.

Acknowledgements The work reported here was generated during the ADB/GEF Rural Development project and the author was engaged as aconsultant by a consortium of GITEC Consult GMBH/Donaev Management Consulting under the direction and guidance of the PMU in Dushanbe. I am indebted to the efforts of my erstwhile colleagues for their inputs into aspects of this work.

References and Further Reading

Banks T (1997) Pastoral land tenure reform and resource management in northern Xinjiang: a new institutional economics perspective: Nomadic peoples, v. 1

Banks T (2003) Property rights reform in rangeland China: dilemmas on the road to the household ranch. World Dev 31(12):2129–2142

Banks TC, Richard PL, Zhaoli Y (2003) Community-based grassland management in Western China. Mountain Res Dev 23(2):132–140

Fernandez-Gimenez ME (2002) Spatial and social boundaries and the paradox of pastoral land tenure: a case study from post socialist Mongolia. Human Ecol 30(1):49–78

Fernandez-Gimenez ME, Allen-Diaz B (1999) Testing a non-equilibrium model of rangeland vegetation dynamics in Mongolia. J Appl Ecol 36:871–885

Fernandez-Gimenez ME, Batbuyan B (2004) Law and disorder: local implementation of Mongolia's land law. Dev Change 35(1):141–165

GRM (2008) Kyrgyz Republic: a study of the impact of land reform on agriculture, poverty reduction, and environment, ADB TA 4408/Project Number: 38079

Hannam I (2012) International perspectives on legislative and administrative reforms as an aid to better land stewardship in Central Asia. In: Squires V (ed) Rangeland stewardship in Central Asia. Springer, Dordrecht, pp 407–430 (Chapter 17, this volume)

IFAD (2008) Improving access to land and tenure security: policy. IFAD, Rome, 2008

Kurbanova B (2012) Constraints and barriers to better land stewardship: analysis of PRAs in Tajikistan. In: Squires V (ed) Rangeland stewardship in Central Asia. Springer, Dordrecht, pp 129–164 (Chapter 7, this volume)

Land Administration Guidelines (1996) With special reference to countries in transition: United Nations, New York and Geneva

Neudert R (2010) The paradox of pastoral land tenure: are solutions with individualized tenure possible? – A case study from Azerbaijan. Greifswald University

Ostrom E (1990) Governing the commons: the evolution of institutions for collective action. Cambridge University Press, Cambridge/New York

Ostrom E (2007) Collective action theory. In: Boix C, Stokes S (eds) The Oxford handbook of comparative politics. Oxford University Press, Oxford

Ostrom E (2009) A general framework for analyzing sustainability of social-ecological systems science, v. 325

Ostrom E, Gardner R, Walker J (1994) Rules, games and common pool resources. University of Michigan Press, Ann Arbor

Oxfam (2009) Reaching tipping point: climate change and poverty in Tajikistan. Oxfam International, Dushanbe, 22 p

Robinson S et al (2012) Pastoral tenure in Central Asia: theme and variation in the five former Soviet republics. In: Squires V (ed) Rangeland stewardship in Central Asia. Springer, Dordrecht, pp 239–275 (Chapter 11, this volume)

Soule M (2000) Land tenure and the adoption of conservation practices. Am J Agric Econ 82:993

Squires V (2012) Better land stewardship: an economic and environmental imperative if there is to be sustainable development. In: Squires V (ed) Rangeland stewardship in Central Asia. Springer, Dordrecht, pp 31–50 (Chapter 2, this volume)

Taylor J (2006) Rangeland policy, privatization and new ecology in Inner Mongolia. Presented at "Survival of the commons: mounting challenges and new realities," the Eleventh Conference of the International Association for the Study of Common Property, Bali, Indonesia

USAID Land Reform Project (2011) Dushanbe, Annual Report February

Part V
Land Protection

Land is the basis of a country's wealth. In Central Asian countries where more than 50% of the people are rural with a high dependency on agriculture and animal husbandry, it is essential to ensure that land protection is a high priority. Land users are the agents for change. The individual decisions made by millions of people on a day-to-day basis determine the longterm sustainability of any given land use strategy.

Chapter 14 In recognition of the fact that land management 'starts and ends with the land user', this chapter considers the steps in engaging with land users in efforts to introduce new management interventions that will foster land stewardship and greater sustainability.

Chapter 15 Many subsistence land users see wildlife as competitors for scarce resources such as forage and water or in some cases as predators on livestock. In other situations, wildlife may be a reservoir of diseases that threaten livestock or even humans. Such attitudes towards wildlife are barriers to implementation of measures to introduce and enforce protection.

Chapter 16 Uzbekistan is a country with rapidly growing human population. This case study examines the role participatory management of desert rangelands to improve food security and sustain the natural resource base.

Chapter 14
Engaging with Land Users: The First Steps on a Long Road

Hua Limin and Degang Zhang

Abstract This chapter is like an anatomy lesson. It dissects the conduct of a major rural development project from the initial site selection through the various steps required to refine the project objectives and adapt them to local conditions and the efforts involved to get stakeholder involvement in project implementation and monitoring and evaluation. The key element is that land users need to be involved right from the start. Farmers are direct land users, and any positive responses on land management approaches can only be achieved through farmers.

Keywords Multifunction • Participatory approach • Community based • PRA • Farmers as land users • Central Asian region • Implementation • Monitoring • Evaluation • M&E • Xinjiang • Gansu • Agro-pastoral • GEF • World Bank • Rural development • Qilian Shan • Tian Shan • Altai Shan • Livelihoods

Key Points

- The livestock production and rangeland management on farmers' land are multifunctional in terms of economical, religious and social purposes. Therefore, the farmers' rangeland management skill and level are the key to both natural resource conservation and successful implementation of rangeland management projects.
- Use of a participatory approach is the basic principle for international assistance projects, and donors demand stakeholder participation in assessment, implementation, monitoring and evaluation. Typically, there are three phases in the participatory approach. These involved using a *simple participatory approach*, *improved participatory approach* and *co-management participatory approach* based on the community's active participation over the 6 years of the project.

H. Limin (✉) • D. Zhang
College of Grassland Science, Gansu Agricultural University, Lanzhou, China
e-mail: hua-lm@263.net

V. Squires (ed.), *Rangeland Stewardship in Central Asia: Balancing Improved Livelihoods, Biodiversity Conservation and Land Protection*,
DOI 10.1007/978-94-007-5367-9_14, © Springer Science+Business Media Dordrecht 2012

1 Introduction

Farmers are direct land users and any responses on land management approaches can only be done through farmers. In China and the developing countries of Central Asian region (Fig. 14.1, see Masseli (2012), Chap. 1 for definition), the rangeland animal husbandry relies on the family operation with the household as the major decision-making unit. This is not the same as the operation pattern in the commercial pasture in advanced counties.

The livestock production and rangeland management on farmers' land are multi-faceted in terms of economical, religious and social purposes. Therefore, the farmers' rangeland management skill and level are the key points for natural resource conservation and successful implementation of rangeland management projects. Moreover, approaches that focus on farmers are a directive of some governments who are seeking long-term stability and more social equity as they strive for national development.

Based on the change from 'how to work for farmers' to 'how to work with farmers', many international projects have achieved much in applying the new approach. There is consensus that the acceptable way is through a participatory approach. However, the participatory approach varies according to different theories. Some pay more attention to the process while others focus on the outcome. Usually though, the participatory approach involves a number of steps – participatory survey, analyses, implementation and appraisal. The final objective of the participatory approach is to manage the natural resource through the power and capacity of farmers at the grassroots level.

China's experience in addressing rangeland degradation and recovery (Squires et al. 2009) has much to offer. China has borders with several Central Asian countries (Kazakhstan, Kirgizstan, Tajikistan and Afghanistan) and has similarities with Mongolia. China has a large area of rangeland with over 3.3 million pastoral households. Most (2.3 million households) are located in the semi-pastoral counties in the

Fig. 14.1 Sketch map of the Central Asian region which according to the UNESCO definition includes Western China (Xinjiang and Tibet) and Afghanistan

north and northwest of China (see map) The rangelands of Northern and Western China are the third largest in the world, supporting the world's largest population of sheep and goats, and fourth largest population of cattle (Squires and Hua 2010). However, rangeland degradation and desertification in China is also one of the most serious of any country in the world.

Land degradation and biodiversity loss are the most critical issues of the ecological environment in the west of China, and they are the main causes for poverty and are constraints to economic development. The Chinese government has paid much attention to these issues, and the significant achievements have been achieved in land degradation control through long-term efforts, especially as a result of Western Development Strategy that was launched in 2000. Despite active efforts aimed at arresting land degradation and conserving biodiversity conservation, the outcome relied on top-down measures and was limited because these efforts were from a single ministry (or bureau), and the efforts were not coordinated across line ministries (World Bank 2003). In 2003, World Bank/GEF initiated a Pastoral Development Project under Operational Program 12 (OP12) in Gansu and Xinjiang, and it was an attempt to arrest and reverse land degradation, conserve biodiversity in the important mountain rangelands in both Gansu (Qilian Shan) and in Xinjiang (Tian Shan and Altai Shan) and improve livelihoods.

The project is the first large-scale operational project of the China/GEF strategy partnership on land degradation in dryland ecosystems under the GEF/OP12. It was funded by the GEF (Global Environment Facility) and linked to the World Bank Gansu and Xinjiang Pastoral Development Project. The global environmental objective of the project was to maintain and nurture rangeland ecosystems to enhance global environmental benefits. More specifically, the project aimed to mitigate land degradation, conserve globally important biodiversity and enhance carbon sequestration.

Like other international cooperation projects, Gansu and Xinjiang GEF carried out the project using a participatory approach. However, various projects use different participatory approaches due to special conditions that apply there both from without (political system, culture, economy, etc.) and within (requirement of project funder, capacity of project management, demand of farmers, etc). The choice of which one should be used for a project implementation depends on the dominance of one or other (or both) sets of conditions. In Gansu/Xinjiang GEF project, the participatory approach experienced three phases during 6 years of the project implementation, which involved *simple participatory approach*, *improved participatory approach* and *co-management participatory approach* based on the community's active participation.

Since the 1980s, using a participatory approach is a basic principle for international assistance projects, and donors demand assessment, implementation, monitoring and evaluation. Almost all projects begin by conducting participatory rural appraisal (PRA). The rationale and methodology for PRA is dealt with more fully in Kurbanova (2012), Chap. 7 and in Adebo (2000). PRA helps communities to:

- Mobilize their human and natural resources.
- Define problems.
- Consider previous successes and failures.

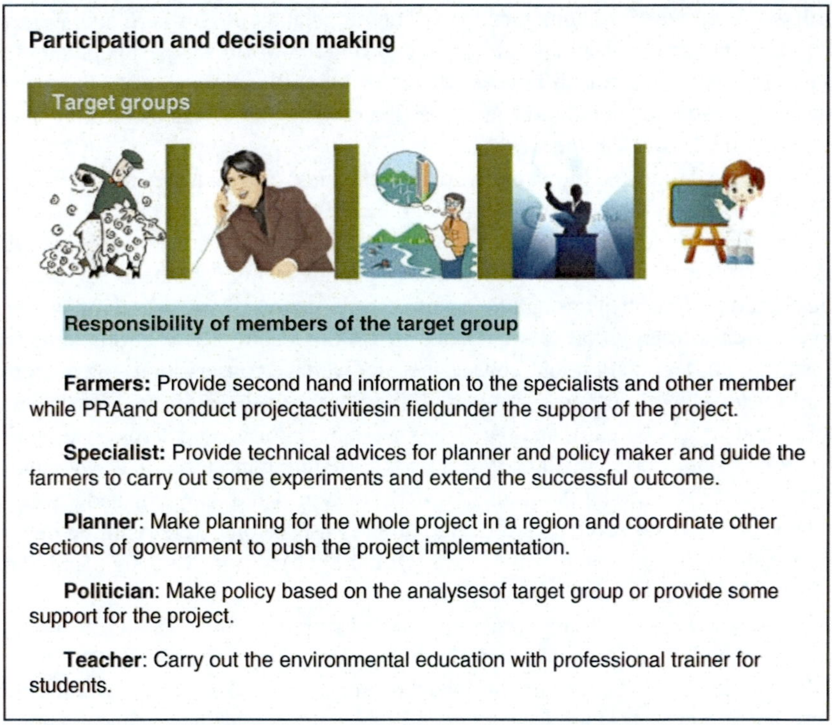

Fig. 14.2 Schematic showing the role of the various actors/stakeholders in the project

- Evaluate priorities and opportunities.
- Prepare a systematic and site-specific plan of action.

A schematic summary of the development of the participatory approach in the GEF project is shown in Fig. 14.2 starting with the project-planning stage (see text below for elaboration). The target groups for the overall GEF project were land users, government agency/extension specialists, planners, politicians/policymakers and teachers and students.

Several steps were involved in the implementation of the participatory approach for Gansu/Xinjiang GEF project starting with the choice of target group through participatory implementation. The steps go through a cycle (Fig. 14.3). Basically, each programme of GEF project in Gansu that was implemented in 9 counties was carried out according to this procedure. More detailed contents of each step will be introduced in the next section.

In the Gansu/Xinjiang GEF project, the PRA approach was used in nine counties, spanning conditions that ranged from the desert margins to areas in the alpine regions (Fig. 14.4).

There was great diversity in the ethnicity of the people (Mongolians, Kazaks, Yugurs, Tibetans, Hui and Han) and in the type of animal husbandry practised

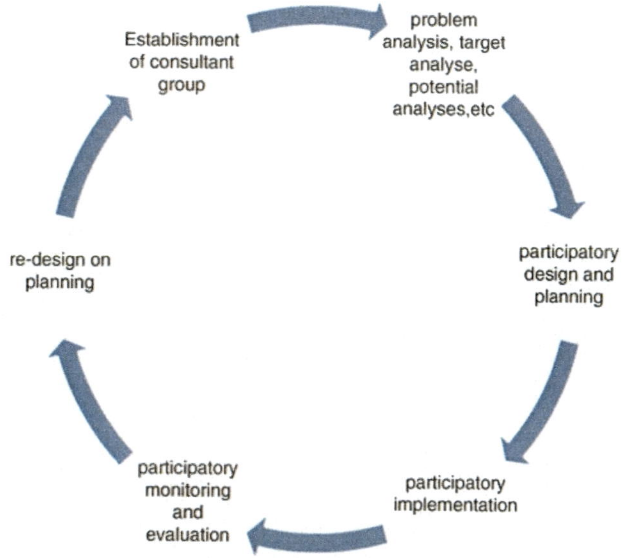

Fig. 14.3 The flow chart of participatory approach in Gansu/Xinjiang GEF project

(sheep, goats, cattle, yaks and horses). The project that is the basis for this chapter was an excellent outdoor laboratory, and by using the many case studies that will be introduced here, the participatory approach was applied step by step in some GEF pilot villages in Gansu/Xinjiang GEF project.

There were eight key steps:

Step 1: Establishment of stakeholder group for preliminary project design
Step 2: Selection of project pilot area and participatory investigation
Step 3: Confirmation of work plan based on participatory analyses and appraisal
Step 4: Participatory implementation
Step 5: Participatory monitoring and evaluation (M&E)
Step 6: Participatory training programme
Step 7: Environmental education

Step 1: Establishment of Stakeholder Group for Preliminary Project Design

The characteristics of the stakeholder group are a key factor in deciding how to proceed with project implementation. Experience gained from a number of international projects suggests that the PRIA group should consist of project agent, local government agent, farmers as well as different specialists with professional background of natural science and social science, such as rangeland specialist, livestock specialist and social development specialist. However, the real purpose of forming the group and conducting the PRA is to know and understand the different stakeholder's opinion on the project itself and their attitude towards environmental conservation versus economic development. Therefore, the project agent should firstly visit the local government and hold workshop or meet with local officials, as well as

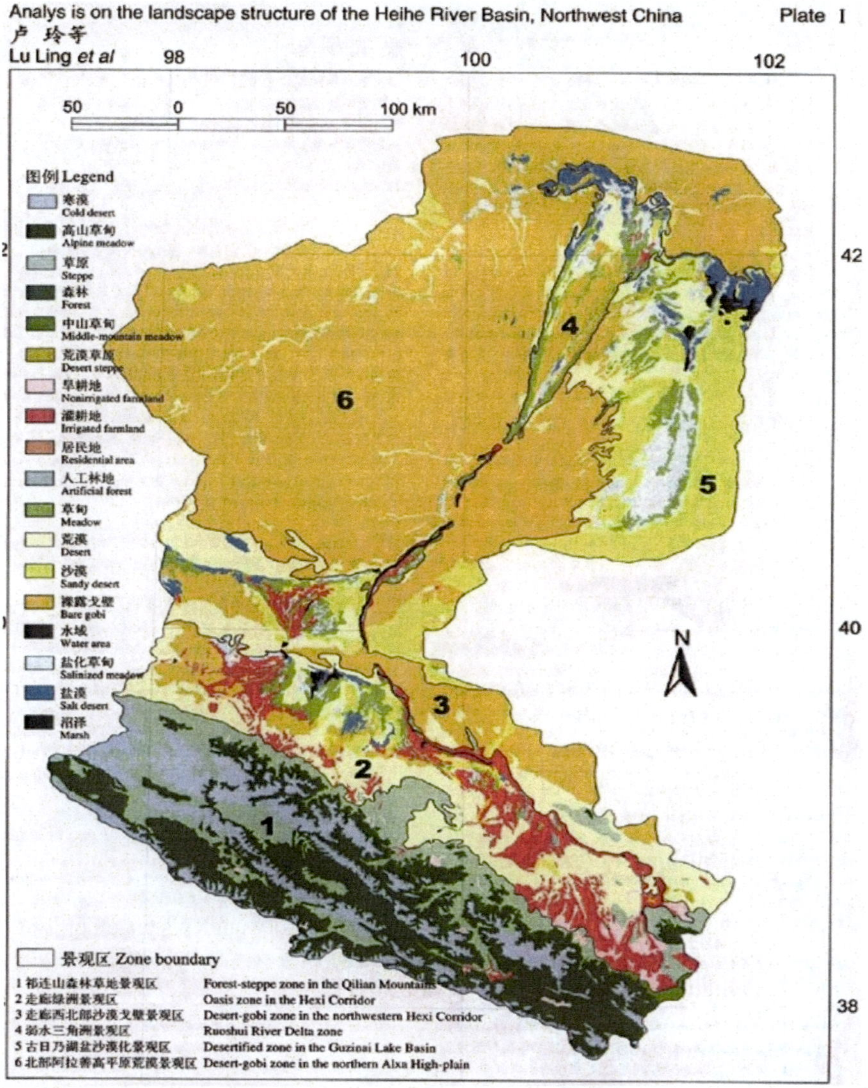

Fig. 14.4 A typical cross section from the alpine regions of the Qilian Mountains (lower left) to the desert margins (upper part). The altitude ranges from >3,000 m in the mountains to about 350 m on the desert edge. This example from the Hei He (Black river) traversing the Hexi corridor is typical of the conditions in which the GEF project was implemented. There are six distinct landscape zones

farmer's representatives. In addition, the project agent should pay more attention to the function of consultant group and use them sparingly at first until there is a clear consensus at local level about the relevance and timing of proposed project interventions.

Initially, the stakeholder group consisted of five to seven members from different government agencies, financial bureau, animal husbandry bureau, forest bureau, environment bureau, as well as land management specialists. In addition, two or three farmers' representatives and one representative from the consultant group were involved in the stakeholder group.

Role of the Consultant Group

In fact, the consultant group is more important than other stakeholders because it is they who are responsible for selection of project implementation area, outlining design of participatory appraisal, conduct of PRA, etc. A highly efficient consultant group can help the project manager to accurately know the real situation before or during the project implementation, and also provide available advice to solve some problem faced in the project implementation. Due to limited financial resource, most clients (project management office) do not like to employ many consultants, especially international consultants. So, '*how to select and establish a consultant group?*' is also a key precursor for project implementation. Normally, the donor will assign some international specialists to participate in the project design. The project counterpart organization selected their specialist from some universities and institutes to join the group. Moreover, the local specialists at county level and township level must be involved in the group because they are familiar with local situation rather than foreign specialists and 'big professor' of university (Box 14.1).

Box 14.1 Composition and Role of Stakeholder Group and Consultant Group

Case Study 1: Stakeholder Group in Gansu/Xinjiang GEF Project
The stakeholder group of Gansu GEF project was the leadership group that consisted of people from Gansu Finance Bureau, Forest Bureau and Animal Husbandry Bureau. The function of the provincial stakeholder group was coordination among different projects to ensure the consistency of provincial strategy objective and provide the counterpart funding for the project as well. The Gansu Animal Husbandry Bureau was the implementing agency in Gansu. At the county level, a stakeholder group consisted of one person from local government, two farmers from pilot village as well as two or three specialists.

Case Study 2: The Consultant Group for PRA of Gansu GEF Project for Yongchang County
In 2004, World Bank employed two specialists to work for GEF project. Under their support, the framework of consultant group was completed. The original consultant group consisted of foreign and domestic specialists from university or institute. However, the consultant group faced the first challenge when they visited the Yongchang County. The interpreter did not know the strong local accent.

(continued)

Box 14.1 (continued)

The foreign specialists had different understanding and opinions on making plan, beneficiary selection, implementation approach, etc., due to different culture and educational background. The local technicians did not like to co-work with consultant group because they thought these specialists were outsiders, which means the specialists do not know the real situation of local rangeland management and livestock production. Finally, the project management office made a decision and recruited two local technicians from grassland workstation at county level and animal husbandry extension station at township level, respectively. The composition of new consultant group evidently improved the work efficiency and accuracy on project area selection, problem analyses, making annual plan, etc. The feedback of project monitoring and evaluation later proved that the useful composition provided the strong support for the project's smooth implementation. On the other hand, the involvement of local technician is also a way of capacity building because they can have more experience and update their knowledge from outsider specialists in the group.

Step 2: Selection of Project Pilot Area and Participatory Investigation

The first mission of consultant group is to select and confirm the project implementation scale. Each project has its own focus and requirement according to investor's target and requirement. Some projects need large ecological scale, such as Ban Grazing Project funded by Chinese government that selected one whole county as its implementation unit. Most of the project activities were carried out on millions of hectares of pastures of two or three places in one county. The guarantees of these projects are enough funds and high-efficiency project management organization, which can get enough support from a strong government. However, most international assistance project lack the guarantees, in particular, in developing counties and undeveloped counties. Based on the understanding and according to the experience of similar project, Gansu/Xinjiang GEF project considered that the scale of project implementation area should not be large, but should focus on a village as the project implementation unit. The reasons for selection were:

1. The village is at the bottom of the political organization. The project management office can easily contact villagers directly and get useful feedback instead of spending lots of time on negotiation and discussion at a higher organizational level.
2. The village is a small unit in the management framework. Most of the villages are separated by natural boundary rather than administrative boundary. So, the natural boundary is easy to work within and convenient for conducting some rangeland management activities, such as rotation grazing and rest grazing. If a project selects a bigger area with several villages for the project site and wants to carry out the similar project activities, the coordination between different villages is a big challenge because the project has to spend more time and more money on it.

3. Farmers living in the same village usually have similar living standards and have common interests, such as livelihood and knowledge of the environment in and around village. This provides a good basis for the investigation. Most international projects focus on environmental protection, poverty alleviation, sustainable agriculture development, etc., and the objectives of the project are always close to the farmers' livelihood. The quick response by farmers when a project conducts a PRA is very useful for information collection and analyses, and it is of benefit to project implementation later.
4. Most farmers living at the same village are relatives and their religion is the same. They trust their neighbours. This point is very important for project demonstration, training and extension. One of the challenges of farmers' training is the choice of a good and believable role model. Selecting a number of farmers for training and designating them as demonstration householders are a useful approach on which to base the training programme.

Based on the analyses above, Gansu/Xinjiang GEF project selected one village in each of eight counties as the project demonstration site. The villages represented several different ecological areas, such as dryland farming area, agro-pastoral area and pastoral area. The entire set of project activities that were implemented was based on the core of village-based farmers' priorities. In addition, the location of the pilot village took into consideration ease of access, reliability of communication, farmers' enthusiasm and natural resources (Boxes 14.2 and 14.3).

Box 14.2 Characteristics of Mayinggou Village of Yongchang County

Case Study 3: Background of Mayinggou Village of Yongchang County in Agro-Pastoral Area

Mayinggou village is located in southeast Yongchang County at an altitude of 2,700 m. It is around 65 km from the county capital. There are 526 householders with a total population of 2060. The area of usable rangeland is 5,067 ha. The area of arable land is 224 ha (mainly growing barley). The forage crops are corn and oats. Over 50% of household income is from livestock production. The village is located in an agro-pastoral area and there were no previous national projects related to animal husbandry development in the village. Although rangeland area is bigger than the cropland, the rangeland was so degraded because of the lack of any rangeland management regulation in this village.

When the consultant group and project officials go to a confirmed village, the first mission is to invite villagers to a meeting for PIA. The members of consultant group should tell villagers the purpose of the PIA and the procedure to be followed. According to the investigation outline prepared in advance, the consultant group uses various tools to conduct investigation. The tools include semi-structured interview, individual interview, key informant interview, group interview, questionnaire,

Box 14.3 Preparations for the PIA

1. Prepare materials in advance

A full list of the materials required for each participatory activity should be drawn up. These materials should be pre-packed before any activity to ensure that anything missing can be sourced. Materials for participatory planning activities typically include:

- Sheets of newsprint
- Copies of topographic maps or GIS maps with overlays for spatial planning
- Copies of regional inventories and other studies from rangeland institutes and rangeland monitoring stations
- Sticky tape
- Coloured pens
- Scissors
- Pencils and/or pens
- Notebooks, pro forma worksheets and planning frameworks for detailed activity planning
- A tape recorder and/or camera

SWOT analysis, factor analysis, direct ranking, matrix ranking, farmer's own record, historical record, seasonal calendar, diary diagram and social and economical mobility. Such information is used to improve the capacity of self-organization, benefit and knowledge sharing.

Which tool should be selected for use in a particular setting depends on the specific local situation in practice. In Gansu/Xinjiang GEF project, the semi-structured interview, individual interview, key informant interview and questionnaire were used widely in participatory investigation. Moreover, during the semi-structured interview, individual interview and key informant interview, the local specialists played the key role because the farmers trust the local specialists more than outside specialists.

Although the members of consultant group expressed the purpose of the PIA and the project to the villagers and emphasized that the GEF project focus was about improving rangeland management and livestock production in the village, the farmers were concerned more with the issues directly related to their life, such as electric power, kid's education and road condition. Under the circumstance, the investigation group could not refuse or ignore the farmers' priorities. The compromise was that the investigation group urged the farmers to change their focus to animal husbandry development. They explained that the improvement of rangeland management and livestock production can be of benefit to solve the other social issues. Finally, the farmers accepted that approach. The ranking of issues at the beginning and end of the PIA are shown in Box 14.4 and indicate how far the villagers were able to go in re-assessing their priorities.

Box 14.4 A Summary of the Villagers' Priorities at the Time of the First Round of PIA and Their Final Set

Case Study 4: Farmers' Initial Concerns and Rankings in Mayinggou Village of Yongchang County at the First Meeting to Discuss Project Implementation

Item	Concerns and difficulties	Rank
1	Lack of fund for life	1
2	Poor infrastructure of village school and bad road	2
3	Lack of techniques and on-site guide form technicians	2
4	Lack of rangeland fence and water supply	3
4	Lack of technique for pen-feeding	4
5	Lack of technique for forage storage	4
6	Productivity of animal is low because of poor bred	5
7	Lack of technical materials	6
8	Difficult to access market information	7
9	Poisonous grasses in rangeland are severe	8
10	Grasshopper threat is severe	8
11	Animal disease (goat abortion, parasite) reduce productivity	9

Farmers' Concerns and Rankings in Mayinggou Village of Yongchang County at the End of First Meeting to Discuss Project Implementation

Item	Concerns and difficulties	Rank
1	Lack of grassland management	1
2	Lack of machines for fodder processing	4
3	Lack of fuel (mainly depend on manure of livestock)	6
4	Lack of high-quality livestock breeds (especially for meat production)	4
5	Lack of warmsheds and severe diarrhea of lamb	5
6	Lack of techniques ration formulation	3
7	Lack of forage seeds	2

Based on the result of semi-structured interview, the investigation group also interviewed some individual householders and a key person to probe the reliability of issues that were obtained from the semi-structured interview. The interview was in the householder's home in a relaxing environment without government officials' interference

Election of participatory project management group at village level

The farmers/herders are main users of rangeland and it is very important to use their local knowledge and involvement in rangeland management. The project supports all GEF villages to establish the participatory rangeland management group (PGMG) based on the vote of farmers during the investigation. Moreover, it is important to sustain the project benefit through community organization in its current phase. The

PGMG members include the members of village committee, rich farmers, poor farmers and females so that the decision made by the group could be easily accepted by the community. According to the local conditions, the group should make policies on participatory rangeland management, such as the animal numbers, utilization rate of forage in rangeland in different seasons, fence maintenance, water point protection and grazing management, and rangeland utilization system, and then, the group will be in charge of executing the policies (Box 14.5).

Box 14.5 Summary of the Role and Function of the Participatory Grassland Management Group

Case Study 5:

Item 1: Name of PGMG

Participatory v management group of Mayinggou Village Yongchang County

Item 2: Members selection and subordinate relationship of PGMG

The members of PGMG are selected by the villagers' congress through PAR tools and led by the villagers' congress (by village committee during adjournment). The GEF project offices of the county and township guide the technical activity.

Item 3: PGMG members

PGMG is constituted by 4 members of village committee and 6 villagers (see the table below).

Members of PGMG in Mayinggou Village

Position	Name	Sex	Age	Nationality	Note
Leader	Zhao Shoulun	M	45	Han	Village leader
Vice leader	Zhao Shibiao	M	31	Han	Village leader
Member	Jiang Xuemei	F	30	Han	Village leader
	Zhao Zongqi	M	28	Han	Villager
	Zhao Zongjun	M	40	Han	Villager
	Cao Hongling	F	25	Han	Villager
	Zhao Zhongnian	M	25	Han	Villager
	Li Lincai	M	36	Han	Villager
	Zhao Shuhong	M	28	Han	Village leader
	Dong Shouchang	M	50	Han	Villager

Item 4: Office place of PGMG

The office of PGMG is arranged in the village committee and PGMG's activity is within the village.

(continued)

Box 14.5 (continued)

Item 5: Law or rule basis of work of PGMG

- Grassland law of People's Republic of China
- Detailed rules for grassland management in Gansu Province
- Temporary rules for grassland contract utilization and management of Yongchang County
- Temporary rules for grassland contract utilization and management of Mayinggou Village

Item 6: Tenet

Follow the national laws and rules, unite villagers to protect and properly utilize grassland and protect ecological environment, wildlife and biodiversity. Make contribution to the sustainable livestock production and provide necessary information for policy-making of government.

Item 7: Basic tasks and obligations

1. Manage the grassland border, grassland vegetation, wild animal, water resource, grassland infrastructure, wildfire control, grassland carrying capacity and stocking rate. Prevent all illegal activities such as trespass grazing, clear vegetation for arable land or building, collect medicine herbs, hunting and overgrazing. If the illegality was found, PGMG has the obligation to provide the suggestion of atonement or punishment to government.
2. PGMG has the right to require villagers to properly utilize the contracted grassland and regulate the reasonable carrying capacity according to the feed supply from natural grassland, artificial grassland and crop residues in order to keep feed balance. The utilization percentage of natural grassland is below 50%. The annual average carrying capacity is below 1.7 sheep units per ha (3.4 sheep units for summer pasture and 0.85 sheep unit for winter pasture). The animals exceeding the carrying capacity should be removed within 2 months. The grazing period for summer pasture is from June 15 to October 19 and from October 20 to next March 19 for winter pasture. The supplements should be given from November and the supplement amount should be gradually increased. Try to feed animals in barn from March 20 to May 19, and semi-barn semi-grazing from May 20 to June 4 to avoid grazing in the critical period of grass growth. If the illegality was found during grassland monitoring, PGMG has the obligation to provide the suggestion of atonement or punishment to government.
3. PGMG has the obligation to encourage villagers to practice the rotational grazing, reseeding, poisonous grass control, fertilization, etc. with the assistance from GEF project office.
4. PGMG has the obligation to guide farmers to mitigate the loss caused by wildfire, snow disaster, freezing disaster, floods disaster and animal disease.
5. PGMG has the obligation to seek assistance from GEF project office or other channel to obtain new techniques for increasing farmers' income.

(continued)

Box 14.5 (continued)

6. PGMG has the obligation to strictly manage the grazing-ban grassland and other infrastructure, to monitor the grassland productivity and provide suggestion for proper utilization approaches (harvest grass or grazing). If the illegality was found during grassland monitoring, PGMG has the obligation to warn the breakers and if the consequence was severe the atonement or punishment suggestion should be submitted to government.
7. PGMG has the obligation to guide farmers to establish artificial grassland, multiple cropping forages, process fodder and forage reservation so that the utilization efficiency of forages could be increased and the grazing pressure on natural grassland could be decreased.

Item 8: Other tasks and duties

1. PGMG should actively organize participatory training activity in terms of new techniques, grassland protection and environment protection under the help of technicians from county.
2. PGMG has the obligation to resolve the grassland tenure confliction between farmers or villages.
3. PGMG should actively search market information to expand the animal product market.
4. PGMG has the obligation to conduct the technique extension work (greenhouse for animal raising, ration formula and disease control) and guide farmers to use recycle energy, such as solar energy, wind energy and biogas for saving energy and decreasing the greenhouse gas emission.
5. Maintain the infrastructure, such as grassland fence and water supply establishment.
6. Collect information in animal production, marketing and livelihood through PAR tools and analyze the problems for new support.

Step 3: Make a work plan for the village based on participatory analyses and appraisal

How to solve the problem that was found during the participatory investigation? A good work plan is another key for a project implementation. The work plan should stand by some principles as follows:

- Aim to the objective of project. Although lots of demands were found and reflected during the participatory investigation, a project cannot solve all problems that exist in the project area because each project has its own target. Hereby, the stakeholder group firstly aims at the problem that can meet the requirements of project support, not the problems whose ranking are No.1 on the table. For example, in Yongchang County, Gansu, No.1 problem is fixing the road because the bad road limits the development of whole village. However, the objective of

Gansu/Xinjiang GEF project is to maintain the rangeland ecosystem and conserve their biodiversity. Perhaps, there is some relationship between fixing the road and rangeland ecosystem conservation, but the limited project funds should be directly used to solve some problems of rangeland management and livestock production. In this point, the stakeholder group, in particular the project official, should explain the objective and requirement to the leader of PGMG. Finally, the project official and leader of PGMG explain to all villagers why the GEF project cannot support fixing the road.

- Development of work plan should involve farmers, project officials as well as specialist (s).

During the participatory planning with its focus on bottom-up, the team encountered many demands from the farmers about their urgent needs. The farmers always focus on short-term benefits; even some wished that the project would pay them money directly. The idea of long-term benefit that might flow from the project was a hard idea to put across when dealing with farmers whose problems are here and now and whose planning horizon is very short., Long-term impact is also another anticipated outcome of the project, and it is hoped that the project interventions will be accepted, replicated and scaled up.

Experience in the early days of project implementation in the Gansu/Xinjiang GEF project led to a change from the simple participatory approach to an improved participatory approach. The project not only considered the farmers' legitimate demand but also specialist's recommendations. This enabled a reconciliation of all points of view and guided the project activities (Fig. 14.5 and Box 14.6).

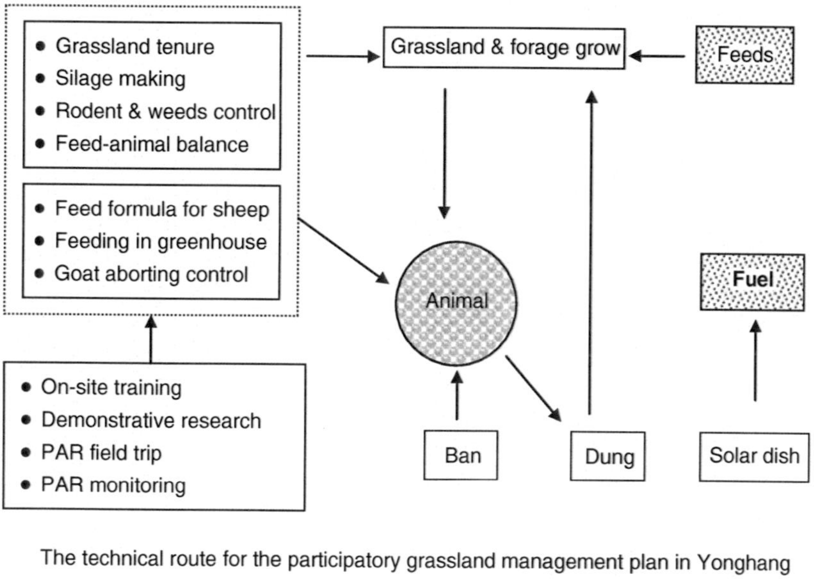

Fig. 14.5 A flow chart showing linkages and interconnectiveness of key factors in Yongchang

Box 14.6 Outline of the Work Plan for Yongchang County

Case Study 6: The Work Plan of GEF Project in Yongchang County

1. **Capacity building**
   Including oversea training for local governor and technicians; national and provincial training for local technicians; county level training and township level training for farmers
2. **Grassland use right reform and natural grassland improvement**
   In project village, 1,333 ha of grassland has been fenced for grassland use right reform. 33 ha of grassland was used for wild forage seeds collection, and 4,000 kg of seeds of *Elymus nutans, Agropyron cristatum* and *Poa pratensis* were collected for reseeding the degraded grassland. 666 ha of land were used for wild forage propagation. 1 set of sowing machine was purchased.
3. **Establishment of grassland management system**
   The village grassland management system was set up by organizing the participatory grassland management group and participatory grassland management plan. The household group contract of grassland tenure was demonstrated in groups 1 and 7; the situation of 'only utilization and no management' was altered.
4. **Grassland rodent controls**
   In northern steppe, 200 stands for eagle attraction were set up, and the controlled area is 13,300 ha.
5. **Grassland monitoring**
   1 set of equipment and 4 monitoring sites were set up.
6. **Forage and fodder production**
   6.6 ha of land was used to grow maize, alfalfa and *Elymus nutans*, in which the maize was for silage making. 60 sets of mini-type and 3 sets of medium-sized hay chopper were purchased for farmers to increase the forage-processing efficiency. A silage pit of 180 m^3 was built and 700 tons of maize silage was made.
7. **Headwater conservation**
   For the objective of improving the water resource conservation function of grassland, 133 ha of grassland in headwater conservation area was reseeded with 4,000 kg of seeds of locally *collected Elymus nutans, Agropyron cristatum* and *Poa pratensis*. 11 water points were built or maintained by purchasing cement and other materials and this ensures the clean water supply for 30,000 sheep and 2,000 yaks.
8. **High-quality sheep introduction**
   30 high-quality breeding bucks, including Suffolk, Dorset, Texel and Borderdale, were introduced to improve the native sheep and more than 100 crossbred sheep have been obtained and the economic efficiency was significantly increased. This activity greatly satisfied the local farmers.
9. **Warm shed construction**
   20 warm sheds, each 50 m^2, were set up in project village. Total construction area was 1,000 m^2.

(continued)

Box 14.6 (continued)

10. **Rural clean energy constructions**
    120 sets of solar dish for farmers and 30 sets of solar panel for herders were purchased. And bricks and cement were purchased for the build of biogas pit. These activities are aimed to reduce the fuel harvesting from natural vegetation and CO_2 emission, and meanwhile, to resolve the electricity supply for reading, TV watching and daily life.

Step 4: Participatory Implementation

Although most projects can pay the money to hire labour to complete the project activities, farmers who are involved in project implementation can also provide labour. Farmers are the master of their village, and they are very familiar with the situation and environment around their village, such as water resource, landscape and soil condition. A successful project not only involves the farmer to take part in the project design but also involves them to implement it. If a project keeps the farmers apart from the project implementation, the farmers will feel the project is not their project; even they do not care about the project implementation. So, inviting the farmers, involving them and respecting their opinion are guarantees of participation (Box 14.7).

Box 14.7 Case Study of Villager Involvement in Yongchang and Jingtai County, Gansu

Case Study 7:
Gansu/Xinjiang GEF project supports herders to fix the grassland fence and build warm sheds in Yongchang and Jingtai County. At the early stage, the project management office paid the money to a construction company to conduct the work. Originally, the project management office thought the quality of works should have been guaranteed because of the contract provisions. However, the result of supervision of the work was not good due to incorrect grassland fence route that hindered the animal road or prevented the drinking route of animals. The project management office figured out the reasons based on the investigation with farmers. Famers did not take part in planning the fence. The contract labour did not know the actual condition of grassland. After that, all civil work on the field must involve the local farmers. And if they provide labour as well, the project paid money to them. The farmers were happy and throw themselves into the project with great enthusiasm.

In another case study in Anding, the project supported the famers in pilot village to conduct livestock improvement based on PRA. The local project office purchased some breeding sheep for the famers. However, the farmers rejected the sheep because they were not good and some farmers even thought that some project officials were corrupt. The project management office changed the tactic on purchase of breeding sheep, which organized the farmers to participate in the whole process of purchase. Finally, the famers accepted the approach.

Step 5: Participatory Monitoring and Evaluation (M&E)

Participatory monitoring and evaluation is very important section of a project implementation. A project will be changed or improved depending on the result of monitoring and evaluation. How to get the right and useful information from monitoring and evaluations? This was a challenge for the project management. Although different ways of participatory monitoring and evaluation are introduced on some manuals, choosing the right one for the project depends on some local conditions, including human resource, technical difficulty and farmers' cooperation. In general, the participatory monitoring and evaluation should obey certain rules.

- Clear and operability indicators of M&E. Normally, the indicators of project M&E have three aspects, which include progress, outcome and impact. Whatever indicators, it should have a simple and clear target. Based on the targets, a feasible work plan for M&E can be made out. However, some M&E indicators of project were designed by some professional people. The M&E team often got confused by it. For example, in Gansu/Xinjiang GEF project, one of M&E indicators was 'Trends in adoption of improved grazing activities'. This indicator was a problem for the M&E team because they did not know what kind of data to reflect it.
- A useful approach is another key to a successful outcome. Many M&E approaches have been developed in recent years and some deliver a good outcome in the project management. Often enough, the M&E team cannot afford the time and patience to get useful information. However, the feedback and attitude of farmers are direct indicators for project M&E, especially on the progress and impact of the project. Many experiences showed the farmers are smart and careful about the minute change in their life or in their village. Trust farmers and make friends with them (Box 14.8).

Box 14.8 On the Spot M&E: An Example from Gansu

Case Study 8:
During the period of M&E for Gansu GEF project implementation, the project officer was in charge of part of project M&E, which were the progress and suitability of the project. The project officer always went to the pilot village and talked with the farmers, ate their food, slept on their bed and even drank some liquor with farmers. The purpose was to let the farmers trust the project officer and the project officer can master the real situation of project in field. A practical advice is to avoid interference the local government officials at this stage while the conversation between the project officer and farmers is ongoing because some farmers do not trust the local government officials. Another tip is that talking to more farmers is useful to get more information; do not only find one or two farmers to finish the M&E.

In Liangzhou, another GEF project county, in the first year, the local project management office conducted one of the project activities, which involved

(continued)

Box 14.8 (continued)

attracting eagles to control rodents in the pasture. Artificial perches were erected throughout the rangeland. During the M&E in the next year, the local farmers reflected to the project officer that the eagle stands were a waste of money because there are trees nearby and it was unnecessary to build an artificial perch for eagles. However, the local project management office had thought the project activity is useful because of some specialist's advice. Finally, the project activity was cancelled based on the careful analyses of the behavior of birds and rodents populations.

Step 6: Participatory Training Programme

The training supported by the project consisted of several kinds: (1) for farmers/herders in the field, (2) technicians at province level and county level and (3) overseas study tour and training. The training at provincial level was aimed at the technicians from county/district and was mainly concerned with the project management and techniques. The training at county level on rangeland protection and livestock production was not only for the members of the PMO and technicians from county bureaus but also included some farmers and herders. The training for farmers/herders in the project area focuses on demonstration and 'learning by doing' in field. The training was not inflexible and was varied according to the training needs evaluation. Farmers and herders were taken to other counties for study tours and for an opportunity to learn by seeing what others had achieved.

The monitoring and evaluation to the training effect was paid more attention in the project management office, based on the results of feedback and participant evaluation. The training approach was continuously improved, and the development of training materials made much progress. At the beginning of the GEF project implementation, the project bought some specialized books as training materials to conduct the technician training; however, the training was monitored and was not deemed to be useful. The major reason was that in the books, the knowledge that the technician wanted could not be found. As an alternative, the project coordinator went down to villages to carry out the evaluation of the needs of farmers/herders and technician as well. Based on the evaluation result, for the phase one of training, seven sorts of technique manuals were compiled for the technician, as well as five sets of multimedia disk, five sets of posters and one set of table calendars. For the farmers, one set of calendars and one set of information booklets were distributed to participants.

The training materials prepared for the first term were welcomed by a great number of technicians and farmers/herders and gave the compiler a good insight into what information should be included and how to present it to them when compiling the local training materials. The training materials in phase two were developed based on the result evaluated again from the training materials of phase one. The training materials of phase two aimed at the requirement of partial technicians want to master the knowledge of advanced livestock feeding and rangeland management, and the requirement of farmer and herder was to handle the simpler and easier-to-understand training materials. Therefore, in phase two, five books of newly developed technique materials (for technician) and ten wall pictures (for farmer and herder) were added.

Step 7: Environmental Education

Environmental education is another approach of capacity building for farmers. At present, the environmental educations focus on the water resource conservation, industrial pollutants emissions and cultivating people's good habits of living. However, the special environmental education for ecosystem conservation in pastoral area had not been tried.

The farmers' children on the rangeland have not experienced the air pollution, automobile exhaust gas, noise pollution, etc., which is happening in the cities. But they do know from their parents and from their own observations the changes happening in their hometown. The children can influence their parents, in particular raise their environmental consciousness and pass on to them better practices of environmental conservation. In addition, children in the pastoral areas have a problem of limited teacher resource. Even when there is a teacher, there are few teaching materials. The children need more understanding of the processes that are leading to changes on their rangeland, their livestock as well as impacts on living conditions in their hometown

There are three quite different aspects of environmental education – 'about', 'in' and 'for' the environment.

The framework 'about', 'in' and 'for' the environment is a popular way of organizing the experiences within an environmental education programme. In the process of the environmental education programme, the cooperation of volunteers, local officials and teachers of the primary and secondary schools specify the tasks of teachers and students, the teaching time, teaching place and group practical.

The programme was rolled out in selected villages in 8 GEF project counties. With the support of the World Bank/GEF project, the Grassland in Hometown and Love Our Hometown segment were completed in elementary and secondary schools in Gansu Province. Demonstration schools were in Sunan Yugur Autonomous County, Shibaocheng township of Subei County, Yinda township of Subei district, Pingshanhu township of Ganzhou district, Wagon village of Liangzhou district, Mayinggou village of Yongchang County, Xindunwan village of Jingtai Country and Xiangquan township of Anding district. The team consisted of an advisory consultant and volunteers from the College of Geography and Environment Science, Northwest Normal University, Lanzhou. The regional textbook and teachers' guide book were compiled and multimedia courseware for environmental knowledge was developed, and the environmental education model of community participation was formulated (Zhao and Squires 2009).

The environmental education is conducted step by step according to the sequence shown in Fig. 14.6 (Box 14.9).

Fig. 14.6 Simple solar kettles such as this were provided to cooperating households as a way to conserve precious fuel wood

Box 14.9 Training Programmes in Three Pilot Villages in Gansu

Case Study 9: Training Programs in GEF Project Pilot Village

Anding: Demonstrative training on altering lambing time for improving lamb survival rate and fattening efficiency, sheep fattening in pen-feeding way, silage making and feeding, ration formulation for sheep fattening, selecting harvest time for improving yield and quality of sainfoin (*Onibrychis* sp.) silage making and using legumes and urea multi-nutrient block on livestock supplementation.

Jingtai: Practical training on forage maize growing, silage making and feeding, roughage ammoniate technology, supplementing tan sheep with locally available feeds resource to determine ration formulation, greenhouse for tan sheep feeding and artificial insemination for tan sheep.

Liangzhou: Demonstration-based training on improving cashmere goat by crossbreeding, goat feeding in modified greenhouse, rotational grazing, degraded rangeland reseeding, land tenure reformat and improving rangeland environment, animal-feeds balance and monitoring on rangeland ecology.

Box 14.9 (continued)

Yongchang: Impact of group contract for rangeland tenure on animal-feeds balance and animal production efficiency and farmers' income, silage making and feeding, rangeland rodent and poisonous plant control, degraded land reseeding, ration formulation for sheep supplementation, goat abortion control, greenhouse-style warm pen for animal feeding and improving sheep productivity.

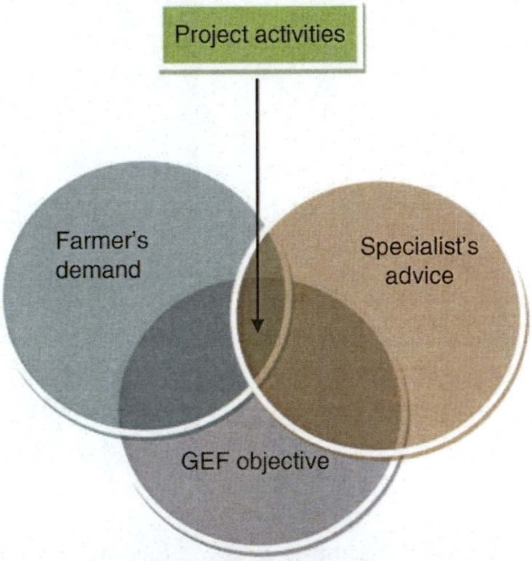

Fig. 14.7 The overlap of the requirements of different project stakeholders

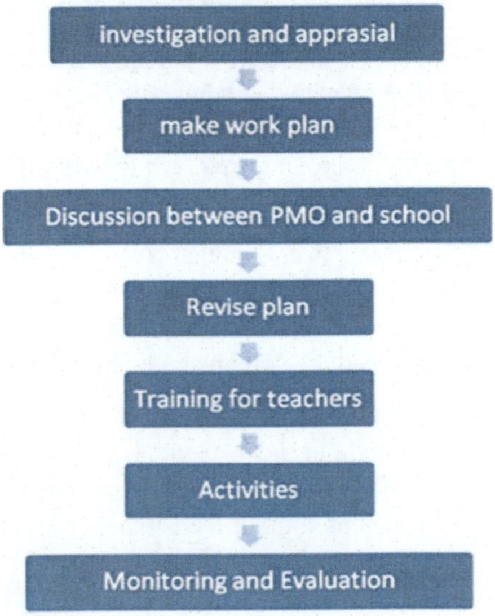

Fig. 14.8 The flow chart of the environmental education programme in Gansu GEF project

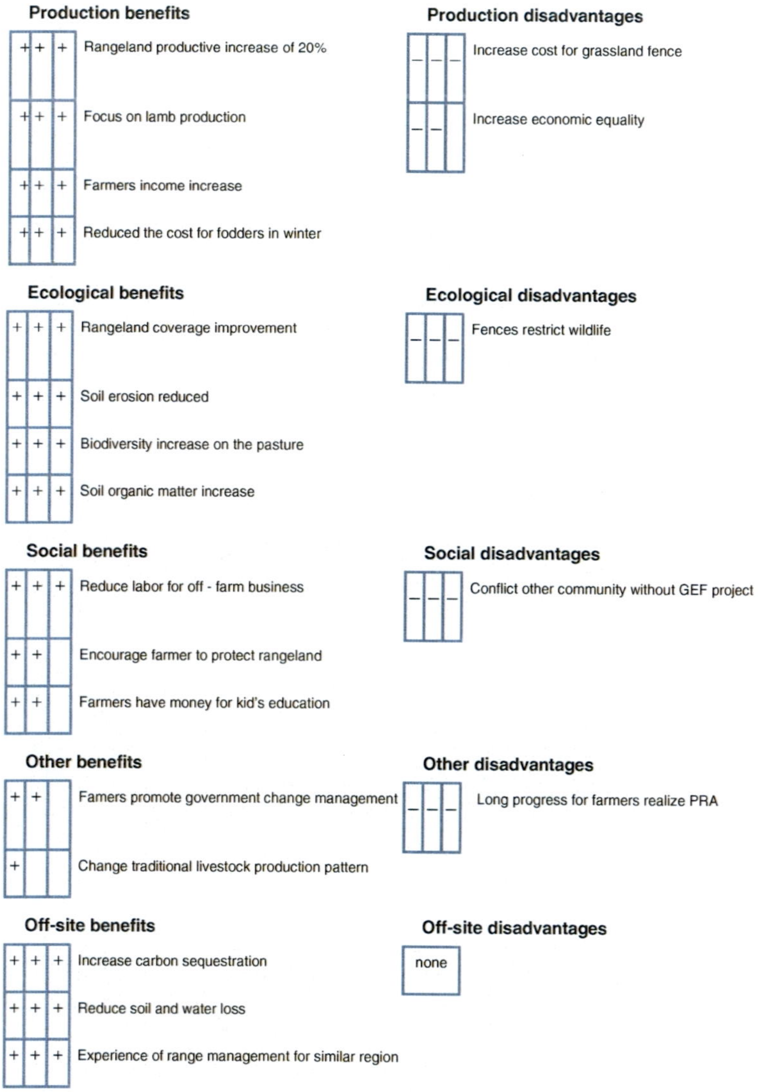

Fig. 14.9 Impact of the participatory approach for GEF project in Yongchang County. *Project benefits according to land user.* Using the format developed by WOCAT for assessment of project interventions, the following summary can be offered under four principal headings: (i) impacts of the technology, (ii) sociocultural benefits, (iii) ecological benefits and (iv) off-site benefits/disadvantages (Fig. 14.7)

References

Adebo S (2000) Participatory rural appraisal. http://www.myfirecommunity.net/discussionimages/NPost8220Attach1.pdf

Kurbanova B (2012) Constraints and barriers to better land stewardship: analysis of PRAs in Tajikistan. In: Squires V (ed) Rangeland stewardship in Central Asia. Springer, Dordrecht, pp 129–164 (Chapter 7, this volume)

Masseli D, Inam-ur-Rahim (2012) Central Asia's bio-physical context: key features, challenges, responses and opportunities. In: Squires V (ed) Rangeland stewardship in Central Asia. Springer, Dordrecht, pp 3–30 (Chapter 1, this volume)

Squires V, Hua LM (2010) Livestock husbandry development and agro-pastoral integration in Gansu and Xinjiang. In: Squires V, Limin H, Degang Z, Guoli L (eds) Towards sustainable use of rangelands in North West China. Springer, Dordrecht, p 384

Squires V, Lu X, Lu Q, Wang T, Yang Y (eds) (2009) Degradation and recovery in China's pastoral lands. CABI, Wallingford, 480 p

WOCAT (2007) Where the Land is greener: case studies and analysis of soil and water initiatives worldwide. In: H Linger, W Critchley (Eds). Co-published by CTA, FAO, UNEP and CDE

World Bank (2003) Gansu and Xinjiang Pastoral Development Project. World Bank, Washington DC. www.worldbank.org

Zhao CZ, Squires VR (2009) Environmental education: a tool for changing the mind-set. In: V. Squires et al, 2010 op. cit

Chapter 15
Fostering Community-Based Stewardship of Wildlife in Central Asia: Transforming Snow Leopards from Pests into Valued Assets

Rodney Jackson

Abstract Addressing human–wildlife conflict is an important requisite to managing rangelands for livestock and wildlife. Despite high altitudes, aridity, and relatively low primary productivity, the rangelands of Central Asia support a rich and diverse biodiversity—including the endangered snow leopard that many herders perceive as a predator to be eliminated. Conserving this and other wildlife species requires carefully crafted interventions aimed at curbing depredation losses and/or reducing competition for forage, along with offering locally sustainable, environmentally friendly income-generating activities for supplementing pastoral household livelihoods. This is best achieved through a combination of incentives designed to foster sound rangeland and wildlife stewardship, along penalties or disincentives targeting herders who violate mutually agreed rules and regulations (including grazing norms and wildlife disturbance or poaching).

When working toward the harmonious coexistence of people and wildlife, conservationists and rangeland practitioners need to seek the cooperation and build goodwill among herders and other stakeholders, including local government and private industry (especially the livestock production, mining, and tourism sectors).

Keywords Gurvan Saikhan National Park • Annapurna National Park • Nepal • Pakistan • India • Mongolia • China • Tibet • Mining • Poaching • PRA • Holistic • Community engagement • Fuel • Habitat fragmentation

R. Jackson (✉)
Snow Leopard Conservancy, 18030 Comstock Ave, Sonoma, CA 95476, USA
e-mail: rodjackson@mountain.org

V. Squires (ed.), *Rangeland Stewardship in Central Asia: Balancing Improved Livelihoods, Biodiversity Conservation and Land Protection*,
DOI 10.1007/978-94-007-5367-9_15, © Springer Science+Business Media Dordrecht 2012

Key Points

- The demise of the Soviet Union in the late 1980s and early 1990s precipitated profound changes to the formerly collectivized livestock and rangeland systems found across large expanses of Central Asia, in tandem with widespread and often catastrophic poaching of wildlife. The collapse of employment in the collectives, state-owned farms, and state-managed factories led to a surge in rural migration and the emergence of new herding households. For example, today, almost every small valley in southern and western Mongolia has a resident family, and this pattern is being repeated throughout Central Asia. These events were accompanied by substantial shifts in livestock spatial distribution patterns following the collapse of the relatively sophisticated, centralized support system which had created and maintained widely spaced groundwater wells and made supplementary feed available during prolonged periods of drought or heavy snowfall. A combination of failing management interventions and the introduction of a free market economy has led to rapid declines in the ecological condition and stability of rangeland resources.
- Central Asia's wildlife are well adapted to living at high elevations, tolerating climatic extremes, dealing with prolonged aridity and severe winters, while sharing scarce resources and minimizing competition through niche specialization, seasonal migrations, and habitat separation. The real long-term challenge lies with moving pastoral communities beyond their harsh and often insecure subsistence livelihood into more economically viable and environmentally friendly activities. Ultimately, local people must be encouraged to perceive snow leopards and other large carnivores as being worth "more alive than dead." They must also assume greater responsibility for protecting their flocks or herds from predators if we are to attain an enduring coexistence between predators and humans in the Central Asian rangelands.
- The real long-term challenge lies with moving pastoral communities beyond their harsh and often insecure subsistence livelihood into more economically viable and environmentally friendly activities. Ultimately, local people must be encouraged to perceive snow leopards and other large carnivores as being worth "more alive than dead." They must also assume greater responsibility for protecting their flocks or herds from predators if we are to attain an enduring coexistence between predators and humans in the Central Asian rangelands.

1 Introduction

The mountains and plateaus of Central Asia are remarkably rich in terms of its wildlife and diversity of habitats. The Chang Tang of Tibet and Mongolia constitutes the last strongholds of Central Asia's hoofed wildlife that includes Tibetan gazelle (*Procapra picticaudata*), Mongolian gazelle (*Procapra gutturosa*), chiru or Tibetan antelope (*Pantholops hodgsonii*), wild yak (*Bos grunniens*), kiang (*Equus kiang*), Asiatic wild ass (*E. hemionus*), Argali sheep (*Ovis ammon*), blue sheep

(*Pseudois nayaur*), ibex (*Capra sibirica*), saiga (*Saiga tatarica*), red deer (*Cervus elaphus*), white-lipped deer (*C. albirostris*), and musk deer (*Moschus* spp.). Carnivores include snow leopard (*Panthera uncia*), lynx (*Lynx lynx*), wolf (*Canis lupus*), dhole (*Cuon alpinus*), and several fox species (*Vulpes* spp.). Of special note are Mongolia's white-tailed gazelle migratory herds which until recently were said to rival the annual ungulate migrations of Africa's Serengeti plains (Mallon and Jiang 2009). The Tibetan antelope migrates across a vast landscape at elevations exceeding 4,500 m and sparsely populated by humans (Schaller 1998).

Central Asia's wildlife are well adapted to living at high elevations, tolerating climatic extremes, dealing with prolonged aridity and severe winters, while sharing scarce resources and minimizing competition through niche specialization, seasonal migrations, and habitat separation (Schaller 1998). Mountain areas support argali sheep, blue sheep, ibex, and snow leopards, while along the northern and western perimeters of Mongolia, one finds red deer, roe deer (*Capreolus capreolus*), and wild boar (*Sus scrofa*). The desert expanses in China, Kazakhstan, or Mongolia harbor the endangered wild camel (*Camelus bactrianus*), Gobi bears (*Ursus arctos*), wild ass, and Saiga antelope. There is a diverse array of small- or medium-sized rodents ranging from voles (*Microtus* and *Alticola* spp.) to jeboas (*Allactaga* spp.), gerbils (*Meriones* spp.), ground squirrels (*Spermophilus* spp.), and marmots (*Marmota* spp.), while hares (*Lepus* spp.) and pikas (*Ochotona* spp.) may constitute important prey items for medium-sized rangeland carnivores such as the red fox (*Vulpes vulpes*), Tibetan fox (*V. ferrilata*), corsac fox (*V. corsac*), Pallas' cat (*Otocolobus manul*), and lynx.

For centuries, humans have coexisted with wildlife practicing nomadic or semi-nomadic pastoralism herding sheep and goat flocks, cattle, horses, yaks, and camels. For example, Mongolia is among the world's leading pastoral nation, with rangelands covering 83% of the country's 1.29 million km^2 (Scharf et al. 2010). Thanks in large measure to the low sparsely distributed human population and their relative economic isolation, Mongolia and Tibet represent the last stronghold in Central Asia for many species of ungulates and carnivores considered rare, endangered, or declining elsewhere in their ranges.

The demise of the Soviet Union in the late 1980s and early 1990s precipitated profound changes to the formerly collectivized livestock and rangeland systems found across large expanses of Central Asia, in tandem with widespread and often catastrophic poaching of wildlife. The collapse of employment in the collectives, state-owned farms, and state-managed factories led to a surge in rural migration and the emergence of new herding households. For example, today, almost every small valley in southern and western Mongolia has a resident family, and the trend is being repeated throughout the uplands of the Central Asian region.

These were accompanied by substantial shifts in livestock spatial distribution patterns following the collapse of the relatively sophisticated, centralized support system which had created and maintained widely spaced groundwater wells and made supplementary feed available during prolonged periods of drought or heavy snowfall. A combination of failing management interventions and the introduction of a free market economy has led to rapid declines in the ecological condition and stability of rangeland resources, especially in Mongolia's southern desert steppes (Behnke 2006). Neglected water sources encouraged herders to concentrate livestock around the few remaining functioning wells and to stay longer than previously at a particular site. Seasonal mobility thus suffered, and many pastures have become notably degraded, including those near administrative centers. As noted by Schmidt (2006), a reversed rural-to-urban migration followed several years of severe winters (known as *dzud*) with a resulting catastrophic loss of livestock. However, the herds are now rapidly rebounding, with Mongolia recording a doubling of its livestock population to some 36 million animals (including sizeable increases in sheep and goats).

Revisions in pasture management laws invested in local political units (*soum* and *bag*), along with ill-defined or contested property rights, economic inequities, and a scarcity of water and winter forage, have all produced mixed results. The poorer households have been marginalized and communal respect for seasonal grazing restrictions and long-held pasture rest-rotational practices severely compromised (Fernandez-Gimenez 1999; Fernández-Giménez and Batbuyan 2004; Scharf et al. 2010). These changes, along with more restricted seasonable movements, are resulting in widespread overuse of fragile pastures due to extensive overstocking that in turn is leading to spreading desertification.

The ever-growing human and livestock population has also affected Central Asia's wildlife populations. The former coexistence between herders and wildlife is on the decline, accelerated by commercialization and increasing hunting pressure in response for rising demands for wildlife products, especially from China (Wingard and Zahler 2006). During the Soviet era, herding and grazing collectives or *negdels*, along with hunting brigades, held sole responsibility for rangeland and wildlife management. In Mongolia, a network of hunting brigades held exclusive rights to harvest wildlife and guarded against illegal hunters. Like the *negdels*, they were administered by local government and were outfitted with vehicles for conducting regular patrols.

Twenty-five thousand people participated nationwide, hunting a variety of furbearing species for export to the Soviet Union and other socialist states. With the shift to a market economy in the early 1990s, the hunting brigades quickly disintegrated from lack of state support. A combination of economic hardship and market access encouraged urban and rural dwellers to trade wildlife as an alternative income sources. For example, the price of marmot skins surged starting in 1995, the same year Mongolia relaxed its gun control laws (Scharf et al. 2010). These authors reported markets in provincial centers were swelled with guns, bullets, traps, furs, meat, antlers, and organs used for medicinal purposes. Wingard and Zahler (2006) estimated that up to 250,000 Mongolians were hunting wildlife for subsistence and the commercial market, with the value of wildlife trade (mostly illegal) estimated at US$100 million. In Mongolia's eastern steppes, Olson et al. (2011) noted the presence of herding households had a negative impact on Mongolian gazelle densities, concluding conservation largely depended on the ability of gazelles to make long-distance, unimpeded nomadic movements, as well as protection from harmful human activities.

In Tajikistan, a civil war followed the Soviet Union's collapse, leading to widespread strife and poverty. Doctors and engineers were forced into subsistence livelihoods, including animal husbandry (Bliss 2006). Along with collective workers, the Kyrgyz of the Pamirs were particularly hard hit, suffering severe winter livestock mortality following the withdrawal of supplementary feed from neighboring Kyrgyzstan. Populations of the famed Marco Polo sheep (*Ovis ammon polii*) plummeted due to widespread poaching, including use of automatic firearms by border guards who were no longer being sufficiently provided for by the state (Jackson, personal communication, 2003). Koshkarev and Vyrypaev (2000) noted widespread poaching of snow leopards in several Central Asia states following the breakup of the USSR.

China's skyrocketing economic growth is also leading to substantial changes across the high mountains and expansive basins of the Tibet Plateau, which is a noted habitat for large migratory herds of wild antelope, gazelle, and bovids. Such socioeconomic changes are fueling a fundamental shift from nomadic or seminomadic pastoralism to a more sedentary rangeland and livestock husbandry practice, in which land privatization and open-range fencing are becoming more prevalent. Besides disrupting wildlife's seasonal movements and its long-distance migratory patterns, such changes are leading to the collapse and replacement of traditional, communally managed, and ecologically adapted grazing management systems (Fox et al. 2009; Richard et al. 2006).

It is against this background that I present the example of the snow leopard, an endangered carnivore widely perceived as a menace to be eliminated by herders. This chapter discusses community-based conservation approaches that have been applied in India, Nepal, and Mongolia which are designed to transform local people's perception of snow leopards from a pest to a species valued more alive than dead.

2 The Setting

Central Asia—the Roof of the World—is characterized by high elevations; a harsh climate with extreme seasonal shifts of temperature, humidity, and precipitation; poor rocky soils; and steep slopes (Maselli, Chap. 1). It is among the world's most inhospitable landscapes where only the hardiest and enterprising of humans are able to survive. There are few roads, the lines of communication are primitive, and most households live in relative isolation for much of the year, making only short visits to nearby administration centers or more distant urban capitals.

It is also a home to the endangered snow leopard, perhaps the world's most elusive and charismatic large felid, that is sparsely distributed across 1.2–1.6 million km^2 of habitat in 12 countries of South and Central Asia, namely, Afghanistan, Bhutan, China, India, Kyrgyzstan, Kazakhstan, Nepal, Mongolia, Pakistan, Russia, Tajikistan, and Uzbekistan (Nowell and Jackson 1996). With an estimated total wild population of 4,500–7,500, snow leopards inhabit mountainous rangelands at elevations of 3,000 to over 5,000 m in the Himalayan and Tibetan Plateau, but as low as 600 m in Russia and Mongolia (Sunquist and Sunquist 2002). This species' habitat is among the least productive of the world's rangelands due to low temperatures, high aridity, extreme seasonal conditions, and a harsh climate. The average peak graminoid biomass has been estimated at 170 kg ha^{-1} (asymmetric 95% CI 128–228 kg ha^{-1} (Mishra 2001)). Consequently, prey population densities are also relatively low, ranging from 6.6 to 10.2 blue sheep per km^2 in productive habitat in Nepal (Oli 1994) to 0.9 ibex per km^2 in marginal habitat of Mongolia (McCarthy et al. 2005). Supplemental prey items include marmots, pikas and other small rodents, and game birds. As noted later, livestock can be an important component of the snow leopards' diet.

Besides its naturally low density, the primary challenges for conservationists are its elusive and cryptic behavior, large home range (in excess of 650 km^2 in Mongolia, unpub. data), and dependence upon ungulate prey populations that are mostly declining due to widespread poaching (Jackson and Ahlborn 1989). With the depletion of its prey, snow leopards may resort to feeding on killing livestock, in turn encouraging herders to resort to retributive killing of predators (Jackson et al. 2010).

The predominant land uses and source of local livelihoods in most of the snow leopard's habitat revolve around traditional pastoralism, with agro-pastoralism at lower elevations, especially in the Himalayan region. According to Mishra et al. (2003), 7 of the 12 range countries contain over 25% of land area in permanent pasture, with 50% or more of the human population engaged in agro-pastoralism. Over 40% of people live below national poverty levels with an average per capita annual income of US$250–400. While relatively few humans reside in snow leopard habitat, their use of the landscape is pervasive, leading to generally escalating levels of human–wildlife conflict, especially within or adjacent to protected areas where resources are more strictly controlled. Perhaps as much as 50–60% of the snow leopard population and sympatric wildlife inhabit areas outside of protected areas, albeit at lower densities (Jackson and Ahlborn 1990). These populations, however, are critical to sustaining genetically viable metapopulations (unpublished data). Since few, if any, of the approximately 100 existing protected areas situated within snow leopard range are free of human influence (Green and Zhimbiev 1997), its survival hinges upon uneasy coexistence with subsistence pastoralists and farmers trying to eke out a living under the same harsh environmental conditions. Thus, innovative and decidedly participatory approaches for engaging pastoralists in snow leopard and biodiversity conservation are needed to ensure the species' long-term survival. This is especially relevant where mountain ranges are more fragmented and habitat less productive, as in Mongolia's South Gobi region (this author, unpublished data).

This brings us to the vexing problem of how best to empower communities and decision-makers to respond effectively and fairly in light of the continually shifting socioeconomic conditions brought about by powerful political and economic changes sweeping the region (Kurbanova, Chap. 7). Of special note are the far-reaching social, economic, and environmental changes associated with the dissolution of the Soviet Union and specially the states of Tajikistan, Uzbekistan, Kyrgyzstan, Kazakhstan, and Mongolia (and even Russia itself) as they emerged in the 1990s from decades of centrally controlled socialism to the open and highly competitive global economic marketplace. The examples of the Pamirs in Tajikistan and steppes of Mongolia have been the focus of a number of academic studies (e.g., Bliss 2006; Kreutzmann 2012).

3 The Conflicts

In order to better understand how communities could become effective custodians of land in a rapidly changing environment, we need to examine the sources and root cause(s) of human-habitat–wildlife conflict (Treves et al. 2006). Here, we are using the example of the endangered snow leopard as a flagship species for how rangeland managers and conservationists could work with pastoral communities to ensure a more harmonious future for wildlife of Central Asia.

3.1 Competition for Forage and Habitat Disturbance

Recent studies suggest the density and diversity of wild herbivores in the cold deserts of the Trans-Himalaya may be depleted through resource limitations imposed by competition with domestic livestock, especially where dietary overlap is extensive (a condition perhaps exacerbated by the low productivity and plant species diversity characteristic of this part of the world).

Bagchi et al. (2004), for example, found that sheep and goat compete with ibex for forage, often excluding ibex from using pastures if flocks were accompanied by shepherds and/or their dogs. These investigators estimated that livestock such as sheep, goat, horse, cattle, and yak removed large amounts of forage from pastures (up to 250 kg of dry matter per day by certain species). Mishra et al. (2001) concluded that most rangelands in Spiti (northern India) were overstocked with domestic herbivores that amounted up to 10 times the biomass of wild herbivores. Livestock herding was possibly implicated in the local extinction of four wild ungulate species: ibex, free-ranging horses, and sheep and goats utilized the same habitat and shared a common diet, thus lacking in forage differentiation necessary for unimpeded coexistence. Supplementary winter feeding helped to elevate livestock numbers, leading to overstocking especially during the critical winter season.

Retzer (2006) confirmed forage competition between livestock and the Mongolian pika (*Ochotona pallasi*), which he concluded was able to harvest forage more closely than domestic stock. Bagchi et al. (2006), working in the Trans-Himalaya, found that small mammalian herbivores served as mediators of plant community dynamics in high-altitude arid rangelands. The policy of eradicating pikas and voles in China may compromise ecosystem functioning and species diversity (Smith and Foggin 1999). However, these authors like others also place blame on changes in traditional pastoral practices and overstocking as the root cause for rangeland degradation, including desertification. Thus, conservation interventions should seek ways for fostering coexistence between humans and livestock (Mishra et al. 2009; du Toit et al. 2009).

3.2 *Livestock Depredation*

Given the prevalence of subsistence agro-pastoralism across most snow leopard range, conflict between herders and cats (including other predators like the wolf) should come as no surprise (Jackson and Nowell 1996). Depredation rates, however, vary widely from under 1% in parts of Mongolia and western China (Schaller et al. 1987, 1994) to over 12% of livestock holdings in hot spots in Nepal (Jackson et al. 1996) and India (Mishra 1997; Bhatnagar et al. 2006), but they typically average 1–3% (Oli et al. 1994; Schaller et al. 1988; Namgail et al. 2007; Bhatnagar et al. 1999). These studies also suggest that depredation tends to be highly site specific, with loss rates varying greatly between successive years and even among nearby settlements over the same time frame. Herders are particularly angered by "surplus killing," or incidents associated where a snow leopard enters a poorly constructed night-time corral and then kills 10–50 (or more) of the sheep and/or goats confined within its walls (Jackson and Wangchuk 2001). In Hemis National Park, India, for example, such incidents amounted to only 14% of all depredation events but accounted for 38% of all livestock lost (Bhatnagar et al. 2006). Such incidents appear to be closely associated with most herder retribution toward the snow leopards (e.g., Jackson and Wangchuk 2001; Hussain 2003), which are perceived as being responsible for wiping the herder's entire "bank account."

Studies to date indicate that the annual economic losses associated with depredation events have ranged from about US$50 to nearly US$300 per household, a significant sum relative to the per capita annual household income of US$250–400 (Oli et al. 1994; Jackson et al. 1996; Mishra 1997; Bhatnagar et al. 2006; Ikeda 2004; Namgail et al. 2007). Typically less than 10% of households suffer disproportionate loss, usually from corralled sheep and goat kills, or when unguarded, but high-valued yaks and horses are killed on the open range (Jackson et al. 1996). Complacent guarding, poorly constructed night-time pens, favorable stalking cover, and insufficient wild prey are cited as the primary factors contributing to livestock depredation.

Snow leopards are capable of killing nearly all types of domestic animals, and while herders typically take measures to reduce the risk of depredation, these are often insufficient. As a result of wildlife poaching and overstocking, both livestock numbers and biomass are usually an order of magnitude greater than that of wild ungulates. In Nepal, for example, livestock biomass may reach 1,700 kg km^2 (Jackson et al. 1996) compared to 330 kg km^2 for blue sheep in the same season (Oli 1994), so the probability of a snow leopard encountering a domestic animal is often high. Bagchi and Mishra (2006) reported higher livestock (58%) proportions in snow leopard diet in an area with more livestock (29.7 heads km^2) and fewer wild ungulates (2.1–3.1 blue sheep km^2) in comparison to an adjoining area stock with less livestock (13.9 km^2) but containing more wild ungulates (4.5–7.8 ibex km^2) and where livestock constituted 40% of the leopard's diet. A more recent food habit study drawing upon genetically confirmed scats indicated that livestock provided up to 70% of the diet of snow leopards in northern Pakistan (Anwar et al. 2011).

These data highlight the importance of livestock as prey, at least for some snow leopard populations. It also confirms the potential role pastoralists could or indeed are unwittingly playing in helping to sustain this highly endangered species by allowing some depredation losses.

The relative abundance of livestock and wild ungulate prey is considered a reasonable predictor of livestock depredation risk (Bagchi and Mishra 2006). Other indicators include the distance to snow leopard travel lanes (e.g., ridges and cliffs) and broken terrain ("depredation hot spots"), along with lax guarding by herders, in part resulting from insufficient manpower or funds for hiring communal shepherds (Jackson et al. 1996). Although losses herders attribute to wild predators are often exaggerated (compared to other sources of mortality like disease and large scale die-offs due to occasional severe winters), it is likely the *perceived* level of depredation mostly drives negative attitudes and subsequent reactions toward wild predators (Oli et al. 1994; Mishra 1997; Jackson and Wangchuk 2001). As shown by Bagchi and Mishra (2006), negative attitudes may have a strong economic basis even in culturally similar areas; thus, communities with more access to alternative income were found to display greater tolerance toward snow leopards, despite losing on average 1.1 heads of livestock per family annually. A nearby community heavily dependent upon animal husbandry, however, held more negative feelings despite losing fewer livestock (0.6 animals). As noted earlier, the snow leopard's habitat tends to be overstocked with livestock, in some cases with herd densities compromising animal production itself (Mishra et al. 2001). The resultant competition for limited forage, along with human disturbance and poaching, is often cited as a cause for natural prey population declines. As livestock numbers increase, so natural prey populations tend to decrease, as illustrated in one study where blue sheep density was 63% lower (2.6 blue sheep per km^2) in a rangeland supporting 30% more livestock than an otherwise comparable area sustaining 7.1 blue sheep per km^2 (Mishra et al. 2004). Range-wide, declining prey populations are a serious and chronic threat to snow leopards, along with the cascading effect of escalating livestock depredation rates leading to intensified herder retribution.

There are a number of other issues which contribute to pastoralist–wildlife conflict and are worth mentioning briefly (with detailed discussion beyond the scope of this chapter):

- Wildlife–livestock disease transmission: governments are particularly concerned about outbreaks of foot and mouth disease, since these can have a major influence on livestock populations as well as marketing opportunities for livestock products. However, it is a little-studied topic, although a number of studies are currently underway in Central Asia (Christine Budke, personal communication).
- The current understanding of snow leopard distribution patterns, abundance, and feeding ecology is quite limited. This is attributed to the technical, demanding logistical, and high financial costs of studying a rare, cryptic carnivore that inhabits some of the most remote and rugged habitat on earth (Jackson and Fox 1997). Hopefully, useful information will be forthcoming from the long-term study currently underway in the South Gobi of Mongolia, although this study site

is rather atypical with respect to habitat productivity, prey abundance and utilization, and genetics and meta-population dynamics at a landscape level. On the other hand, conservationists can ill-afford to delay action, given the rapidity with which wildlife is disappearing across vast swaths of Central Asia, notably in the Pamir Mountains and steppes of southern, western, and eastern Mongolia (Wingard and Zahler 2006; Scharf et al. 2010). Rather, rangeland and wildlife managers need to make prudent decisions based on the best practices and the available information, implementing these under a framework which includes regularized monitoring and adaptive management.

- The remoteness of most areas from consumer markets, along with high transportation cost and daunting logistics, means that those herders located close to administrative centers have distinct competitive advantages over herders living in more remote areas (and which are more likely to constitute core snow leopard habitat).
- Mineral and oil exploration reaches far into formerly remote areas, as governments and the private industry respond to the rapidly growing national and international demand for energy and mineral resources like coal, copper, uranium, and rare earth minerals. While the rapidly expanding network of roads and even railroads facilitates economic development, improved access also allows more outsiders to poach wildlife.
- As noted above, urgent issues relating to land tenure and resource rights need to be addressed (Robinson, Chap. 11, Halimova, Chap. 13, and Leake, Chap. 18).

Salafsky and Margolius (1999) described a threat-based model for conceptualizing key underlying threats affecting species, habitats, or ecosystems, establishing desirable conservation targets and conditions, along with necessary interventions for achieving these goals. Figure 19.2 from Jackson et al. (2010) summarizes the key threats and interventions for ensuring a viable population of snow leopards and prey species, including local support and protection from poaching and illegal trade. It is reproduced here as Fig. 15.1.

Clearly, managing human–wildlife conflict is critical, requiring reduction of livestock depredation levels through better animal husbandry and wild prey population and habitat restoration as well as devising sustainable means for offsetting or sharing in the economic cost of coexisting with a large carnivore. Given the snow leopard's endangered and elevated protected status, traditional predator control strategies do not appear an option.

Rather, incentive programs for garnering local community support are imperative; it is largely when tangible economic returns are realized that rural communities are willing, indeed able, to assume their role as conservation partners and become effective stewards of their surrounding environment (Western et al. 1994; Jackson and Wangchuk 2001; Mishra et al 2003).

I will next focus on participatory approaches to community-based stewardship of snow leopard and other wildlife, drawing upon my experience in Tibet's (China) Qomolangma Nature Preserve (Jackson 1998, 2001) and work in the Trans-Himalayan region. Here, I supervised the implementation of a grassroots pilot program in Ladakh (Jackson and Wangchuk 2001) with the goals of (1) reducing livestock loss by

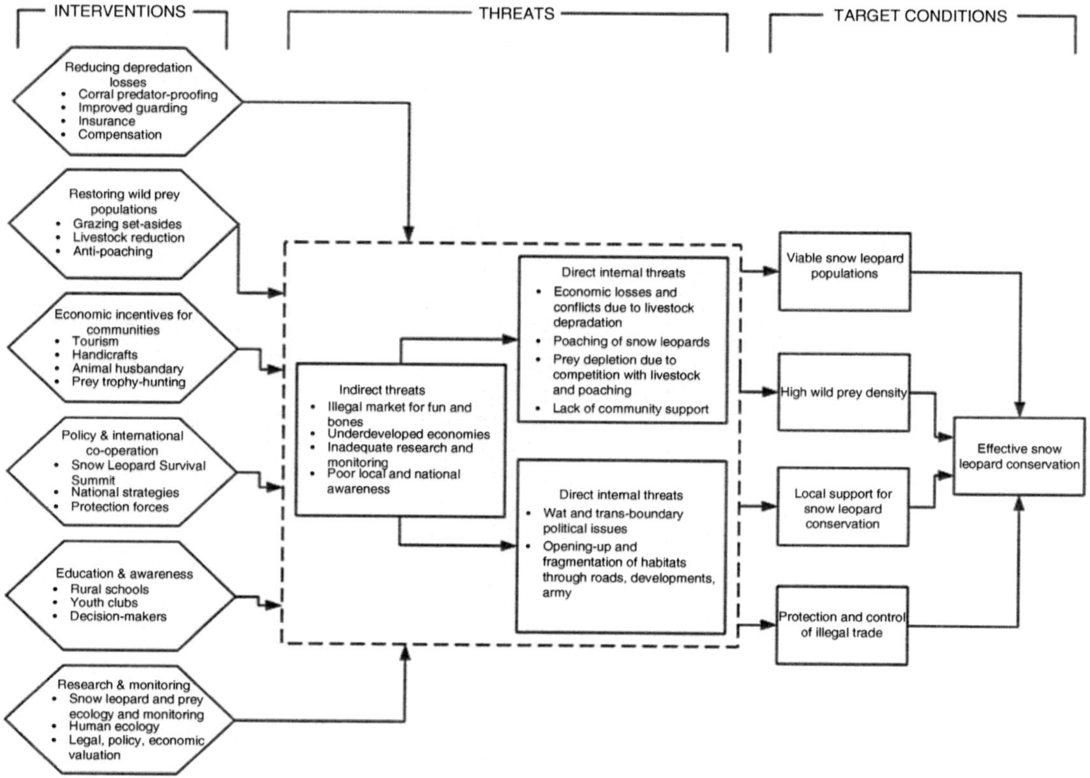

Fig. 15.1 A conceptual threat-based model for snow leopard conservation (Figure 19.2 from Jackson et al. (2010)

predator-proofing night-time pens and improving daytime guarding practices, (2) enhancing rangeland habitat and prey populations through community-based stewardship and sustainable resource management, and (3) increasing household incomes to help offset unavoidable depredation losses.

4 Community Engagement

The process of community engagement and planning is outlined below (see also Kurbanova, Chap. 7 and Hua, Chap. 14 for other examples of community participation).

4.1 APPA: A Holistic Process for Engaging Pastoralists in Rangeland Management and Biodiversity Conservation

Communities were engaged using a highly participatory process pioneered by the Mountain Institute in Tibet and Sikkim in the late 1990s and known as Appreciative Participatory Planning and Action or *APPA* for short (Jackson 1998; Jackson and Wangchuk 2001). *APPA* builds upon the community's interest, motivation, and available resources for improving resource management and enhancing income generation through an integrated set of low-cost, locally appropriate interventions. Using simple Participatory Rural Appraisal or PRA tools, the facilitator "leads" villagers and other stakeholders through a sequential four-step reiterative process termed the *4-D's*, namely, *Discover, Dream, Design*, and *Delivery*. It takes advantage of local people's wealth of traditional knowledge relating to animal husbandry, predator occurrence, habitat conditions, and animal behavior and then blends this with information derived through more structured, rigorous, and tightly focused scientific or "problem-solving" approaches, outlined below:

Step 1 Discovery—Learning About Livestock Herding Practices and Identify the Community's Strengths and Valued Resources

The most effective remedial actions emerge from a sound understanding of the root causes for depredation, which in turn requires knowledge of how people manage their domestic herds and their rationale for adopting favored strategies, including the role played by local institutions and the supporting decision-making procedure(s).

Common participatory rural appraisal or *PRA* tools were deployed to encourage learning from pastoralists for a better understanding of the root cause(s) for livestock depredation, livestock guarding practices, and other driving factors behind the generally negative perception of large predators among herders. These tools, which do not require special educational grounding, enabled planners and villagers to gather and then evaluate diverse information on existing conditions affecting rangeland use, herd productivity, and depredation risk. Table 15.1 provides some

Table 15.1 Examples of *PRA* tools used for appraising livestock depredation and animal husbandry patterns

Map of valued natural resources and village assets
Map of seasonal pastures and depredation "hotspots"
Calendar of seasonal livestock movements and daily herding cycle
Calendar of seasonal depredation losses (shows peak depredation periods)
Pair-wise matrix ranking of major sources of livestock mortality
Pasture ranking with respect to depredation and other losses
Ranking of different guarding measures
Income and livelihood ranking matrix
Semi-structured interviews to assess predation causes and patterns, along with possible remedial actions
Venn diagram showing village institutions affecting livestock production and management
Village or pasturage transects to obtain firsthand understanding of livestock management practices and issues

examples of the type of information generated and that helped forge mutual understanding vital to coming up with remedial measures which livestock owners and the community could implement and manage with minimal outside technical or financial inputs; the reader is referred to the online manual for more information (http://www.snowleopardconservancy.org/pdf/community_engagement_handbook.pdf).

In nearly all cases, these exercises conducted implicated poorly constructed livestock pens and lax daytime guarding practices as the primary cause of depredation. Livestock were allowed to forage, poorly or completely unguarded, in areas with well-broken terrain and cliffs that constitute prime habitat for the snow leopard (e.g., Jackson et al. 1996). Historically, there was better emphasis on daytime guarding, and problem predators were controlled by trapping or other traditional control methods no longer permitted. With children at school and the youths increasingly reluctant to assume animal husbandry, even highly vulnerable small-bodied livestock were left unguarded for extended periods of time.

Step 2 The Dream—Envision the Community's Short- and Long-Term Futures Were Necessary Resources Suitably Mobilized and It Acted in Concert

In the second phase, participants envisioned (i.e., visualized) what their settlement could realistically "look like" over the short-term (1–3 years) and long-term (5–10 years) time frames if a majority of its members acted collaboratively to reduce predation loss, protect snow leopards and other wildlife, and successfully enhance household incomes. Collective discussions broadened the framework upon which stakeholders could help devise interventions for achieving the preferred objectives, including harmonious coexistence of people and wildlife. The overwhelming majority of herders agreed that existing corrals were inadequate, having low, flimsy walls that easily allowed snow leopards to enter the structure. The highest ranked interventions included predator proofing of night-time corrals to prevent future multiple attacks, followed by protection of the natural prey base and herder education for improving daytime guarding practices. With predation on the open range being virtually impossible to eliminate, project sponsors focused on ways of enhancing income generation.

Step 3 Design—Design an Action Plan to Guide the Community's Development and Encourage Nature Protection While Minimizing Long-Term Dependency upon Outside Financial Sources and/or Technical Expertise

Stakeholders were asked to follow "best practices" design guidelines for better ensuring that remedial actions are (1) environmentally responsible, (2) economically sustainable within the local context, (3) socially responsible (i.e., build on tradition and cultural values compatible with the protection of nature), and (4) implemented under a mutually agreeable and communally signed "action plan" that sets forth specific responsibilities, contributions, and obligations of each partner (Jackson and Wangchuk 2001) (Table 15.2).

These guidelines encourage stakeholders to blend traditional knowledge with external expertise and scientific information so that interventions comply with protected area regulations, yet are also integrated with the longstanding pastoral system. Drawings and plans prepared in the field encourage all livestock-owning households to design and then build secure and cost-effective livestock pens, especially known as depredation "hotspots." The typical improved sheep or goat corral measures 6×10–15 m with a 2.5-m stone drywall and 10-×110-cm wire-mesh cover over the roof and supported by wooden cross-poles for preventing a snow leopard from jumping into the enclosure. Each structure is provided with suitably protected (or small) air vents and a well-fitting wooden door which can be securely closed at night. These materials (wire mesh, cable fasteners, poles, door, and door frames) typically cost between $400 and $1,500 (USD) depending upon the dimensions of the structure and transportation to typically remote site via horse or yak. The beneficiary community contributes by providing all mud and stones for drywall construction, along with needed labor for constructing the corral and then maintaining it in good repair.

Stakeholder consensus on setting realistic targets for alleviating conflict is also important. Given limited availability of labor and costs involved with hiring shepherds, there is no easy way to avoid depredation on the open range. Large-bodied stock such as horses, yaks, and cattle crossbreeds must roam widely when foraging and are only rarely tended by a shepherd. These animals are especially prone to snow leopard predation in the winter when they are the weakest due to poor nutrition and when escape can be impeded by deep snow. Herders are offered other incentives, such as training revenue-generating skills to address this type of human–wildlife conflict.

Step 4 Motivating the Participants—To Initiate Improvements Immediately and Largely on Their Own Rather than Waiting Until an Undefined Time in the Future

Exploring how wildlife and biodiversity conservation can directly or indirectly benefit local people is a key objective of the community planning session. Posters which visually illustrate alternative rangeland management and livelihood opportunities, including tourism, help foster debate and build ownership among stakeholders who focus on improving existing activities rather than trying to establish unfamiliar activities or businesses. For example, villagers could be encouraged

Table 15.2 Conditionality and best practices for community-based wildlife conservation interventions

(a) Project conditionality

Projects are most likely to succeed if its sponsor and beneficiary communities endorse and include the following conditionality within the project's implementation and operational framework:

Ensure Tangible Conservation Results and Benefits for People, Rangelands, and Wildlife—Project activities should be implicitly linked with snow leopard and mountain biodiversity conservation (i.e., must have positive impact and not adversely impair species, habitat or rangeland resources). Specific and clearly defined actions are needed to benefit snow leopards, their prey, and habitat as well as helping to improve local people's livelihoods and income-generating opportunities

R*equire Reciprocal Investment by Participating Organizations or Institutions*—Each stakeholder (whether villager, NGO, or government) must make a reciprocal (cofinancing) contribution, within their means, in support of the agreed-to project actions or activities. This may be in the form of cash or in-kind services like materials and labor, which are valued using existing market rates and prices

Full Participation by All Involved Parties—There must be strong commitment to active and equitable participation from each involved stakeholder group throughout the life of the project (from planning to implementation, monitoring, and evaluation and reporting). In addition, project-supported activities should benefit as many households as possible (and especially those that are more marginalized)

Responsibility for Project Facilities—The beneficiary community must be willing to assume all or a significant responsibility for repairing and maintaining any infrastructural improvements (e.g., predator-proofed corral) that may be provided by the project

Monitoring and Compliance Performance—Stakeholders should be willing to employ their own simple but realistic indicators for measuring project performance and impact, according to an approved community monitoring and evaluation plan. Similarly, the project donors or implementing agency will monitor project activities, outputs, and results

(b) Design and operational best practices (Adapted from Jackson 2001)

Biodiversity conservation and livelihood enhancement actions and their associated activities should comply with the following general "best practices" guidelines:

Environmentally sound—Measures should result in no or only very minimal harm to plant or animal species, habitats, or ecosystems and preferably increase the abundance of key or endangered species, restore degraded habitats, and sustain natural ecological and biodiversity-sustaining processes

Economically sustainable—Control actions should be affordable, contain cost-sharing mechanisms, and be capable of being sustained with minimal outside cost and technical input

Socially responsible—Measures should build upon proven traditional customs and good animal husbandry or pastoral practices

Based on good science and adaptive management—Projects should benefit from the blending of modern science and traditional knowledge while being driven by principles of adaptive management

Imbedded with clear responsibilities and a transparent budget—Implemented based upon a signed agreement and *action plan* that clearly sets forth the responsibilities and contributions of each partner, in accordance with a mutually agreed-to work plan and budget. *The work plan should specify details such as where (location), who (responsible party), what (inputs/activities), how much (quantity), when (scheduling), how implemented (method), and how monitored (indicator and process to be used)* to improve their capacity for capturing more of the revenue from existing tourism activities, but without unduly increasing dependency upon this fickle sector. Improved marketing of livestock products, the production of high-quality cashmere wool, and the production of handicrafts represent other income-generating options that could be explored.

Examples of Community-Based Conservation Actions: In this section, I present examples of community-managed conflict alleviation and incentive conservation programs.

The most important steps toward addressing human–snow leopard conflict involve reducing livestock depredation and offsetting costs (real or perceived) of resulting losses. Better antipredator livestock management represents an important means of reducing livestock depredation by wild carnivores. Predator-proofing corrals have been shown to significantly reduce losses to snow leopards (Jackson and Wangchuk 2001). This simple and effective conservation partnership of providing predator-proofing materials to local virtually eliminates multiple depredation incidents, removes long-held herder animosity toward this large cat, and helped lower livestock in its diet from over 30% to around 11%. To date, no depredation losses have resulted from properly improved livestock pens, although there have been a number of visitations and attempts at entry by snow leopards. The community's sense of ownership and satisfaction with this approach contrasts with outside agencies who built similar pens in Hemis National Park without seeking significant community involvement during site selection or design. One such corral was situated at the base of a cliff where snow leopards could gain easy access, and not surprisingly, this corral remains unused by the community that it was intended to benefit. The importance of engaging communities on equal terms as planners cannot be overemphasized.

Reducing livestock losses on the open range is far more challenging. Vigilant daytime herding by communally hired shepherds is one option, especially if infused with cash rewards for herders demonstrating better shepherding with fewer depredation losses in their pastures. As noted, low wild ungulate and high livestock numbers may lead to increased livestock depredation by snow leopards and sympatric carnivores. Besides antipoaching initiatives, ungulate recovery can be facilitated by establishing grazing set-asides on community-used pastures, where grazing and resource extraction are curtailed seasonally or on a more prolonged basis. This approach enabled the blue sheep population to double in one area in India (Charadutt Mishra, unpublished data) that together with improved herding practices has led to livestock depredation losses declining from 12 to 4% of holdings annually (Mishra 1997).

However, balancing predator–prey populations is not necessarily simple, as indicated by events following the natural reoccupation of Mt. Everest National Park by snow leopards following nearly 30 years of their absence from the ecosystem. Here, researchers observed the number of Himalayan tahr (*Hemitragus jemlahicus*) declined by nearly 70% within a few years of the leopard's return, which was attributed to heavy and sustained kid predation (Lovari et al. 2009). This suggests that elevated snow leopard numbers and/or use of the area follow better nature protection, leading to intensified predation upon wild prey and possibly eventually followed by increased livestock depredation losses over the long run.

An ungulate trophy hunting program in Pakistan generated substantial economic benefit, with 80% of the US$25,000 trophy fee being distributed to local communities and amounting to as much as US$150 per household. Concurrently, the number of markhor (*Capra falconeri*) increased due to stringent community-imposed curbs on poaching. However, a negative outcome of such protection occurred when one beneficiary community demanded monetary compensation for snow leopards preying "their valuable wild ungulate" (Hussain 2003). These two examples underscore the importance for establishing multipronged conflict management programs along with emphasizing wild prey population recovery should be accompanied by more holistic range management and better livestock protection.

Livestock are inherently vulnerable to depredation given their reduced antipredatory abilities: depredation loss may ensue even given vigilant herding practices and high wild prey densities. Therefore, addressing human–carnivore conflict by offsetting or sharing depredation-related economic impacts may be a more promising long-term strategy. Compensation, an accepted predator conservation strategy in many parts of the world, faces numerous challenges in this remote and heavily rural region. Low compensation amounts (offsetting only 3–35% of the financial losses), false claims, corrupt disbursement officials, difficulty in authenticating claims, and bureaucratic apathy often associated with state-run programs (Mishra 1997; Jackson and Wangchuk 2004) have inadvertently elevated local resentment toward protected areas harboring snow leopards. A variant approach whereby the communities themselves manage funds, verify claims, and disburse payments appears more promising (Hussain 2000; Mishra et al. 2003). With local management by village-level committees and family-paid premiums, ownership is strengthened along with leveraging internal peer pressure against corruption or false claims. Nonetheless, fund capitalization remains the most important obstacle, as exemplified by a conservation organization in Nepal that had to invest $60,000 to establish a livestock insurance-compensation fund to serve some 45 households in a protected area (Gurung et al. 2011).

Livestock losses to disease (often exceeding 50%) typically exceed the loss to snow leopard. In Chitral (Pakistan), for example, depredation loss amounts to less that 0.5% of the average herd, so the Snow Leopard Trust launched a pilot livestock vaccination program in 2003. In exchange for tolerating depredation, some 1,500 livestock were vaccinated against common diseases which declined by 90%. Participants agreed to cease persecution of snow leopards, reduce their livestock holdings, and improve fodder handling methods in order to increase forage availability for wild herbivores. This program hopes to create sufficient economic incentive by increasing livestock survival and productivity in order that sales of excess animals will bring each family US $400 or more income per annum.

The most promising avenues for offsetting the economic burden from livestock depredation rests with providing herder households the necessary resources and skills training to enable them to generate alternative or supplementary income from activities like tourism or sale of handicrafts. A prime example is the largely self-sustaining *Himalayan Homestays* incentive program in India which builds upon existing tourism and trekking to enhance livelihoods for local people and garner support for snow leopard conservation (Jackson and Wangchuk 2004). This UNESCO-supported initiative provides villagers training in homestays and nature guiding, with bookings facilitated by local travel agents. Individual households, operating through women's groups, accrue revenue from "bed and breakfast" stays by tourists in village homes (rotated among households), catering and handicraft sales at tented cafes on trekking routes, and nature guiding.

The Snow Leopard Conservancy established the first traditional homestays in 2003 in Hemis National Park, premier snow leopard protected area. Currently, some 100 households in 20 communities participate, with those operators situated in prime snow leopard habitat earning US$100–650 (average US$230) during the 4-month tourist season (unpublished data). Tourist visitation increased from 37 in 2001 to >700 by 2006. Client satisfaction exceeded 85% and tourists welcome the cultural interaction. Another US$400 in sales from cafes is shared among 4–8 families. Approximately 10–15% of homestay profit goes into a village conservation fund which has supported tree planting, garbage management, and recently the establishment of a village wildlife reserve for the threatened Tibetan argali (*Ovis ammon hodgsonii*). One community constructed predator-proof corrals, another paid a full-time herder to guard livestock in high summer pastures, and a third insured large-bodied, high-valued livestock like yak through a national livestock insurance program.

Mongolia's Snow Leopard Enterprises program generates income for herder–artisans through handicrafts sales, in exchange for community support of snow leopard conservation (Mishra et al. 2003). Artisans are trained and given simple tools to develop culturally appropriate woolen products which are marketed by an international NGO in the west. In exchange, communities sign conservation contracts stipulating a moratorium on snow leopard and wild ungulate poaching; contract compliance provides herders with a 20% bonus above the agreed price of their products and which is shared between artisans and the community's conservation fund. Violation of this contract results in the entire community losing the bonus. Compliance is monitored by nearby protected area rangers and law enforcement agencies. This program is currently benefiting 29 communities and over 400 herder families across snow leopard range in Mongolia and is being expanded to Kyrgyzstan and Pakistan.

5 Conclusion: The Next Steps

Pastoralism in Central Asia has undergone profound changes in recent decades. Recent attention has been directed at mineral exploration and extraction, driven in large part by China's phenomenal growth. In Mongolia's eastern and central grasslands, the Nature Conservancy (2011) recently evaluated areas of potential conflict between this country's conservation portfolio and the areas leased for mining or petroleum development in order to determine how development impacts to biodiversity could best be offset. Areas with relative conservation value in the highest 30th percentile were designated as places to avoid development; the remaining conflict areas were removed from the portfolio and replaced with sites of similar biotic composition and ecological condition located outside of existing leases or projected development sites.

Besides this "Development by Design" approach, community stewardship (i.e., custodian responsibility) of wildlife and rangelands should be encouraged across the region. This type of initiatives needs to be designed and implemented within a highly participatory framework that ensures involvement from all stakeholders at the very onset of the project, including incorporating traditional knowledge among the planning tools and process (Berkes et al. 2000; Fernández-Giménez 2000). An open process valuing information exchange and the sharing of ideas and experience is critical to establishing resource norms and management guidelines that will best address local and regional environmental, economic, and social conditions (Schmidt et al. 2002). For example, herder training in the Gurvan Saikhan National Park of Mongolia's South Gobi (Bedunah and Schmidt 2004) where training focused on (1) improving nature conservation, (2) maintaining mobility of herder households, (3) improving skills in collaborating between households and local government, and (4) instilling skills for resolving conflicts, including actions at sharing and strengthening resource governance between herders and local government at the *soum* and *bag* level. Critical pasture management issues included rotational grazing, mutually agreeable moving dates, reservation of winter pastures, negotiating agreements with outside users, collaboration in transportation between seasonal camps, forage and winter risk management, and commitment to alternative fuels and fuel efficiency in order to reduce the pressure upon shrubs.

Strategies aimed at establishing ecologically appropriate rangeland utilization systems may need to include poverty alleviation since a relatively high proportion of herders in Central Asia are poor. Herders in colder climes and capable of producing higher-quality cashmere wool may earn substantial income, with many showing such indicators of wealth as solar-operated TVs, satellite dishes, and 4-wheel vehicles. Access to micro-credit is a crucial step in breaking the cycle of poverty in remote areas. When paying back the loan in the spring, after combing cashmere, traders demand payback in cash if prices are low and in cashmere when prices are high (Schmidt 2006). Recruiting and providing basic financing for volunteer rangers to assist herders in developing pasture protection and management plans is another option meriting further consideration: the designation of *Community Conserved Areas* offers the possibility of extending biodiversity conservation beyond established protected areas, thereby connecting fragmented habitat into more suitable units for the maintenance of viable predator and prey populations.

Facilitating protection of carnivores like snow leopards, which are highly valued globally but viewed negatively by local people, presents a special challenge. A promising avenue may rest with special incentive packages designed to increase household incomes in environmentally friendly and socially acceptable ways and fitting within the framework of "Payments for Ecological Services" or *PES* (Pagiola and Platais 2007). Financial instruments that translate the snow leopard's global values into tangible local values large enough to drive conservation "on the ground" need to be expanded to meet site-specific conditions and made available for implementation by rangeland managers (Dickman et al. 2011). While these could evolve from the examples provided above and be structured to offer conservation payments, compensation, or insurance mitigation or meet local development needs, the relevant cost, constraints, and capacity requirements (e.g., skills training, product quality control, and marketing) should be taken into account under a cost–benefits framework (e.g., Figure 1, Dickman et al. 2011).

Monitoring of conservation outputs is required whether payments for biodiversity protection are on the project's actions (e.g., predator-proofing corrals, proper guarding of livestock while grazing, antipoaching patrols) or its results (e.g., number of snow leopards and/or prey species present). Fortunately, the emergence of noninvasive genetic sampling from scats provides a means for determining the number of snow leopard present, so tracking populations over time has become easier (Janečka et al. 2008).

The ultimate challenge lies with moving pastoral communities beyond their harsh and often insecure subsistence livelihood into more economically stable and environmentally friendly activities. With sufficient incentives, and by embracing proven pastoral practices like seasonal or rotational grazing and winter forage banking, livestock production could be better sustained on these arid or semiarid rangelands. Transforming herder perceptions of snow leopards from a pest to an asset also depends upon pastoralists assuming greater responsibility for protecting their flocks or herds from predators.

5.1 *Take-Home Lessons*

Regularly emphasize the rationale for providing such incentives, the benefits they bring, and the community's vested responsibility for protecting snow leopards, its prey, and habitat:

- Clearly articulate each stakeholder's conservation responsibilities, the arrangements for reciprocal financing or in-kind contributions, and provide efficient tools enabling participatory planning and action along with supporting collaborative compliance monitoring and project evaluation.
- Value traditional ecological knowledge and blend it with findings from science-based approaches which address the key threats to wildlife and plant biodiversity as well as rangeland/ecosystem management.
- Incentive-driven, income-generating programs typically require substantial financial and human resource investment in order to sustain competitive market-based enterprises like handicrafts production, traditional homestays, or nature guiding. Many donors fail to appreciate that significant returns on community-based programs may not be forthcoming for 5–10 years, while implementing agencies are hard-pressed to demonstrate tangible results within the typical 2–5-year time frame expected by donors.

- Central Asia's varied topography, climate, biodiversity, economic geography, and cultures demand locally crafted approaches that address the particular set of environmental, social, and political conditions.
- Success is closed tied to the effectiveness with which project sponsors and stakeholders alike manage conflict, alleviate local threats to biodiversity and rangeland productivity, and yet meet fundamental aspirations of the targeted community. This, in turn, requires a well-informed, networked group of beneficiaries and project instigators—a process which is greatly enhanced with equitable participation and empowering decision-making.

References and Further Reading

Anwar MB, Jackson RM, Sajid NM, Janečka JE, Hussain S, Beg MA, Muhammad G, Qayyum M (2011) Food habits of the snow leopard, *Panthera uncia* (Schreber, 1775) in Baltistan, Northern Pakistan. Eur J Wildlife Res 57:1077–1083

Bagchi S, Mishra C, Bhatnagar YV (2004) Conflicts between traditional pastoralism and conservation of Himalayan ibex (*Capra sibirica*) in the Trans-Himalayan mountains. Anim Conserv 7:121–128

Bagchi S and Mishra C (2006) Living with large carnivores: predation by snow leopard Uncia uncia Conservation Biology 17 1q512–1523

Bagchi S and Mishra C (2006). Living with large carnivores: predation on livestock by the snow leopard (Uncia uncia). J Zoo 268:217–224.

Bedunah DJ, Schmidt SM (2004) Pastoralism and protected area management in Mongolia's Gobi Gurvan Saikhan National Park. Dev Chang 35(1):167–191

Behnke R (2006) The socio-economic causes and consequences of desertification in Central Asia. Springer, Bishkek

Berkes F, Colding J and Folke C (2000) Rediscovery of traditional knowledge as adaptive management. Ecol App 10:1251–1262

Bhatnagar YV, Wangchuk R and Jackson R (1999) A survey of depredation and the related human-wildlife conflicts in Hemis National Park, Ladakh, Jammu and Kashmir. International Snow Leopard Trust, WA, USA (Project Report)

Bhatnagar YV, Wangchuk R, Prins HHT, van Wieren SE, Mishra C (2006) Perceived conflicts between pastoralism and conservation of the kiang *Equus kiang* in the Ladakh Trans-Himalaya, India. Environ Manage 38:934–941

Bliss F (2006) Social and economic change in the Pamirs (Gorno-Badakhshan, Tajikistan). Routledge, Taylor and Francis Group, London, 378 pages

Dickman AJ, Macdonald EA, Macdonald DW (2011) A review of financial instruments to pay for predator conservation and encourage human–carnivore coexistence. PNAS 2011 108(34):13937–13944

du Toit J, Kock R, Deutsch JC (eds) (2009) Wild rangelands: conserving wildlife while maintaining livestock in semi-arid ecosystems. Wiley, London

Fernández-Giménez ME (2000) The role of Mongolian nomadic pastoralists' ecological knowledge in rangeland management. Ecol Appl 10:1318–1326

Fernández-Giménez ME, Batbuyan B (2004) Law and disorder: local implementation of Mongolia's land law. Dev Chang 35:141–165

Fox JL, Dhondup K, Dorji T (2009) Tibetan antelope *Pantholops hodgsonii* conservation and new rangeland management policies in the western Chang Tang Nature Reserve, Tibet: is fencing creating an impasse? Oryx 43:183–190

Green MJB, Zhimbiev B (1997) Transboundary protected areas and snow leopard conservation. In: Jackson R, Ahmad A (eds) Proceedings of the eighth international snow leopard symposium. WWF Pakistan and International Snow Leopard Trust, Seattle, pp 194–203

Gurung GS, Thapa K, Kunkel K, Thapa GJ, Kollmar M, Mueller Boeker U (2011) Enhancing herders' livelihood and conserving the snow leopard in Nepal. Cat News 55:17–21

Halimova N (2012) Land tenure reform in Tajikistan: implications for land stewardship and social sustainability: a case study. In: Squires V (ed) Rangeland stewardship in Central Asia. Springer, Dordrecht, pp 305–332 (Chapter 13, this volume)

Hua L and Zhang D (2012) Engaging with land users; the first steps on a long road. In: Squires V (ed) Rangeland stewardship in Central Asia. Springer, Dordrecht, pp 333–356 (Chapter 14, this volume)

Hussain S (2000) Protecting snow leopards and enhancing the farmer's livelihoods: a pilot insurance scheme in Baltistan. Mt Res Dev 20:224–229

Hussain S (2003) The status of the snow leopard in Pakistan and its conflict with local farmer livelihoods. Oryx 37(1):26–33

Ikeda N (2004) Economic impacts of livestock depredation by snow leopard *Uncia uncia* in the Kanchenjunga Conservation Area Nepal Himalaya. Environ Conserv 31:322–330

Jackson R (1998) People-wildlife conflict management in the Qomolangma Nature Preserve, Tibet. In: Ning W, Miller D, Lhu Zhu J (eds) Tibet's biodiversity: conservation and management. Springer, WWF, Washington, DC, pp 40–46

Jackson R (2000) Managing people-wildlife conflict on alpine pastures in the Himalaya. In: Richard C, Basnet K, Shah JP, Raut Y (eds) Proceedings of regional grasslands workshop, volume 3. Technical and status papers on grasslands of protected areas. ICIMOD (International Centre for Integrated Mountain Development), Kathmandu

Jackson R (2001) Managing people-wildlife conflict in Tibet's Qomolangma National Nature Preserve. In: Field R, Warren RJ, Okarma H, Sievert PR (eds) Wildlife, land and people: priorities for the 21st century. Proceedings of the second international wildlife management congress. The Wildlife Society, Bethesda, pp 188–191, 399 pages

Jackson R, Ahlborn G (1989) Snow leopards (*Panthera uncia*) in Nepal: home range and movements. Nat Geogr Res 5(2):161–175

Jackson R, Ahlborn G (1990) The role of protected areas in Nepal in maintaining viable populations of snow leopards. In: Blomqvist L (ed) International pedigree book of snow leopards, volume 6. Helsinki, Finland, pp 51–69

Jackson R, Fox JL (1997) Snow leopard conservation: accomplishments and research priorities. In: Jackson R, Ahmad A (eds) Proceedings of the 8th international snow leopard symposium Islamabad, November 1995. International Snow Leopard Trust, Seattle and WWF-Pakistan, Lahore, pp 128–145

Jackson P, Nowell K (1996) Problems and possible solutions in management of felid predators. J Wildlife Res 1(3):304–314

Jackson R, Wangchuk R (2001) Linking snow leopard conservation and people-wildlife conflict resolution: grassroots measures to protect the endangered snow leopard from herder retribution. Endangered Species UPDATE 18(4):138–141

Jackson RM, Wangchuk R (2004) A community-based approach to mitigating livestock depredation by snow leopards. Hum Dimens Wildl 9:307–315

Jackson R, Ahlborn GA, Gurung M, Ale S (1996) Reducing livestock depredation losses in the Nepalese Himalaya. In: Timm RM, Crab AC (eds) Proceedings of the 17th vertebrate pest conference. University of California, Davis, pp 241–247

Jackson RM, Mishra C, McCarthy TM, Ale SB (2010) Snow leopards: conflict and conservation, Chapter 18. In: Macdonald DW, Loveridge AJ (eds) Biology and conservation of wild felids. Oxford University Press, Oxford, pp 417–430, 762 pages

Janečka JE, Jackson R, Yuguang Z, Diqiang L, Munkhtsog B, Buckley-Beason V, Murphy WJ (2008) Population monitoring of snow leopards using noninvasive collection of scat samples: a pilot study. Anim Conserv 11(5):401–411

Koshkarev EP, Vyrypaev V (2000) The snow leopard after the break-up of the Soviet Union. Cat News 32:9–11

Kreutzmann H (2012) Pastoral practices in High Asia: Agency of 'development' effected by modernisation, resettlement and transformation. Springer, Dordrecht

Kurbanov B (2012) Constraints and barriers to better land stewardship: analysis of PRAs in Tajikistan, pp 129–164 (Chapter 7, this volume)

Leake JE (2012) Conclusions and a way forward. In: Squires V (ed) Rangeland stewardship in Central Asia. Springer, Dordrecht, pp 431–442 (Chapter 18, this volume)

Lovari S, Boesi R, Minder I, Mucci N, Randi E, Dematteis A, Ale SB (2009) Restoring a keystone predator may endanger a prey species in a human-altered ecosystem: the return of the snow leopard to Sagarmatha National Park. Anim Conserv 12:559–570

Mallon DP, Jiang Z (2009) Grazers on the plains: challenges and prospects for large herbivores in Central Asia. J Appl Ecol 46:516–519

Maselli D, Inam-ur-Rahim (2012) Setting the stage: key features of the present-day Central Asian region. In: Squires V (ed) Rangeland stewardship in Central Asia. Springer, Dordrecht, pp 3–30 (Chapter 1, this volume)

McCarthy TM, Fuller TK, Munkhtsog B (2005) Movements and activities of snow leopards in Southwestern Mongolia. Biol Conserv 124:527–537

Mishra C (1997) Livestock depredation by large carnivores in the Indian trans-Himalaya: conflict perceptions and conservation prospects. Environ Conserv 24:338–343

Mishra C (2001) High altitude survival: conflicts between pastoralism and wildlife in the Trans-Himalaya. PhD thesis, Wageningen University, The Netherlands, 131 pages

Mishra C, Prins HHT, van Wieren SE (2001) Overstocking in the Trans-Himalayan rangelands of India. Environ Conserv 28:279–283

Mishra C, Allen P, McCarthy T, Madhusudan MD, Bayarjargal A, Prins HHT (2003) The role of incentive programs in conserving the snow leopard. Conserv Biol 17:1512–1523

Mishra C, van Wieren SE, Ketner P, Heitkonig IMA, Prins HHT (2004) Competition between livestock and bharal, *Pseudois nayaur* in the Indian Trans-Himalaya. J Appl Ecol 41:344–354

Mishra C, Bagchi S, Namgail T, Bhatnagar YV (2009) Multiple use of Trans-Himalayan rangelands: reconciling human livelihoods with wildlife conservation. In: du Toit J, Kock R, Deutsch J (eds) Wild rangelands: conserving wildlife while maintaining livestock in semi-arid ecosystems. Blackwell Publishing, London, pp 291–311

Namgail T, Fox JL, Bhatnagar YV (2007) Carnivore-caused livestock mortality in Trans-Himalaya. Environ Manage 39:490–496

Nowell K, Jackson P (1996) Wild cats: a status survey and conservation action plan. IUCN, Gland

Oli MK (1994) Snow leopards and blue sheep in Nepal: densities and predator: prey ratio. J Mammal 75:998–1004

Oli MK, Taylor IR, Rogers ME (1994) Snow leopard *Panthera uncia* predation of livestock: an assessment of local perceptions in the Annapurna Conservation Area Nepal. Biol Conserv 68:63–68

Olson KO, Mueller T, Kerby J, Bolortsetseg S, Leimgruber P, Nicholson C, Fuller T (2011) Death by a thousand huts: effects of household presence on density and distribution of Mongolian gazelles. Conserv Lett 4(4):304–312

Pagiola S, Platais G (2007) Payments for environmental services: from theory to practice. The World Bank, Washington, DC

Retzer V (2006) Forage competition between livestock and Mongolian Pika (*Ochotona pallasi*) in Southern Mongolian mountain steppes. Basic Appl Ecol 8:147–157

Richard C, Zhaoli Y, Guozhen D (2006) The paradox of the individual household responsibility system in the grasslands of the Tibetan Plateau, China. In: Bedunah DJ, McArthur ED, Fernandez-Gimenez M (eds) Rangelands of Central Asia: Proceedings of the conference on transformations, issues, and future challenges. 2004 January 27; Salt Lake City, UT. Proceeding RMRS-P-39. Fort Collins, CO: U.S. Department of Agriculture, Forest Service, Rocky Mountain Research Station, pp 83–92

Robinson et al. (2012) Pastoral Tenure in Central Asia: theme and variation in the five former Soviet republics In: Squires V (ed) Rangeland Stewardship in Central Asia Springer, Dordrecht, pp 239–274 (Chapter 11, this volume)
Salafsky N, Margoluis R (1999) Threat reduction assessment: a practical and cost-effective approach to evaluating conservation and development projects. Conserv Biol 13:830–841
Schaller GB (1998) Wildlife of the Tibetan Steppe. University of Chicago Press, Chicago, 373 pages
Schaller GB, Li H, Ren JR, Qiu MJ, Wang HB (1987) Status of large mammals in the Taxkorgan Reserve Xinjiang China. Biol Conserv 42:53–71
Schaller GB, Jungang R, Mingjiang Q (1988) Status of snow leopard *Panthera uncia* in Qinghai and Gansu provinces China. Biol Conserv 45:179–194
Schaller GB, Tserendeleg L, Amarasanaa G (1994) Observations on snow leopards in Mongolia. In: Fox JL, Jizeng D (eds) Proceedings of the seventh international snow leopard symposium. Northwest Plateau Institute of Biology and International Snow Leopard Trust, Seattle, pp 33–46
Scharf KM, Fernández-Giménez ME, Batbuyan B, Enhkbold S (2010) Herders and hunters in a transitional economy: the challenge of wildlife and rangeland management in Post-socialist Mongolia. In: du Toit JT, Kock R, Deutsch JC (eds) Wild rangelands: conserving wildlife while maintaining livestock in semi-arid ecosystems. Wiley, London, pp 312–339
Schmidt SM (2006) Pastoral Community Organization, livelihoods and biodiversity conservation in Mongolia's Southern Gobi Region. In: Bedunah DJ, McArthur ED, Fernandez-Gimenez M (eds) Rangelands of Central Asia: Proceedings of the conference on transformations, issues, rangelands of Central Asia: Proceedings of the conference on transformations, issues, and future challenges. 2004 January 27; Salt Lake City, UT. Proceeding RMRS-P-39. Fort Collins, CO: U.S. Department of Agriculture, Forest Service, Rocky Mountain Research Station, pp 18–29
Schmidt S, Gongor G, Kar K, Swenson K (2002) Community organizing – a key step towards sustainable livelihoods and co-management of natural resources in Mongolia. Policy Matters September 2002:71–74
Smith AT, Foggin JM (1999) The Plateau Pika (*Ochotona curzoniae*) is a keystone species for biodiversity on the Tibetan plateau. Anim Conserv 2:235–240
Sunquist M, Sunquist F (2002) Wild cats of the world. University of Chicago Press Chicago. IL USA
Treves A, Wallace RB, Naughton-Treves L, Morales A (2006) Co-managing human-wildlife conflicts: a review. Hum Dimens Wildl 11:383–396
Western D, Wright DH, Strum S (1994) Natural connections: perspectives in community-based conservation. Island Press, Washington, DC
Wingard JR, Zahler P (2006) *Silent steppe: the illegal wildlife trade crisis in Mongolia*. Discussion Paper, East Asia and Pacific Environmental and Social Development. Department World Bank, Mongolia/Washington, DC

Chapter 16
Participatory Management of Desert Rangelands to Improve Food Security and Sustain the Natural Resource Base in Uzbekistan

Makhmud Shaumarov, Kristina N. Toderich, E.V. Shuyskaya, Shoaib Ismail, T.F. Radjabov, and Osamu Kozan

Abstract This chapter presents an analysis of contemporary status of rangelands, vegetation, and land use in Uzbekistan. Constraints to development include the climate and other physical conditions, but the socioeconomic and the policy environment are also barriers to adoption of better land management. The large rural population in remote areas is in subsistence mode, and there is little opportunity to improve household income in the face of the lack of access to grazing lands, lack of secure land tenure, and limited rural financial services. Measures to introduce and implement sustainable land management concepts and practices are outlined.

Keywords Aral Sea • Kyzyl Kum Desert • Land tenure reform • Policy • Subsistence • Poverty • Climate change • CACILM • Socioeconomics • Post-Soviet era • Rural credit • Gender

---

M. Shaumarov
Institute of Agricultural Economics and Social Sciences in the Tropics and Subtropics, University of Hohenheim, Stuttgart, Germany
e-mail: makhmud.shaumarov@uni-hohenheim.de

K.N. Toderich (✉)
International Center for Biosaline Agriculture for Central Asia and Caucasus (ICBA-CAC), 6, Osiyo street, 700000 Tashkent, Uzbekistan
e-mail: ktoderich@yahoo.com

E.V. Shuyskaya
K.A. Timiryazev Institute of Plant Physiology Russian Academy of Sciences, Moscow, Russia

S. Ismail
International Center for Biosaline Agriculture (ICBA), Al-Ruwayya, Dubai, United Arab Emirates
e-mail: s.ismail@biosaline.org.ae

T.F. Radjabov
Samarkand State University, 15, Universitetskii Bulvar, Samarkand, Uzbekistan

O. Kozan
University of Kyoto, 46 Shimoadachi-cho, Yoshida Sakyo-ku, Kyoto, Japan

V. Squires (ed.), *Rangeland Stewardship in Central Asia: Balancing Improved Livelihoods, Biodiversity Conservation and Land Protection*,
DOI 10.1007/978-94-007-5367-9_16,

Key Points

- The rangelands are a national asset of high economic, environmental, and cultural value, for which the whole Uzbek community needs to take responsibility. Therefore, a national rangelands strategy (NRS) is needed to be developed because the rangelands under current climate change face a wide range of problems, which cannot be easily resolved with current legislation and institutions, present knowledge, and existing dispute resolution procedures.
- Land degradation in Uzbekistan has two ultimate effects – ecosystem instability and poverty. In fact, these two effects are inextricably linked and each can cause the other. They are also part of a closed loop whereby degraded land leads to ecosystem instability and poverty forcing desert communities to further stress the fragile desert environment, thus degrading the land even further and exacerbating the ecosystem instability and poverty.
- Agricultural policy of Uzbekistan in the early transition period was primarily concentrated in support of staple crop production in arable land areas and in strengthening of the other sectors of economy in order to avoid deepening of the crisis and food insecurity (UNDP 2010b). However, national policies toward smooth institutional transition and development of livestock sector in dryland areas have lagged behind of the country priorities (UNDP 2010c). Since independence in 1991, the government took nationwide agricultural reform: state-induced farm restructuring, state ownership of land, reforms to transfer land from collective farms to private farming use (IAMO 2008), and continuation of area-based state quota for cotton and wheat production at fixed output prices.
- Implementation of better sustainable land management (SLM) practices in Uzbekistan requires the development of new capacities of people involved in the management of pastureland. New skills and knowledge are required for officers working in related institutions – particularly at the regional and local levels – but also for communities that are using pastureland for their livelihoods. Related with the need for new skills and knowledge, there is also the need to strengthen procedures – particularly for assessing pastureland – in order to provide pastureland managers with better information for assessment and to develop an integrated land use planning systems.
- The arable land per capita in Uzbekistan is much below the global average (0.183 ha/capita vs. 0.26 ha/capita), and high population growth, desertification, and salinity will further decrease this relation. Income-generating strategies need to be developed for rainfed desert and semidesert steppes. Better utilization of low quality water/mineralized water and reclamation of marginal resources would also help.

1 Introduction: The Geographic and Socioeconomic Setting

1.1 The Physical Setting

The Republic of Uzbekistan is a double land-locked country, centrally situated in the heart of Central Asia (CA) within the Aral Sea basin (Fig. 16.1). Almost 80% of land areas of the country are comprised of deserts and semideserts, including the

Fig. 16.1 Administrative map of Uzbekistan Showing boundaries with neighboring Central Asian countries

Kyzyl Kum, the largest desert of CA. As a whole, the territory of the country, located in the arid zone of Asia, is highly susceptible to land degradation, desertification, and climate change.

It is estimated that more than 52% of the arable lands and 73% of rangelands are presently undergoing accelerated land degradation (CACILM 2006). There is a broad agreement [IFPRI (ADB 2009)] that Uzbekistan is among those countries most vulnerable to climate change due to a high sensitivity of its arid arable lands (Fig. 16.2), high density of population, and growing food insecurity (UNFCCC 2008). Average rates of warming since 1950 along the territory of the republic have been increased by 0.29°C per decade, which is more than twice the world average.

1.2 Climate

Uzbekistan being situated in the middle of the Asian continent is a subject to the cold influence of the Arctic and Siberian fronts that may reach the southern border near Afghanistan. Uzbek climate features a strong continental influence. In some years, winters are characterized by an unusually thick layer of ice on top of snowfall (locally known as "*jut*") that may occur and which are strongly detrimental to wildlife, especially ungulates population, preventing them from accessing snow-covered feed (Bekenov et al. 1998). Summers can be hot and dry (Gintzburger et al. 2003).

Fig. 16.2 Topographic sketch map of Uzbekistan with the large Kyzyl Kum Desert and the Aral Sea

The appearance, or not, of optimal distribution of spring precipitation makes the difference between the good vegetation (rangeland grazing year) and a poor season as a whole. Cycles of drought, which occurred in 2007–2008 and 2010–2011 seasons, negatively affected the vegetation of the rangelands, especially in Central Kyzyl Kum (Shuyskaya et al. 2012 in press).

1.3 Socioeconomic Situation

The GDP per capita in 2011 was $3,300 (CIA 2012). The level of low-income population in the country has decreased from 27.5% (2001) to 25.8% by 2005, but in rural areas it remains still high. Uzbekistan's HDI is 0.641, which gives the country a rank of 115 out of 187 countries with comparable data. The HDI of Europe and CA as a region increased from 0.644 in 1980 to 0.751 today, placing Uzbekistan below the regional average. About 64% from a total population of 27 million live in rural areas, and they directly or indirectly depend on irrigated agriculture. The arable land per capita in Uzbekistan is much below the global average (0.183 ha/capita vs. 0.26 ha/capita), and high population growth, desertification, and salinity will further decrease this relation. Source: (IMF 2008; IMF 2008a)

Population in CA is going to increase from 60.6 million (2008) to 79.9 million by 2050 mainly by the forecasted growth in Uzbekistan from 27.8 to 38.4 million. Already, now there is food insecurity in the poorer strata and increasing numbers of rural poor. Climate change has significant impact on rural poor, as poor people have little resistance to drought and have no savings to recover afterwards.

The crisis in public finance following the breakup of the Soviet Union prompted a dual process of demonetization and re-agrarianization in rural Uzbekistan, increasing reliance on household and subsidiary plots for self-subsistence and on off-farm and nonfarm informal income-generation activities significantly. The shift from collective farms to joint-stock shareholding companies *(shirkat*s) has resulted in a process of labor retrenchment that has affected women significantly. The liquidation of collective farms in favor of independent farms has consolidated farm management as a male occupation. While the actual labor input of women into farming activities on household plots and private subsidiary plots and in cotton production has remained extremely high, they are increasingly incorporated into the workforce either as unpaid family laborers or as casual laborers earning piece-wage rates.

In the 1990s, rural households were allocated land plots on the basis of what had been personal plots on state and collective farms during the Soviet period. Some 82% of Uzbekistan's households benefited from these plots, which allowed for a minimal standard of living. These plots (average size of which is 0.12 ha) have since become the basis for small family or *dehkan* farms. During the same period, the vast majority of state and collective farms were converted into *shirkats* (cooperative farms based on household contracting). Since the late 1990s, private commercial farms have also been introduced. In contrast to the *dehkan* farmers, the private farmers are allowed to use hired labor and to have larger plots. Since 2003, the *shirkats* in irrigated croplands have been gradually transformed into private commercial farms, and in the near future, the share of commercial private farms is forecasted to increase to up to 85% of total sown land and dehkan farms to 12%. However, the major parts of desert rangelands are still formally belonging to the state owned by Karakul production *shirkats*, which remain functional until today. Land is likely to continue to be owned by the state (Robinson, Chap. 11), but nonetheless, private farmers can rent it from the state for 49 years and peasant dehkans for livelong period with inheritance rights.

Still, without strong private sector development in the countryside, much of the rural population still relies on employment in *dehkan* farms, which in most cases amounts to little more than subsistence agriculture. While these workers are classified as employed, their actual income-generating opportunities are very limited, due to poor access to markets, credit, equipment, etc. The transformation of *shirkats* into private farms is also further reducing employment since the more efficient private farms have higher labor productivity rates and therefore employ fewer workers.

Rural areas in Uzbekistan were always disadvantaged compared to cities even in the Soviet period. However, the post-1995 economic policies have probably reinforced these disadvantages, producing rising poverty levels in remote countryside and more disparities between rural and urban areas. Demographic trends have meant rapid growth in the supply of (mainly low-skilled) labor and poor employment opportunities in rural areas and small towns. These trends are pushing growing numbers of rural residents into small subsistence farming, onto *mardikhors* (informal labor markets), or into seasonal migration. In human terms, this translates into more vulnerability for rural residents with growing numbers of children being brought up in incomplete families.

2 Poverty, Inequality, and the Character of Economic Growth

Independence and the onset of the transition in Uzbekistan coincided with a number of severe shocks. To respond to these shocks and to transform its economic system, Uzbekistan followed a policy model different from that adopted in most other transition economies. This approach has been "unorthodox" both during the macroeconomic stabilization phase of 1991–1995 and during the recovery phase of 1996–2003. However, on the basis of this "home-grown" approach, the country managed to outperform the rest of the region during the stabilization phase and in the recovery phase sustained moderate but acceptable annual GDP growth rates of 3.5–4.0%, while in the last 5 years, the Uzbek economy has been developing at higher growth rates – 7.3% on average (IMF 2008). By 2001 it was the only country of the former Soviet Union to have surpassed its estimated 1989 level of GDP.

These relatively encouraging outcomes have, however, been marred by a number of persistent problems, which have prompted analysts and more recently the government to recognize that some policy changes might be in order. While poverty increased gradually – and, to some extent, unavoidably – during the contraction-stabilization years of 1991–1995, it then stagnated during the subsequent recovery and was accompanied by growing inequality. Indeed, income inequality has escalated markedly since 1995–1996, due largely to the policy bias that favored urban-based, capital-intensive, medium- and large-size enterprises. The capital-intensive nature of the import-substitution model followed since 1995 has meant that both urban and rural labor markets were unable to absorb the rapidly growing working-age population as well as the labor being shed in agriculture and the state-owned enterprises. As a result, while open unemployment has remained low, the number of underemployed engaged in low-productivity, low-wage, or part-time jobs is high. This can largely explain the continued high poverty levels despite relatively good growth rates.

2.1 The Profile of Poverty in Uzbekistan

It is estimated that 27.5% of the population or 6.8 million people were living below the poverty line in 2001 and that 9% were living in extreme poverty Source: (IMF 2008a). Within the framework of the Poverty Reduction Strategy Papers (PRSP), the government has updated the poverty estimates for 2008 using the same methodology and estimates that the overall level of poverty fell slightly to 26.2%. The household budget survey (HBS) results suggest that around 70% of the poor population lives in rural areas. However, other surveys have also suggested that the poverty risk is high for residents of small towns, where employment opportunities have decreased (due to, e.g., the nonfunctioning of large employers) and where the access to land is more limited. In 2001, households with an unemployed head of household were more likely to be poor, but over 50% of the poor live in households where the head is actually employed. This suggests that poverty risk is linked not just to unemployment but also to underemployment and low wages. Low salaries in the public sector (where wages are 60% of the national average) and agriculture (50% of the national average) contribute to poverty among the employed. Poorer households were larger than average and had both more children (four or more) and adults Source: (IMF 2008; IMF 2008a).

The Gini coefficient[1] for Uzbekistan is estimated at 0.35 and has decreased since the mid-1990s. This indicates that, as in other countries in the region, there are average-to-high levels of income inequality. However, there are doubts about the household budget survey's ability to capture the upper-income population. This would lead to the underreporting of the largest incomes and the underestimation of inequality. Differentials in average wages between sectors also suggest high and growing disparities, particularly between agriculture and other sectors.

Income-generating strategies need to be developed for rainfed desert and semidesert steppes. Better utilization of low quality water/mineralized water and reclamation of marginal resources would also help. The situation is aggravated by the fact that local farmers have limited experience and knowledge of advanced soil and water conservation technologies for promoting the best practices of SLM.

3 Rangelands of Uzbekistan

Pastureland represents 22 Mha in Uzbekistan or about 50% of the total area of the country (Gintzburger et al. 2005). It is mostly desert land that is being degraded over time due to overgrazing, wood overharvesting, and unsustainable agricultural practices; the result is an increasing instability of desert ecosystems and poverty for populations living in these areas. Owing to its geographical and climatic characteristics, Uzbekistan is highly susceptible to environmental degradation, in particular its arid ecosystems (Fig. 16.3). The most serious ecological problems threatening the country's natural resources are incremental soil and water salinization, wind and water erosion, overgrazing and deforestation, loss of biodiversity, and the reduction in productive potential of arable land and pastures.

A major distinguishing feature of land degradation in Uzbekistan is loose sand, and according to the Forestry Department, some 2.3 million ha in Bukhara Oblast

[1] Gini coefficient – a measure of income inequality.

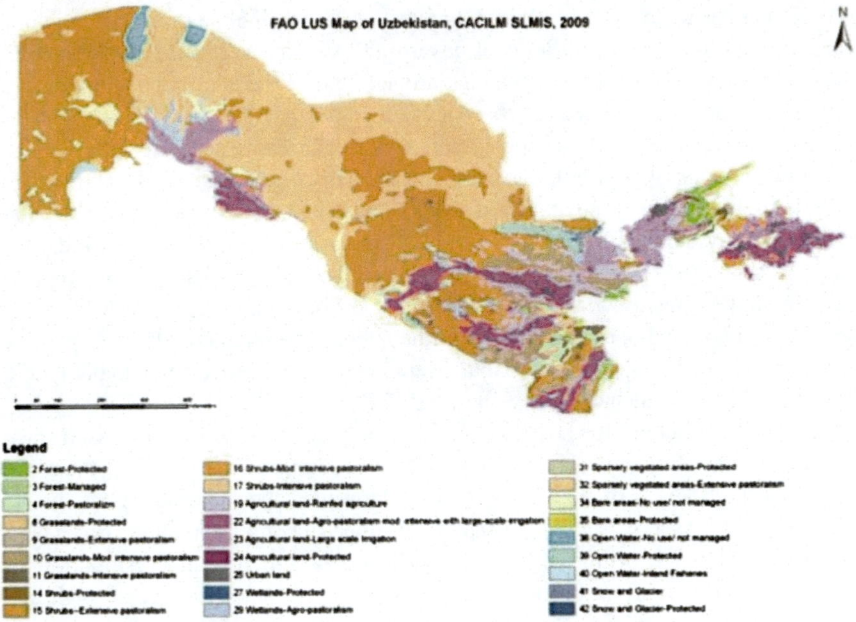

Fig. 16.3 Land use categories in Uzbekistan (*Source*: FAO, CACILM SLMIS 2009)

and 4.5 million ha in Karakalpakstan are affected by wind erosion. Sand is blown around by the wind with negative effects ranging from impact on people's health and well-being, reduced agricultural productivity, impact on roads and other infrastructure, contamination of water resources and environmental pollution, and morphological changes to land. In the case of the exposed Aral seabed, the sandy substrate has not had the chance to develop a surface fertile substrate, which would prevent most of the wind erosion that is responsible for the raising of 70 million tonnes of sand and dust per year into the atmosphere. It has been estimated by local experts that up to five million people in Uzbekistan are affected directly, and a lot more indirectly, by wind-blown sand and dust. Some have given up and have migrated away from the desert to cities in Uzbekistan or even other CA countries – communities are breaking up because of the impact of wind-blown sand.

Land degradation in Uzbekistan has two ultimate effects – ecosystem instability and poverty. In fact, these two effects are inextricably linked and each can cause the other (Holland 2010). They are also part of a closed loop whereby degraded land leads to ecosystem instability and poverty forcing desert communities to further stress the fragile desert environment, thus degrading the land even further and exacerbating the ecosystem instability and poverty.

Soil salinity is a serious problem in some of the locations of agricultural irrigated lands throughout the country due to groundwater, resulting from excess irrigation during cultivation of cotton and other crops, high evaporation during hot summer, and/or poor drainage function and maintenance. Due to the gradual increase in

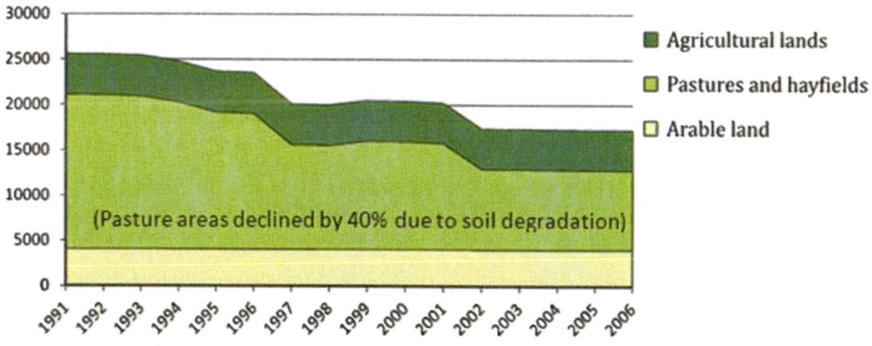

Fig. 16.4 Dynamics of land degradation in Uzbekistan (1,000 ha) (*Source*: Modified from UNDP 2010c)

salinization of soils and rising of water table, the majority of lands in the cropping farms have been gradually removed from cultivation of traditional agricultural crops. Data on salinity of soils and surface water from various sources (open canal, drinking water, and collector-drainage water) indicates a significant increase in the salinity levels. A direct correlation exists between the water table levels and the spread of salinization over different irrigated lands. In most of the farms, the farmers are forced to use the drainage water (4,000–5,000 ppm, 6–7 dS/m) when irrigation water is not available. The development of the new strategy of the sustainable use or rehabilitation of desert pasturelands (which currently declined by 40% due to soil degradation as is shown in Fig. 16.4) and adjoining agricultural lands in present conditions requires nontraditional and more complex approach to the assessment of the naturally ecological state of these anthropogenically transformed ecosystems. Nowadays, it is crucially difficult to rehabilitate and sustainably develop the salt-affected and waterlogged lands that are roughly estimated to make up about 60–70% of territories in the dry bottom of the Aral Sea, patchily throughout the Kyzyl Kum Desert, Aydarkul-Arnasay Lakes systems (AALS), and saline prone areas frequently found in the old irrigated agricultural lands of Uzbekistan.

The principal threats to land degradation affecting desert and semidesert ecosystems of Uzbekistan are as follows:

Overgrazing. Land vulnerability is exacerbated by local residents who overgraze available pastures by domestic stock, in an effort to survive. As the situation worsens, farmers tend to move further into marginal areas and to replace sheep with goats, which unfortunately complete the total denudation of land leaving it susceptible to wind action. Overgrazing of marginal land is particularly concentrated in the vicinity of settlements and around wells. In these areas, the land is not only denuded of all vegetation, but it is also prevented from forming the surface "skin" that is necessary to prevent wind erosion and begin the process of soil development.

Wood overharvesting. Local population cuts down trees and shrubs for wood fuel (Holland 2010). Unlike former times, when population in deserts was mostly nomad and the population number was not large, contemporary settlements require an extensive amount of wood fuel for cooking and dwelling heating. Obviously the settlers tend to cut any wood available instantly around the settlements in the first place. Besides, availability of motor vehicles provides an opportunity to harvest wood from more distant areas when wood resources around a settlement are exhausted. Country's desert and semidesert forest ecosystems are composed of a complex of trees, shrubs, and grass communities. Ecosystem diversity of desert and semideserts is low in comparison with other types of ecosystems, which makes them more vulnerable to any kind of outside interventions. Overharvesting of trees and shrubs by local population significantly lowers resilience of the ecosystem and results in its serious degradation.

Unsustainable agricultural practices. Because of their vulnerability, rainfed, unirrigated lands demand special attention in terms of their utilization for agricultural purposes. Inappropriate patterns of land use severely affect ecosystem stability and add to existing problems of land degradation. One of the factors that have contributed to environmental degradation in many parts of the world is that different agencies have had responsibilities for different aspects of the management of what is, essentially, one system. This means that even when there is effective planning, the potential benefits that could arise from integrated planning are not secured, and more commonly one sector impacts on others, frequently due to nothing more than a lack of awareness. However, integrated land use planning (ILUP) may be a simple concept, but it is hard to deliver; it is also a relatively new concept in Uzbekistan and will need considerable support if it is to become effective. There is considerable scope for ILUP to be a foundation for better land stewardship.

Lack of Enabling environment for pasture management. For policies to be successful, they need to be enabling and inclusive. At the local and national level, all relevant stakeholders need to involve, where possible, from the design stage. Their involvement enhances the chances that policies will be effective. The active participation of pastoral communities and district-level decision-makers is required for local policies and projects, such as integrated land use planning or community-based pasture use plans, to be practically effective and successful. It is crucial that the national and international agencies make a joint effort to seek development of enabling environment for pasture management of dryland areas in Uzbekistan. It includes the strengthening of the legislative framework and the related institutions and the capacity development of officials and land users in local communities.

Climate change impacts. Central and western arid regions of Uzbekistan are highly vulnerable to climate change (UNECE 2010) given that agriculture is a key factor concerning GNP and even more so with regard to food security, employment, and social stability. Increasing aridity and frequent droughts, combined with ineffective land management policies, also cause degradation and desertification of pastures. Predicted climate variability can lead to regular drought seasons in the downstream areas of the Amu Dar'ya River delta in western regions (SNC 2009). The climatic effects might become even more severe in the region of Karakalpakstan with current shrinking conditions of afflicted Aral Sea basin. It is expected that the composition of plant communities in the rangeland used for pasture will be significantly altered, with consequences for forage production, cattle breeding, and sheep rearing. Implementing appropriate land and water management strategies that increase soil quality and productivity in drought affected areas is crucial.

Since practices that sequester soil carbon often enhance land productivity (FAO 2010), effective institutional policies designed to encourage soil carbon sequestering management practices could lead to potential dividends in greater crop production and enhanced producer income. There is an increasing global interest in soil carbon finance mechanisms by international agencies, investors, and national governments, e.g., through the nationally appropriate mitigation (NAM) mechanism. Since soil carbon sequestration potential is very high, it is most likely that this trend will increase further and it would create additional economic incentives for the government of Uzbekistan to invest and to support SLM practices (see below) in drylands. Practices that sequester carbon in soil stocks also tend to enhance resilience in the face of climate variability and likely to enhance long-term adaptation to changing climates.

Currently, there is a significant potential to increase carbon sequestration in rangelands of Uzbekistan by improving management of vegetation resources under the moderate grazing of livestock and ecologically friendly rangeland improvement (reseeding technique and others). However, mismanagement of rangelands and increasing livestock number lead to deterioration and overexploitation of vegetation resources in desert and semidesert zones. Increased livestock number decreases directly the biomass production of vegetation and indirectly the processes of carbon storage capacity in rangelands. Learning from best practices of international pilot projects in soil carbon storage and accessing available finance mechanisms could benefit both public agencies and rural livelihoods in drylands and help to enhance grassland productivity in the long run.

Hand harvesting of range forages for feed and fuel wood and concentration of livestock around populated areas and active wells are the main concern of imbalance between fodder resources, livestock, and human population. The dramatic decreasing of rangeland's grazing capacity and lack of winter forage and fuel wood lead to overgrazing and uprooting of shrubs that will devastate forage resources for a long time. Many of valuable forage species disappeared or are replaced by less palatable plants. Distant pastures are not utilized any longer, since availability of water resources in remote seasonal pastures is limited due to high levels of salinization or obsolete infrastructure. Most of the remote water wells maintained during the former Soviet period are no longer available. During the Soviet era, local pastoral communities have lost their skillful shepherds and vast of traditional knowledge accumulated since many centuries in CA. Thus, it is impossible to recover this unique knowledge and experience at present and either to use distant rangeland zones without detailed assessment of the water cycle combined with vegetative seasonality (UNDP/GM 2007). These processes induce further land degradation and desertification process from plains into the foothills and mountain areas, consequently – food insecurity in the region.

The livestock inventory is high in CA: sheep (43,923.000), horses (1,837.000), and goats (7,463.000). They are fed in the steppes and other rangelands for about 6–7 months a year. The dramatic increase in the number of goats (2007–2011) caused overgrazing that seriously diminished vegetation and transformed foothill steppes (dominated by *Artemisia)* into desert. For improving grazing capacity of rangelands, it is most important to increase the total plant biomass per unit area by ensuring a sustainable level of harvest or uprooting perennials as winter forages. This would allow development of more forage-based production for livestock feeding and would prevent further desertification (which can deplete the soil C stock, if left unchecked), impoverishment, and migration of the poorest strata. For the poorest strata, sustainable forage production on marginal, unused land might become an additional or even alternative source of income.

The dry grasslands and shrubland have long been neglected by policymakers because they were misperceived as being degraded marginal areas, offering poor returns on development investment. There is a lack of current information of the behavior of agropastoralists to the response of rangeland degradation and reduction of livestock production and livelihood income. Despite the existence of a large database on natural resources (soil, botanic diversity, bioproductivity of rangelands, livestock sector, etc.), most of these data, however, are not in digital form and are at risk of being lost due to rapid institutional changes. These databases have never been linked for the purpose of indication of potential role of desert grasslands for carbon sequestration, ensuring food security for the poor and making ecosystems more resilient to further climate change.

Application of contemporary research methods with the combination of available data is important for providing the relevant information for sustainable rangeland management. A lack of resources and experience inhibits the application of geographical information systems (GIS) and remote sensing (RS) technologies to establish the digital databases for assessment of vegetation resources of natural rangelands. However, nowadays, availability of RS data with high spatial and temporal resolution enables monitoring and assessing the current condition of rangeland vegetation and its long-term changes under the anthropogenic and ecological disturbances. Using vegetation indices derived from satellite data, seasonal and annual changes of vegetation cover and the trends of degradation processes can be analyzed.

Rangelands of Uzbekistan were once productive with quite diverse plant communities and well adapted to sustain grazing pressure from wild and domestic animals. It was due to unique traditional knowledge and local skills on seasonal pasture management accumulated by nomadic pastoralists since many centuries in CA drylands. It was based on practical knowledge of spatial and temporal variability of climate, water cycle, and vegetative seasonality (UNDP/GM 2007). Nowadays, many rangeland areas in the region, however, are not properly managed resulting in feed deficits, soil erosion, loss of plant biodiversity, and expanding desert margins. Undesired/unpalatable annual species is increased under intense and continuous grazing, and proportion of such species in vegetation composition consisted of 42% in an area affected by heavy grazing (Rajabov 2011). Increased numbers of unpalatable plants in vegetation composition led to a decrease in the qualitative values of the wrangelands.

Presence of *Peganum harmala* can be estimated as a sign of beginning of severe changes in vegetation structure under the intense grazing.

In addition, the collapse of the Soviet Union has led to disruptions of migratory patterns between summer and winter pastures and across national borders. In Uzbekistan, however, the tradition of nomadic pastoralist production was disrupted during the "modernization" effort during the 1950s, which resulted in sedentarization and large-scale conversion of rangelands to croplands. State-owned farms, specialized on Karakul production, were forced to develop larger areas of rangelands in new desert territories. Cropping on unsuitable lands was a contributor to accelerated land degradation mainly due to inefficient use of water resources, soil mineralization, and overapplication of chemical fertilizers. However, considerable field experiments by Soviet scientists contributed to combat land degradation and desertification with practically feasible methods and sound techniques. Nonetheless, sedentarization had also led to patterns of overutilization of rangelands near settlements and watering sources, resulting in declining of rangeland's carrying capacity, a clear gap in feed supply, especially in the autumn-winter period. But in late 1950s, wide introduction of winter fodder depository system in seasonal pastures had increased livestock productivity in state farms.

Increasing human population and expanding agricultural areas have resulted in heavier grazing pressures in rangelands in spite of the increasing availability of crop residues and grains and their role in livestock production. As a result of erratic cropping in low-rainfall zones, overgrazing of the good rangelands, and cutting of shrubs by local population for firewood, the natural vegetation of these desert areas is under pressure from anthropogenic degradation factors. This leads to the eradication of useful, endemic, or rare wild animals and desert plant species and to the reduction of rangeland productivity. Instead of the valuable fodder herbs, subshrubs, and bushes, plants of low palatability to livestock have appeared. The desert forests and subshrubs have completely disappeared from large areas. The all-year-round pastures, that were rich earlier, now have low productivity of fodder plants. The fast growth of cattle numbers urgently requires the increase of productivity of the pastures. The radical improvement of such pastures is possible only by the implementation of the large-scale phyto-melioration[2] and careful intervention. All of which involves a huge investment.

Climate change accelerates the rangeland degradation caused both by intense storms, soil erosion, fluctuations of rainfall, depletion of groundwater, as well as overgrazing and human activities such as gathering fuel wood and medicinal plants. More research and monitoring is needed to assess pasture carrying grazing capacity and improvement productivity, ascertain optimal times for grazing, and assist in improving regulation for pastureland in the context of climate change. To cope with climate changes, we need a better understanding of the current carrying grazing capacity of different types of rangelands and livestock controlled grazing system

[2] Plant-based land reclamation/restoration.

and use of integrated approaches on water harvesting and oasis development programs by using mineralized groundwater. Livestock production in rangelands of Uzbekistan depends mainly on access to groundwater sources since availability of rainfall is very limited, especially in desert and semidesert drylands. Introduction of biosaline agriculture production systems has proven to be an effective way to overcome the limitation of rising salinity levels for livestock production in drylands (Toderich et al. 2009).

3.1 Institutional Settings Around Grasslands

The national agricultural strategy of Uzbekistan in the early transition period after independence was mainly concentrated on support of staple crop production in arable land areas and in strengthening of the other sectors of economy in order to avoid deepening of the crisis and food insecurity (UNDP 2010b). However, national policies toward smooth institutional transition and development of livestock sector in dryland areas have lagged behind of the country priorities (UNDP 2010c). Since early independence, the government took nationwide agricultural reform: state-induced farm restructuring, state ownership of land, and reforms to transfer land from collective farms to private farming use (IAMO 2008). The land reform consisted of land distribution to individual households and small-scale producers in combination with farm restructuring programs whereby not only the land but all state-owned and collectively operated production units were distributed to the new market-oriented agricultural producers, including smallholders and private farmers (Robinson, Chap. 11, World Bank 1999).

The Ministry of Agriculture and Water Resources (MAWR), supervised by the Cabinet of Ministers of Uzbekistan, holds a leading position in development and execution of agricultural policy, including institutional governance of agriculture production and management of national land and water resources (SNC 2009). Regional- and district-level municipalities, as well as local branches of the MAWR, are responsible for planning and allocation of resources on meso-levels. The State Committee for Land Resources, Geodesy, Cartography and State Cadastre (Goskomzem) participates in its function of land use planner and repository for land use information. Goskomzem reviews land use policies and land use legislation and develops an interagency land use planning and management system. This is basically the main national institution to govern land registration, monitoring of land use change which has responsibility to regularly assess the state of soil quality in agricultural and nonagricultural land plots.

Large-scale pastoral land areas in central and northwestern parts of Uzbekistan are allocated to Karakul farm enterprises called "shirkats." All shirkat farms specialized in Karakul fur and wool production belong to the state-owned corporation "Uzbek Korakuli." Today it comprises of 106 Karakul shirkats, and it is the main responsible Uzbek institution for production, processing, and export of Karakul produce as well as for overall management of rangeland territories allocated for pastoral livestock management. A nationwide network of community-based traditional institutions – "Mahalla" – plays a vital role in execution of district municipality orders for management, distribution, and monitoring of resource among rural communities in grassroot levels. Practices of development projects show that although National Land Code has been established in 1998 as the major legal framework for management of land resources, property rights and institutional arrangements over pasture land resources are yet weakly defined. Monitoring and enforcement capacities of district-level public authorities are limited in reaching distant communities and rangelands due to regular shortages in budget allocations, obsolete infrastructure and equipment, reduced number of administrative staff, and lack of economic incentives for them (SNC 2009). This is also supplemented by emerging conflict of interests over utilization of pastoral resources and informal land use on local level by community livestock herds and formal shirkat farm managers (IAMO 2008, Robinson, Chap. 11).

New for the post-Soviet state, input supply markets, insurance system, and private banking sector institutions were introduced in early years of independence in order to create basic support mechanisms for development of private farming and to strengthen agricultural production (UNDP 2010c). However, many small-scale farmers faced constraints in receiving these services due to lack of farming experience, insufficient collateral, and credit histories (IAMO 2008). It has to be noted as well that legal institutions and governance structures toward planning and utilization of pasture resources still undergo transitional changes and trial-error processes. It is not surprising, therefore, that informal rent-seeking solutions and patronage are widely used by producer groups, service providers, and market suppliers as a common practice to minimize risks of market failures and contractual uncertainties and to reduce transaction costs.

3.2 Pastoral Governance Challenges and Institutional Capacity Gaps

The present enabling environment for pastureland management in Uzbekistan is relatively weak. It does not provide a comprehensive and cohesive approach for the sustainable management of pastureland. The laws and resolutions often reflect the narrow institutional goal of a particular organization with no attempt to coordinate or integrate the effort for a comprehensive approach to the sustainable management of pastureland. Moreover, most of the legislation and institutions involved in the agricultural sector focus on arable land and particularly on irrigated land and on forests, but none are strictly focused on pastures. Furthermore, pastureland is not classified as a type of land use within the current land classification system used in Uzbekistan.

This weak enabling environment for the sustainable management of pastureland leads to two major limiting factors for good pastureland management:

- *Complex land ownership:* There are different types of pastureland users (*shirkat, dehkan,* private farm, and household livestock), and each one has its own set of rules and regulations. Additionally, households do not have formal access rights to pastures for their livestock husbandry activities, despite owning a large portion of livestock grazing in these areas. This land tenure complexity is hampering a comprehensive and effective approach for pastureland planning and management. In the context of sustainable pastureland management, these land tenure issues need to be addressed over the long term if a long-term sustainable and economically productive system is to be put in place (Robinson, Chap. 11).

- *Lack of comprehensive and integrated planning and management approach for pastureland:* Despite some existing elements of pasture management, there is no comprehensive integrated planning process to manage pastureland in Uzbekistan. There is a need for putting the existing elements together in a comprehensive integrated approach, including a stronger participation of stakeholders (see below).

Little knowledge exists on the value of pastureland ecosystems, and according to national priorities, there is little chance for more national budget allocation to this sector in the short term. Decision-makers need to analyze and as much as possible quantify the importance of ecosystems to human well-being in order to make better decisions regarding the sustainable use and management of these ecosystem services. Better knowledge of the value of pastureland would support the mobilization of more national resources to protect and conserve these pasturelands. Implementation of better SLM practices in Uzbekistan requires the development of new capacities of people involved in the management of pastureland. New skills and knowledge are required not only for officers working in related institutions – particularly at the regional and local levels – but also for communities, *dehkans,* and *shirkats* that are using pastureland for their livelihoods. Related with the need for new skills and knowledge, there is also the need to strengthen procedures – particularly for assessing pastureland – in order to provide pastureland managers with better information for assessment and to develop an integrated land use planning systems.

The mandates of institutions involved in the management of pastureland need to be strengthened. The current mechanism of coordination in place among many institutions involved in managing pastureland is not sufficient. Additionally, no institution has the clear mandate to coordinate an integrated planning and management approach for pastureland. Moving forward will necessitate the clarification of roles and responsibilities of each institution and a clear coordination mechanism to ensure an applicable integrated planning and management approach for pastureland. Water points are key elements for managing pastures in dryland areas. Yet, it is not clear which institution is in charge of managing and maintaining water wells in dryland areas, and little legislation exists specifically for this type of water wells. It is estimated that the total number of wells is more than 3,000 scattered through dryland areas of Uzbekistan; however, these water points are in a mediocre state of repair. Many water wells are not functioning, preventing the pastureland areas around the non-functioning wells to be fully exploited. Overall there is a capacity gap between the enabling environment and the institutional framework (mandates, procedures, and mechanisms) and the reality in the field. Laws and institutions exist to assess, monitor, and manage pastureland in Uzbekistan. However, there are discrepancies between the way it is supposed to work – including the laws, policies, and institutions – and the reality in the field; numerous capacity gaps exist (Squires, Chap. 12).

3.3 Options and Potential Strategies for Sustainable Land Management

The rangelands are a national asset of high economic, environmental, and cultural value, for which the whole Uzbek community needs to take responsibility. Therefore, a *national rangelands strategy* (NRS) is needed to be developed because the rangelands under current climate change face a wide range of problems, which cannot be easily resolved with current legislation and institutions, present knowledge, and existing dispute resolution procedures.

One of the key elements in this respect is to develop an integrated participatory planning and management methodology for pastureland. It is central for the long-term sustainability of the desert communities. Hence, the development of NRS and future implementation of the national land management plan should continue to emphasize a participatory approach inclusive of all key stakeholders. The emphasis on participatory processes maximizes ownership of project activities by stakeholders who in turn should become the ultimate custodians of planned achievements. This approach contributes also to maximizing the long-term sustainability of planned land management activities. The need for a strong stakeholder participatory process is particularly true for the development of a methodology for an integrated pastureland planning process. One suggested approach is to set up small working groups of key scientists and national stakeholders to review economic valuation of national grassland resources. This process will contribute to a greater understanding about the nature and extent of land degradation impacts on national economy and environment and among producers, consumers, and policymakers and would raise ownership of NRS and future planned achievements by national stakeholders.

Another key option is to strengthen the legislation and policy framework for pastureland management since the current legislative framework is a major impediment for good pastureland management in Uzbekistan. It has to include (Bellamy 2010):

- Revision of the Land Code and to include pastureland as a type of land with specific legal considerations; using international land type classification standards
- Improvement, simplification, and clarification of the process to become pastureland user such as revising the terms upon which shirkat farms operate and introducing clear pasture user rights for households livestock owners
- Legal recognition of an integrated planning and management approach for pastureland
- Reviewing the livestock policy and to approve a specific law on pastureland

The next option for SLM is to develop an integrated participatory planning and management methodology for pastureland with templates and manuals (Bellamy 2010). Based on the lessons learned and international best practices, it is recommended to develop a participative methodology to develop a participatory land use plan for district-level municipalities and national-level agencies, including a spatial approach using GIS. It is important that this methodology be developed with a strong participation/ownership of key government organizations, which should be involved in the development of these plans.

Considering that pastureland represents about 50% of the total area of Uzbekistan and a significant percentage of the population uses pastures as their main livelihood, an important option would be to conduct a national socioeconomic study in dryland areas to assess the value of this type of ecosystems. The study would identify the services provided by this type of ecosystem, the contribution to the local economy, the contribution to the livelihood of local communities, and more generally the importance of this type of ecosystem for human well-being. In addition to this, another option would be to conduct a study of pastureland assessments conducted by Soviet scientists over the last 30–40 years and to assess the data collected by the Soil Institute through pastureland assessments conducted over the past decades. Using an appropriate methodology, the study would try to collate this information and assess the desertification trends over time (if any) for both dryland areas in Karakalpakstan region and the Kyzyl Kum Desert. As part of this exercise, the study would also identify vegetation capacities of grasslands versus capability toward livestock production, the monitoring indicators used to monitor land degradation in these areas, and assess how the WOCAT[3] indicators and the LADA[4] methodology (international standards) could be used to strengthen the monitoring of pastureland.

As mentioned above, a clear land ownership and tenure system have been a major challenge for improving pastureland management over the long term in Uzbekistan. Land ownership in Uzbekistan is a complex issue with different types of pastureland users including shirkats, dehkan, private farms, and household livestock owners. Each type of pastureland users has its own set of laws and regulations, which leads to a competing use of pastures and, in some instances, a skewed taxation system whereby a shirkat has to pay land taxes for an area that is also used by other pasture users. However, it was also noted that the government is encouraging the development of private enterprises through more and more allocation of certificates/licenses to private enterprises to produce Karakul sheep. This is a sign of privatizing the production of Karakul; however, it may add pressure on pastureland if the land ownership is not clarified.

4 Likely Impact of Climate Change and Its Implications

As a whole, the territory of the country, located in the arid zone of Asia, is highly susceptible to land degradation, desertification, and climate change (Chub 2007). Recent research by ICARDA has revealed the limitations in current knowledge on the

[3] World Overview of Conservation Approaches and Technologies (WOCAT). www.wocat.org.

[4] Land Degradation Assessment in Drylands (LADA). http://www.fao.org/nr/lada.

impacts of climate change on dryland forests and people. Dryland forest adaptation studies are relatively recent, and only a few have documented evidence of success in the implementation of adaptation strategies. Given the diversity of dryland forests, more precise regional and local climate change projections are required. Much more research is especially needed on the forest-related social and economic impacts of climate change. Poverty may push people to invade protected areas to use wild products, possibly unsustainably, while greater wealth may lead to even more exploitation of natural resources. It is necessary, therefore, to continue to support research that will reduce uncertainty about the climate change impacts on forests and improve knowledge about management and policy measures that will promote successful adaptation. Migration and urbanization are particular challenges. Protected virgin rangeland areas potentially can provide benefits in the form of genetic resources to the pharmaceutical, biotechnology, agrochemical, seed, horticulture, cosmetic, and phyto-medical markets, but these different markets give rise to a wide range of approaches to benefit-sharing.

The Fourth Assessment Report of the IPCC identifies soil carbon sequestration potential source of mitigation from agricultural sector – from both a technical and economic viewpoint (FAO 2010). International climate change negotiations have not been yet successful in development of fully functional soil carbon finance scheme yet. However, there is currently a soil carbon finance scheme available from voluntary carbon market. Voluntary market trades emission reductions known as verified emission reductions that cannot be used for regulatory compliance. Although the voluntary market is tiny compared to compliance market, the future potential of soil carbon finance from large-scale grasslands can be utilized by Uzbekistan as well. However, this future development requires efficient contributions and efforts from both international carbon market actors and national governments.

There are major government and donor programs for environmental protection and natural resource management in CA (see below), and Uzbekistan as a case study has sought to support agroforestry. Over the last few decades, a series of low-cost technologies the efficiency of nonconventional water use in agri-silvi-horticultural and silvi-pastoral systems to meet the food and feed demands and develop adaptation strategies for vulnerable communities to climate change and water resource shortage were tested and adopted (Toderich et al. 2008). Alternatives of land use contributing to income generation such as the reclamation of marginal lands by using nonconventional water for irrigation (drainage, takyr surface rainfall water, artificial pods with slight salinity, underground water) were implemented in Central Kyzyl Kum and Priaralie and in rainfed desert and semidesert foothills of Navoi region and Fergana Valley. Furthermore, introduction of drip deep small-scale irrigation technologies significantly assisted farmer and agropastoral communities. Various low-cost biosaline technologies for crop diversification including conservation tillage (or no-tillage intervention), efficient water (marginal quality) use, feed and livestock production, and rangelands management were evaluated with participatory work of rural communities, especially women groups (Toderich unpublished data). Rural-based communities were engaged through organizing of learning alliance between farmers, animal breeders, agropastoralists, and householders.

4.1 *Initiatives Under the Aegis of the UN*

Land resources in CA are severely affected by desertification caused by degradation of vegetative cover, sand drifting, water and wind erosion, salinization of arable lands, human-induced desertification, and soil contamination and water pollution with industrial and domestic wastes. The Sub-regional Action Programme (SRAP) of the UNCCD entails support for a regional network of stations to monitor desertification and creation of a regional mechanism to fight desertification, including afforestation of the dry bed of the Aral Sea and adjacent areas. The SRAP also includes measures to reduce poverty, including improvement of degraded pastures and arable lands, restoration of the irrigation network, diversifying agriculture and livestock breeding, and promotion of traditional and new methods for income generation, including agriculture, handicrafts, and ecotourism.

In the mountain ecosystems in eastern Uzbekistan, land degradation and overgrazing, coupled with the replacement of useful plant species by weeds, have led to reduced biodiversity, desertification, and reduction in ecosystem stability. The SRAP entails the development of a regional system for management of CA mountain ecosystems and protection from natural hazards. Several actions are proposed to improve social and economic conditions for the inhabitants of mountain territories (e.g., alternative energy sources, recreational activities, infrastructure, biodiversity protection, and promotion of eco- and agrotourism).

The SRAP provides a number of means to enhance public participation in regional decision-making related to sustainable development. These include strategies for ensuring independent analysis of information on CA projects and environmental conditions. Farmers, pastoralists, animal herders, and householders, especially women groups, were trained and involved in the activities. Accordingly, our recent studies began to involve the farming and agropastoral communities through participatory decision-making. The research approach, on-farm testing and verification and other measures, led to faster dissemination of sustainable technologies.

4.2 *Central Asian Countries Initiative for Land Management (CACILM)*

CACILM is a multicountry and multidonor, long-term (2006–2016) program in the spirit of UNCCD aimed at restoring, maintaining, and enhancing productive functions of land in the five countries of the CA while preserving its ecological functions. Its ultimate goal is to increase economic and social well-being of the population who depends on the land resources. In each participating country, National Programming Framework on SLM forms its strategic basis. CACILM is implementing the programs at national and multicountry levels, through the Support Project of CACILM Multicountry Partnership Framework.

4.2.1 Recent CACILM Activities

The overall goal of the National Focal Point to the UNCCD of Uzbekistan is to combat land degradation through the strengthening and mainstreaming of SLM approaches among all land management stakeholders. Attaining this goal will result in stabilized/improved ecological integrity and better living standards of rural population affected by dryland degradation and desertification. The goals and projects of the NPF group on the seven priority program areas identified are (a) capacity building for creating an enabling environment and for the integration of SLM into policy and planning, (b) adaptation of agriculture to climate change, (c) sustainable management of forests and renewable energy, (d) sustainable management of pastures and carbon sequestration, (e) applied research, (f) integrated resource management, and (g) restoration of vulnerable ecosystem in the Aral crisis region.

Issues and challenges of land degradation have impacts which are not limited to the national borders, but which affect ecosystems of neighboring countries and are of global importance. To assist the CA countries in their efforts on implementation of the UNCCD, the Global Mechanism of UNCCD has defined the Strategic Partnership Application (SPA) of the UNCCD in the CA countries. The primary goals of the SPA are originally connected with development of coordinated, integrated, and certain donor responses for CA countries' assistance in implementation of UNCCD. The SPA has supported CACILM at the partnership forum held in Tashkent, Uzbekistan, in June 2003. The forum participants (donors and representatives of the countries) have adopted the agreement to step forward (i) to integrate basic issues of SLM both in the field of sustainable development planning and into development frameworks of external cooperation of the countries-partners, (ii) to promote inter-sectoral coordination for harmonized operation of SLM initiatives, (iii) as a strategy of resource mobilization to take advantage of the GEF financing programs to combat land degradation, and (iv) to establish in each country of the CA the UNCCD National Working Group on partnership development for implementation of the UNCCD.

5 Barriers and Challenges to Implementation of Sustainable Land Management

As with other CA countries, Uzbekistan faces several major problems – accelerated land degradation and poverty. The majority of rural dwellers face uncertainty about land tenure. Most households are involved in subsistence farming or animal husbandry. Lack of collateral, poor rural credit facilities, and lack of knowledge are formidable barriers to improved incomes. Lack of knowledge about ecological concepts and the reality of living from day to day mean that there will be some time before the concept of land stewardship can take root and be translated into better land management practices.

The steps toward the adoption of SLM are well known, and programs are in place in Uzbekistan to put these into practice. Participatory planning and the support for the formation of land and water user groups may eventually lead to better land management. In the meantime, government policy on land tenure reform, support for rural finance facilities, and the development of the national rangelands strategy will be steps in the right direction.

Related to the need for new skills and knowledge, there is also the need to strengthen procedures – particularly for assessing pastureland – in order to provide pastureland managers with better information for assessment and to develop an integrated land use planning systems.

Acknowledgment We are grateful to Dr. Victor Squires (Australia) for his assistance with the preparation of this manuscript.

References and Further Reading

ADB, CACILM (2007) CMPF Support Project – SLMIS project document (http://cacilm.adb.org)

ADB, CACILM (2009) Final report on field monitoring and assessment studies conducted at selected sites. NSIU, Tashkent

Amelin IS (1944) Rangeland rotations in Karakul Sheep Husbandry of Central Asia (in Russian). All-Union Scientific Research Institute of Karakul Sheep Breeding, Samarkand, 108 p

Ashurmetov OA, Rakhimova T, Hasanov O, Shomurodov X (1998) Recommendations on improvement of desert rangelands of Uzbekistan. J Desert Dev 1:87–90

Bekenov B, Grachev I, Milner-Gulland EJ (1998) The ecology and management of the Saiga antelope in Kazakhstan. Mammal Rev 28(1):1–52

Bekmirzaeva I, Yusupov S, Rabbimov A (2010) Livestock and pasture management plans for Kazakhdarya and Kyzyl Ravat region. Materials of the UNDP project "Achieving ecosystem stability in degraded land in Karakalpakstan and Kyzylkum desert". UNDP CO, Tashkent

Bellamy J (2010) Assessment report and recommendations for institutional capacity development. Materials of the UNDP project "Achieving ecosystem stability in degraded land in Karakalpakstan and Kyzylkum desert". UNDP CO, Tashkent

CACILM (2006) National programming framework of the Republic Uzbekistan, Tashkent, 48 p

CIA – Central Intelligence Agency, The World Factbook, www.cia.gov, accessed in May 2012

Chub VE (2007) Climate change and its impact on the hydro-meteorological processes, agricultural and water resources of the Republic of Uzbekistan. Uzhydromett, NIGMI, Tashkent, 132 p

FAO (2010) Carbon sequestration in dryland soils. World Soil Resources Report, 102, Rome

UNFCCC (2008) Fourth Assessment Report of IFCC Synthesis Report, Adaptation and Mitigation Strategies. WMO and UNEP

Gaevskaya LS (1971) Karakul sheep breeding rangelands of Central Asia. Fan, Tashkent, 321 p

Gaevskaya LS, Krasnopolin FS (1957) Rangelands of Samarkand Province and its use in Karakul sheep husbandry. In: Agricultural Issues of Zerafshan Basin. AN USSR, Tashkent

Gaevskaya LS, Salmanov NS (1975) Rangelands of deserts and semi deserts of Uzbekistan. FAN, Tashkent, 140 p

Gintzburger G, Toderich K, Mardonov B, Mahmudov M (2003) Rangelands of the arid and semi arid zones in Uzbekistan. CIRAD/ICARDA, Montpellier, 420 p

Gintzburger G, Le Hourou HN, Toderich K (2005) The Steppe of Middle Asia: Post 1991 agricultural and rangeland adjustment. J Arid Land Res Manage (ALRM) 19:19–43

Holland M (2010) Mid-term SLM project evaluation report. Materials of the UNDP project "Achieving ecosystem stability in degraded land in Karakalpakstan and Kyzylkum desert". UNDP CO, Tashkent

IAMO (2008) Continuity and change: land and water use reforms in Rural Uzbekistan. Socio-economic and legal analyses for the northern region Khorezm. Available at: www.iamo.de/dok/sr_vol43.pdf

IMF (2008) International monetary fund, welfare improvement strategy of Uzbekistan: full strategy paper for 2008–2010. IMF, Washington, DC

IMF (2008a) International monetary fund. Republic of Uzbekistan: Poverty Reduction Strategy Paper (PRSP). Jan, 2008. IMF, Washington, DC. Accessed online in Sep 2012: http://www.imf.org/external/pubs/ft/scr/2008/cr0834.pdf

Lal R (2007) Soil and environmental degradation in Central Asia. In: Lal R, Suleimenov M, Stewart B, Hansen D, Doraiswamy P (eds) Climate changes and terrestrial sequestration in Central Asia. Taylor & Francis/Balkema Publishers, Leiden, pp 127–137

Nazaruk LA (1968) Biologicheskaya harakteristika nekotorih kormovih vidov odnoletnih solyanok v svyazi s ylychsheniem pastbits na adirah Nishanskoi "stepi". Avtoreferat dissertatzii na soiskanie ychenoi stepeni kandidata biologicheskih nayk, Tashkent [In Russian]

Rajabov TF (2010) Some peculiarities of spatio-temporal changes of vegetation of Karnabchul rangelands under different condition of grazing. J Agro Ilm 4:31–32

Rajabov TF (2011) Spatio-temporal changes of vegetation cover of semi desert along the grazing gradient (in case of Karnabchul). PhD thesis. Samarkand Division of Academy of Sciences of Uzbekistan

Robinson S, Michel S, Wiedemann C, Zhumabayev Y, Singh N (2012) Pastoral tenure in Central Asia: theme and variation in the five former Soviet republics, Chapter 11. In: Squires V (ed) Rangeland stewardship in Central Asia. Springer, Dordrecht, pp 239–274 (Chapter 11, this volume)

Salmanov NS (1996) About sustainable utilization of desert and semi desert rangelands. Uzbek Biol J 3:51–54

Sergeeva GA (1951) Improvement of Karakul sheep breeding rangelands of Uzbekistan. Karakulevodstvo i zverovodstvo, no 4, pp 77–78

Shamsutdinov ZS (1975) Establishment of perennial pastures in arid zones of Central Asia. FAN, Tashkent, 176 p

SNC (2009) Second national communication of Uzbekistan on UNCCC, Tashkent

Toderich KN, Black CC, Ekaterina J, Osamu K, Tolib M (2007) C3/C4 plants in the vegetation of Central Asia, geographical distribution and environmental adaptation in relation to climate. In: Lal R, Suleimenov M, Stewart B, Hansen D, Doraiswamy P (eds) Climate changes and terrestrial sequestration in Central Asia. Taylor & Francis/Balkema Publishers, Leiden, pp 33–65

Toderich KN, Ismail S, Juylova EA, Rabbimov AR, Bekchanov BB, Shyuskaya EV, Gismatullina LG, Kozan O, Radjabov T (2008) New approaches for Biosaline Agriculture development, management and conservation of sandy desert ecosystems. In: Chedly A, Munir O, Muhamad A, Claude G (eds) Biosaline agriculture and salinity tolerance in plant. Birkhauser Verlag, Basel, pp 247–264

Toderich KN, Shuyskaya EV, Ismail S, Gismatullina L, Radjabov T, Bekhchanov BB, Aralova D (2009) Phytogenic resources of halophytes of Central Asia and their role for rehabilitation of sandy desert degraded rangelands. J Land Degrad Dev 20(4):386–396

UNDP (2010a) In: Olhoff A, Schaer C (eds) Screening tools and guidelines to support the mainstreaming of climate change adaptation into development assistance – a stocktaking report. UNDP, New York

UNDP (2010b) Food security in Uzbekistan. UNDP, Tashkent

UNDP (2010c) Livestock production in Uzbekistan: current state, challenges and prospects. Review in the context of agricultural sector development trend. UNDP, Tashkent

UNDP/GM (2007) Traditional land management knowledge in Central Asia Joint publication of the UNDP and Global Mechanism of the UNCCD. UNDP, Almaty

UNECE (2010) Environmental performance review series nr 29. Uzbekistan: 2nd report. UN, New York & Geneva

World Bank (1999) Uzbekistan: social and structural policy review, Report No 19626, Washington, DC

Yeo A (1998) Molecular biology of salt tolerance in the context of whole-plant physiology. J Exp Bot 49:915–929

Yusupov SU (2003) Interaction between livestock and the desert environment in Uzbekistan. In: Schrader F, Alibekov L, Toderich K (eds) Proceedings of NATO Advanced Research Workshop "Desertification problems in Central Asia and its regional strategic development", Samarkand, pp 93–96

Yusupov SY (2010) The management plan for livestock/cattle-breeding and rangelands improvement in Kizil Ravat (southeast Kyzylkum), Tashkent 2010, p 45

Part VI
Where to From Here?

Key questions have been answered in the chapters that precede this section such as:

- Where are we now in our quest for sustainable development?
- What role can land stewardship and heightened awareness of intergenerational equity play?
- What is the importance of creating an enabling environment?

The questions remain:

- Where do we want to be in the near, midterm and long-term future?
- How do we get there?

Chapter 17 reviews international perspectives on legislative and administrative reforms as an aid to better land stewardship in Central Asia. Comparisons of the approach adopted in China and in Mongolia with that in the other countries of the Central Asian region provide useful ideas for consideration of those who are devising new land tenure legislation.

Chapter 18 draws conclusions based on the chapters in this book and from the author's extensive experience in Central Asia and in other former Soviet countries. A possible way forward is outlined.

Chapter 17
International Perspectives on Legislative and Administrative Reforms as an Aid to Better Land Stewardship

Ian Hannam

Abstract This chapter discusses the international and national environmental law for rangelands and outlines frameworks for legislative reforms to achieve sustainable use of rangelands. International and national legal instruments and institutional systems play a significant role in rangeland conservation. At the international level, it discusses a number of multilateral agreements that could be better used to promote the sustainable use of rangeland. Two national level approaches to reform environmental law for rangeland are presented, for Mongolia and PRC respectively, and may offer useful guidelines for other countries in the Asian region to follow in environmental law reform for rangeland management.

Keywords International conventions • IEM • Soil conservation • Land degradation • Rangeland • Biodiversity • Land tenure • Mongolia • China • Kyrgyzstan • Tajikistan • Africa • Land code • Climate change

Key Points

- Since the early 1900s, over 200 multilateral environmental treaties, agreements and protocols have been developed to manage and protect the world's natural environments and natural resources. A number of these instruments contain elements that can contribute to rangeland conservation and management. However, none of them is sufficient on its own.
- Legislation has been used for years in many countries, usually in a piecemeal fashion, to manage specific types of problems on grazing land (e.g. soil erosion), to control grazing activity which causes land degradation problems (e.g. overgrazing of cattle and sheep), and to indirectly control land management problems (e.g. through environmental planning and land use allocation). There is room to improve the capacity of both international environmental law and national environmental law frameworks concerning rangeland management.

I. Hannam (✉)
Centre for Agriculture and Law, University of New England, Armidale, Australia
e-mail: ian.hannam@ozemail.com.au

V. Squires (ed.), *Rangeland Stewardship in Central Asia: Balancing Improved Livelihoods, Biodiversity Conservation and Land Protection*,
DOI 10.1007/978-94-007-5367-9_17, © Springer Science+Business Media Dordrecht 2012

- Two principal framework structures that an international environmental law instrument for rangeland may take are in the form of either a legally binding instrument or a non-legally binding instrument. As a binding instrument, this could take the form of either a specific treaty with all the essential legal elements for rangeland, a framework treaty which identifies the rangeland elements in existing treaties and links them through a separate binding instrument or a protocol to an existing treaty that creates specific rules for rangeland. The development of a binding instrument under one of the three global treaties discussed earlier in this chapter, with key rules and guidelines for rangeland management, is an option that would benefit all rangelands of the world. As a non-binding option, this could take the form of an international charter for rangeland or a declaration for rangeland.

1 Introduction

Rangelands, as defined earlier in this volume, occupy about 50% of the world's land area and cover a large proportion of many Asian countries. Of significance with rangeland management has been the relationship between land user and individual governments. Different countries demonstrate considerable diversity in the preferred uses of rangeland and the administrative and legal processes for its use, in particular the form of land tenure (Nori et al 2005; Fernandez-Gimenez 2006; Taylor 2006). Various political, social and geographical factors influence the different approaches to the land use and land tenure issues. Traditionally, in a nation with an abundance of rangeland, fewer land use restrictions have been applied, and land users have enjoyed a great deal of freedom with few limitations on individual rights. Rangelands have many functions that need to be properly recognised by legislation. Environmentally, the most important one is that it provides a vegetation cover and thus protection for the soil and water, which also ensures sustainable economic production of animal fodder, firewood and other indirect benefits. Rangelands are a product of environmental factors, but they also contribute to both the local and global environment (Friedel et al 2000; Neely et al. 2009).

1.1 Rangeland Law

Globally, while rangelands are considered significant in terms of environmental issues, they are inadequately recognised in international environmental law. This is an attitude of ignorance because they play a major role in climate change with far-reaching consequences for the environment. Although there is no specific category of law recognised as "rangeland law", for the purposes of this chapter, "rangeland law" is considered as that area of law which includes specific laws, regulations and legal instruments that govern the use of rangelands. Other legal approaches to control the use of rangeland, especially the conversion of pastoral land to non-pastoral uses, are also relevant. In some countries, pastoral law can regulate land consolidation or the procedure of land reallocation. In addition to these laws, other laws and regulations affect rangeland, often significantly, and in some cases have played a greater role in the management of rangeland than traditional pastoral land law which may be limited in scope with administrative role over leases or access conditions. Land protection laws have been important for keeping pastoral land productive; these laws have focused on soil productivity, prevention of erosion and protection of land from other environmental damage (Grossman and Brussaard 1992; Chalifour et al. 2007).

In Central Asia, the situation is still fluid, for example, in the Republic of Tajikistan, there are no differentiated legal provisions between pastureland[1] and arable land. There is a lack of policy and special regulations for pastureland use, and the land reform designed for agricultural land prevailed in the distribution of pastureland and the emergence of current modes of tenure for pastureland. A small proportion of the pastureland belongs to few state entities, and some pastureland that is included in the Forest Fund is leased (Halimova, Chap. 13). In contrast, in the Kyrgyz Republic, pasture management, its improvement and utilization are regulated by the Land Code of the Kyrgyz Republic and the Law of the Kyrgyz Republic on Pastures 2009, as well as corresponding legislation and other regulatory acts of the Kyrgyz Republic.

This chapter discusses the international and national environmental law for rangelands and outlines frameworks for legislative reforms to achieve sustainable use of rangelands. International and national legal instruments and institutional systems play a significant role in rangeland conservation. At the international level, it discusses a number of multilateral agreements that could be better used to promote the sustainable use of rangeland. Two national level approaches to reform environmental law for rangeland are presented, for Mongolia and PRC respectively, and may offer useful guidelines for other countries in the Asian region to follow in environmental law reform for rangeland management.

2 Multilateral Environmental Treaties

Since the early 1900s, over 200 multilateral environmental treaties, agreements and protocols have been developed to manage and protect the world's natural environments and natural resources. A number of these instruments contain elements that can contribute to rangeland conservation and management. However, none of them is sufficient on its own. While some of the existing instruments could assist by promoting the management of the activities that can maintain a sustainable rangeland environment, this role is not readily apparent except for those that include provisions specifically directed to rangeland ecosystems. Since UNCED 1992, the global policy environment has changed considerably, with the adoption of the Millennium Development Goals, the outcomes of the World Summit on Sustainable Development in 2002, increased support to the least-developed countries, stronger commitment for climate change mitigation and adaptation and prospects of global agricultural trade liberalisation. As a result, international environmental law is being called upon

[1] In this chapter, and throughout the book, rangelands and pasturelands are used interchangeably.

to provide a wider application to global environmental issues (Chalifour et al. 2007), and rangeland conservation should receive more attention. The scientific environment has also evolved with the work of the Millennium Assessment on dry land ecosystems, which has contributed to improved understanding of the biophysical and socio-economic trends relating to land degradation in global dry lands and their impacts on human and ecosystem well-being (White et al. 2002).

The following is a brief account of the main instruments.

2.1 The World Charter for Nature

The World Charter for Nature (UN 1982) called on states to cooperate in the conservation of nature, establish methods to assess the adverse effects on nature and implement international legal provisions for the conservation and the protection of the environment. The charter states that the productivity of land shall be maintained or enhanced by the use of measures that safeguard their long-term fertility, the processes of organic decomposition, and safeguard against all forms of degradation.

2.2 The Rio Declaration and Agenda 21

The Rio Declaration on Environment and Development of 1992 established the goal of a new and equitable global partnership through the creation of new levels of cooperation among states, key sectors of society and individuals. It emphasised the need for states to work towards international agreements to protect the integrity of the global environmental system and to enact effective environmental legislation (Principle 11) (UN 1992a). The other key instrument of the same time, Agenda 21 (UN 1992b), discusses international environmental law processes that can assist in the global management of rangeland (Chapters 8, 38 and 39 of Agenda 21 discuss international legal instruments and mechanisms).

2.3 Convention to Combat Desertification

The United Nations Convention to Combat Desertification (CCD) (UN 1995a) addresses land degradation and the methods to protect and manage soil and water resources of rangeland areas. Desertification and drought are problems of global dimension, affecting all rangeland areas of the world, and joint action of the international community is often called upon to combat these problems (Neely et al. 2009). Under Article 1 of the CCD, "desertification" means land degradation in arid, semi-arid and dry sub-humid areas resulting from various factors, including climatic variations and human activities. The CCD acknowledges that arid, semi-arid and dry sub-humid areas together account for a significant proportion of the earth's land area and is the habitat and source of livelihood for a large segment of its population. The objective of the CCD is to prevent and reduce land degradation, rehabilitate partly degraded land and reclaim desertified land, particularly in countries that experience serious drought. In this regard, it recognises the high concentration of developing countries, notably the least-developed countries, which are among those experiencing serious drought and/or desertification. It recognises that there must be support by international cooperation and partnership agreements which contribute to sustainable development. This will involve long-term integrated strategies that focus on the rehabilitation, conservation and sustainable management of pastoral land and water resources (Neely et al. 2009).

The six program areas of Chapter 12 *(Managing Fragile Ecosystems: Combatting Desertification and Drought)* of *Agenda 21* are acknowledged by the CCD, and it is considered that they provide a useful basis to establish an approach for combating desertification in rangelands. A principal feature of the CCD relevant to rangeland management is that it outlines how countries can approach the development of national action plans and obtain scientific and technical cooperation (Articles 9–18) and supporting measures (Articles 19–21). These plans can address many important land degradation issues and the appropriate methods to protect and manage soil and water resources of rangeland areas (Secretariat CCD 2000; European Commission 2000). Although the CCD does not have the specific elements that recognise rangeland as an individual ecological element, it does contain many elements that cover legal principles and processes for rangeland conservation.

2.3.1 UNCCD 10-Year Strategy

The CCD 10-Year Strategy 2008–2018 was adopted in 2007 and provides a global framework to support the development and implementation of national and regional policies, programmes and measures to prevent, control and reverse desertification/ land degradation and mitigate the effects of drought through scientific and technological excellence, raising public awareness, standard setting, advocacy and resource mobilisation. The strategic objectives are directly relevant to management of rangelands and were developed to guide the actions of CCD stakeholders and partners in the period 2008–2018, including raising political will. They include improvement of living conditions of affected populations, improvement of affected ecosystems, generating global benefits, mitigating climate change and mobilising resources to support the implementation of the CCD. It is advocated that the operational objectives of the CCD 10-Year Strategy could be enshrined in national legislative frameworks for rangeland management as they focus on advocacy, awareness, education, policy development, science and technology, capacity building and financing and technology transfer (Secretariat CCD 2007).

2.4 Convention on Biological Diversity

Fundamental to the Convention on Biological Diversity (CBD) (UN 1995b) is the concern that biological diversity is being significantly reduced by human activities (e.g. overgrazing and biomass harvesting), and this obviously includes ecological processes in rangeland environments. The CBD stresses the importance of, and the need to promote, international, regional and global cooperation among countries and intergovernmental organisations and the non-governmental sector for conservation of biological diversity and the sustainable use of its components (see Article 16) and for nations to prepare strategies to implement the CBD.

The objective of the CBD is relevant to rangeland management as it includes the conservation of biological diversity, encouraging the sustainable use of its components and the fair and equitable sharing of the benefits arising out of the utilization of genetic resources, including by access to genetic resources and by transfer of relevant technologies. It takes into account various rights over those resources (Article 1). The CBD also recognises that nations have a responsibility for conserving their biological diversity and for using their biological resources in a sustainable manner. In this context, the objective of rangeland conservation is implicit within the definitions of "biological diversity" and "biological resources" in the CBD.

The CBD acknowledges that substantial investments are required to conserve biological diversity and that there is a broad range of environmental, economic and social benefits from those investments (see Articles 6–10). It stresses the importance of, and the need to promote, international, regional and global cooperation among countries and intergovernmental organisations and the non-governmental sector for conservation of biological diversity and the sustainable use of its components (see Article 16) and for nations to prepare strategies to implement the CBD (Miller and Lanou 1995). For the CBD to take on an expanded, more precise role in the sustainable use of rangelands, for example, specific provisions could be drafted for rangeland and included as a protocol to the CBD. The rules should focus on the ecological functions of rangeland that are essential for the conservation of biodiversity and the maintenance of human life in these areas.

2.5 *Convention on Climate Change*

The United Nations Framework Convention for Climate Change (FCCC) (UN 1995c) recognises the role of terrestrial ecosystems as a sink and reservoir for potential greenhouse gases and is concerned that human activities have been substantially increasing the atmospheric concentrations of greenhouse gases. Two of the principal sources of greenhouse gases are changes in land use cover and land use. Scientists have established that rangeland ecosystems provide a significant reservoir of carbon (Squires and Glenn 1997; Squires 1998) and that pastoral activities play a role in emissions of greenhouse gases and initiate or exacerbate soil and vegetation degradation. In particular, these include biomass burning, cultivation, using organic manure, applying nitrogenous fertilisers and livestock grazing (Neely et al. 2009). Excessive vegetation clearance, a principal cause of land degradation, is one of the key concerns of the FCCC. Land degradation exacerbates the emission of gases from terrestrial ecosystems to the atmosphere. Accelerated wind and water erosion, on a global scale, is the principal soil degradation process. Some 1,643 million ha are affected worldwide, of which 250 million are affected by strong or extreme forms of soil erosion. While the FCCC does provide for changes to the terrestrial environment, it is not considered to be the most appropriate international legal vehicle to address rangeland conservation because it presently has a primary focus on changes in the industrial sector rather than the non-industrial and agricultural land use sectors.

The Kyoto Protocol (Kyoto 1997) under the FCCC includes a responsibility to promote sustainable forms of land use in the light of climate change characteristics. It specifically recognises the need to expand and preserve soil carbon sinks and improve agricultural practices in countries where a significant proportion of the emissions are related to the clearing of vegetation for agriculture.

2.5.1 UNFCCC Adaptation and Mitigation

In recent years, the FCCC process has made significant progress in providing a mechanism for developing countries to access climate change funds to implement adaptation and mitigation activities. In this regard, the development of an effective legal, policy and institutional framework would be an essential component of a national strategy to manage the climate change impacts on rangeland ecosystems. Adaptation strategies can be based on reducing land degradation, improving livestock management and improving human livelihood. Mitigation actions could consider policy development, monitoring and reporting methodology, economic assessment and capacity building (Wilkes et al 2011).

2.6 Draft International Covenant on Environment and Development

The IUCN Draft Covenant has been prepared as an umbrella agreement to knit together the principles reflected in the sectoral treaties that impact upon the environment and development (IUCN 2000). The Draft Covenant has many articles that are relevant to the protection and management of rangeland. In particular, Article 20 relates to natural systems and calls on parties to take appropriate measures to conserve and where necessary and possible restore natural systems which support life in all its diversity, including biological diversity, and to maintain and restore the ecological functions of these systems as an essential basis for sustainable development.

2.7 Regional Instruments

A number of regional instruments include elements that countries (other than those within in the specific jurisdiction of the instrument) can adopt in framing national rangeland legislation. Of particular relevance is the African Convention on the

Conservation of Nature and Natural Resources which was adopted in 1968 and revised in 2003 (African Convention 2003). Instead of taking a purely utilitarian approach to natural resources conservation, it acknowledges the principle of common responsibility for environmental management and calls for conservation and rational use of natural resources for the benefit of present and future generations. After 25 years, developments in international environmental law made it necessary to revise this treaty, update its provisions and enlarge its scope, in particular, to provide for the establishment of institutional structures to facilitate compliance and enforcement. Many aspects of the African Convention are useful for other countries to follow, especially its objectives which correspond to key elements of a sustainable approach for rangeland conservation: the achievement of ecologically rational, economically sound and socially acceptable development policies and programmes. Importantly, the African Convention embodies a comprehensive and integrated regional approach to environmental protection and sustainable development. It reflects a renewed perception of resource management that reconciles nature and culture. This instrument advocates an integrated approach to resource management and provides international legal principles and best practices that are relevant to rangeland management. In this context, the African Convention provides a useful approach to regional natural resources management, and a similar concept could be applied to rangeland conservation for the Central Asian region.

3 National Legal and Institutional Frameworks

Legislation has been used for years in many countries, usually in a piecemeal fashion, to manage specific types of problems on grazing land (e.g. soil erosion), to control grazing activity which causes land degradation problems (e.g. overgrazing of cattle and sheep) and to indirectly control land management problems (e.g. through environmental planning and land use allocation) (Kurucz 1993). An analysis of national legislation associated with rangeland conservation indicates that individual states have used a variety of approaches to frame domestic legislation and to deal with protection and management of rangeland. This is generally reflected by the broad structural features of the legislation, as well as in the variety of specific mechanisms used to protect and manage rangeland (Hannam with Boer 2002). In some countries, pastoral land law was an early form of legislation used to manage rural land. Although its primary role has been land administration, as against environmental management and land conservation, this type of law often included mechanisms that could achieve sustainable use of land, including enforceable conditions on land leases for grazing management, vegetation, soil and water management and monitoring rangeland condition and pasture quality (Hannam 2007).

In the past, the main type of legislation aimed at the control of degradation of rangeland has generally been the law associated with "soil conservation". The legislation had a land utilization focus, which is no longer adequate to effectively protect and manage the world's pastoral degradation problems. As a rule, as the area of land affected by degradation grew, practical land conservation techniques were developed and applied in conjunction with expanding agricultural activities. The conservation capabilities of the legislation were overshadowed by the objective of agricultural production, price support schemes for domestic and export needs and land settlement and development schemes. Soil conservation legislation was introduced in the first half of the nineteenth century primarily to control the effects of soil erosion by wind and water on pastoral and cultivation lands (Grossman and Brussaard 1992). Over time, a variety of laws have been developed for rangeland and pastoral land use, and some examples appear in Box 17.1.

Box 17.1 Examples of National Laws Relating to Pastoral Management

- United States of America – *Public Rangeland Improvement Law 1978, Forest and Range Renewable Resources Planning Law 1974*
- Canada – *Agricultural Land Commission Law 1979* (British Columbia)
- Australia – *South Australia Pastoral Land Management and Conservation Law 1989*
- New Zealand – *Resource Management Law 1991*
- Iceland – *Bill of Legislation on Soil Conservation 2002*
- People's Republic of China – *Grassland Law 2002*
- Mongolia – *Pastureland Law* (2007 draft)
- Kyrgyzstan – *Law of the Kyrgyz Republic on Pastures 2009*

Some observations of the characteristics of laws used for rangeland management are summarised below (Hannam 2007):

- Various laws have been used to rectify land management problems caused by poor land use planning or inappropriate land use, as against the inherent ecological characteristics of land being used as the premise for land use decision-making.
- Many national laws associated with rangeland are still very much overshadowed by the physical problems of land use, mainly grazing and agriculture.
- The primary land functions of rangelands are not well represented, and few laws refer to the ecological features or needs of rangeland.
- The legislation does not acknowledge rangeland as having a central role in terrestrial ecology, the conservation of biodiversity and maintenance of environmental amenity.
- Many laws are not clear on the purpose or objectives for rangeland.
- There is often not a logical development of legal elements, and many laws do not include the elements necessary to protect rangeland.
- There is inconsistent use of terminology, and often there is an absence of definitions, inadequate or poorly stated definitions.
- The structure of some laws indicates that they are a reaction to a political or institutional issue rather than designed to effectively manage rangeland.

- Some states have developed a framework of legislation to manage specific land use management problems of which pastoral land management is one type – but they generally lack a linking or coordinating mechanism.

3.1 Other Legal Regimes with a Role in Administration of Rangeland

Other areas of environmental law which together make up a framework of law that can be effective in administration of rangeland include (Grossman and Brussaard 1992) the following.

Land Administration Law
There are many laws that provide for the administration and management of rangeland. These have been employed to control the use of land and its mismanagement. This legislative regime includes various forms of land tenancy and leasehold regimes, with provisions to assess land and regulate conditions of occupancy, use, sale, lease and reservation. There may also be provisions for forfeiture of holdings, alteration of conditions of use and protection of land dedicated for public use.

Biodiversity Law
There is a body of law that has a general objective to protect the environment and conserve biological diversity and ecosystems. This area of law assists in the management of rangeland through its promotion of ecologically sustainable development and through the conservation of biological diversity in general. Biodiversity legislation generally does not apply directly to the control of agricultural land uses, but it is applied to achieve more effective conservation and management of protected and reserved areas in agricultural landscapes (De Klemm 1993).

Vegetation Conservation Law
This area of law is important because it focuses on the conservation and sustainable management of vegetation and pastures and can control their destruction. It promotes vegetation and pasture management in consideration of social, economic and environmental parameters. It sets rules for the ecological assessment of vegetation (its biodiversity, habitat values, flora and fauna values, regional patterns and threatened species) and rules for issue of permits.

Forest Law
Forest laws have been created for both public and private forestry with provisions to develop and apply land management guidelines (Tarasofsky 1999). Plantation establishment and reafforestation law have been applied in some countries to promote reafforestation of land and establishment of shrub land in rangeland badly affected by land degradation (water and wind erosion and salinity) from overgrazing and cultivation.

Environmental Protection Law
The main purpose of this legislation is to protect, restore and enhance the quality of the environment by reducing the risks to human health and preventing the degradation of the environment by pollution, waste management, discharge of harmful substances and point-source pollution. National environmental protection measures can cover any activity that may impact or has impacted on rangeland and affected by pollution, waste dumping, chemicals or other substances that may impact on the pastoral environment.

Environmental Planning Law
The objective of environmental planning legislation is to protect the human and natural environment through the preparation of land use or zoning plans, prescribing environmental assessment standards for land use and determining significant environmental impact from land use change. This type of legislation has the potential to be more widely applied to benefit rangeland by preserving its natural character, protecting outstanding natural features and landscapes, efficient use and development of rangeland resources, maintenance and enhancement of amenity values and heritage protection.

Climate Change Law
A global review for evidence on rangeland systems and climate change indicates that greater recognition and support are needed for sustainable pastoral and agro-pastoral systems in view of their contributions to climate change adaptation and mitigation, disaster risk management, biodiversity protection and sustainable agriculture and rural development. Targeted support by governments, civil society organisations, development agencies and community donors and researchers is needed to harness this opportunity (Neely et al. 2009). Given the important role of rangelands in the management of climate change impacts, it is considered that more attention needs to be paid in national environmental law reform to provide for adaptation and mitigation actions on rangeland.

4 Approaches

The following two examples outline two different approaches to legislative reform to manage environmental problems in rangeland areas. Mongolia and the People's Republic of China (PRC) share a similar ecological gradient and have both been undergoing socio-institutional and ecological transformations: from collective to market economies, increasing market integration and adverse climate impacts on vegetation. However, the two areas are different in their demographic, ethnic, cultural, economic and political contexts, and the respective governments have quite different approaches to rangeland management. These features of the two areas make for a comparative analysis of their legislative response to rangeland management, to enhance an understanding of the respective rangeland socio-ecological systems and to identify solutions towards more sustainable governance of the rangelands.

The Mongolian example discusses a single law approach, where the Pastureland Law has been drafted to manage a range of environmental, social and economic problems associated with grassland management. The PRC example discusses a comprehensive and integrated law approach to manage a complex range of environmental, social and economic issues associated with land degradation in the dry land ecosystems of western PRC. The legislative approaches taken with each of the two examples are vastly different and partly reflect the complexity of problems associated with rangeland usage including the existing political environment. In the Mongolian example, at the time of its drafting in 2007, the Pastureland Law was not considered within the context of the existing national environmental law framework, whereas the PRC approach was considered within a complex of nine related national environmental law areas that apply to the western region.

4.1 Mongolian Pastureland Law (Single Law Approach)

Mongolian rangeland is degraded because herders are unable to apply sustainable grazing practices. Mongolian grassland is not valued so its regulation and management have been avoided in the past. Herders continue to graze their livestock on public land unrestrained, where there is high competition for good pasture. They use public pasture and water free of charge and without initiative to protect and properly use it (Swift 2007). Grasslands and arid grasslands cover a large proportion of the country (112 million ha or 71%) but around 122 million ha of Mongolia is devoted to nomadic pastoralism: 4.6% of this lies in the alpine zone, 22.9% in the forest-steppe zone, 28% in the steppe zone, 23.3% in the semi-desert zone and 16.2% in the desert. As relatively intact terrestrial ecosystems, the Mongolian pasturelands play a significant role in sequestrating atmospheric carbon dioxide, conserving biodiversity and especially in providing livelihood benefits to local herders. However, over the past 50 years, pastureland degradation has undermined the ecosystem services they generate. Under the Mongolian Land Law 2002, "pastureland" (which equates with rangeland) means *rural agricultural land covered with natural and cultivated vegetation for grazing of livestock and animals* (Article 3.1.6 Land Law 2002).

There are many reasons why Mongolia drafted the Pastureland Law (Mongolian Government 2007). The main aim was to develop a legal framework to provide land tenure for livestock producers, herder groups, cooperatives and herding households; to establish remote pasture reserves; and for the operation of pasture use committees as a basis to overcome many of the problems that stem from lack of land ownership (Fernandez-Gimenez 2006). It also aims to develop a legal framework to exempt herding households from individual income tax and to introduce pasture use fees differentially based on economic and ecological assessments and to consider a livestock husbandry risk fund. The *State Policy on Herders*, introduced in 2009, will provide strategic support to the Pastureland Law. The policy aims to create a favourable legal, economic and business environment which enables development of better living conditions for herders, prevention of poverty in herders and employment and social security (Mongolian Government 2009).

4.1.1 Legal Concept

The draft Pastureland Law 2007 has been through an extensive public and parliamentary discussion process. Although at the time of writing this chapter it had still not been promulgated, the government and community are working through a complex of issues to ensure effective legislation finally results. The main purpose is to provide a legal method for the transition from an unplanned and unregulated grassland user system to a system characterised by secure possession of pastureland for herders and legal entities; a pastureland planning and management system; to improve the development and management of pastureland information; to distinguish the functions, duties and responsibilities between the different levels of administration; and to improve the system to identify and manage problems associated with land degradation and the effects of global climate change. Importantly, it includes procedures to classify pastureland on an ecological basis and provide for the agricultural and economic needs of traditional herding communities and the livestock husbandry industry. In this regard, the new administrative, implementation and operational procedures of the Pastureland Law will support economically productive pastoral agriculture while managing pressures on the ecological environment from climate change, desertification and natural disasters (Mongolian Government 2007, 2010).

The draft Mongolian Pastureland Law includes a procedure to allocate land for grazing and for its management and protection, which constitutes the basis of land tenure, and this will help overcome many problems that stem from the traditional pasture usage system (Fernandez-Gimenez 2006). The procedure includes the identification and classification of pastureland, a request (application) for pastureland possession for the purpose of livestock husbandry and the issue of a certificate for possession. It will also engage the communities in the land tenure process. Global experience shows that land use systems that enable stakeholders to formally participate in the decision-making process generally provide a more satisfactory and balanced outcome for all parties. This procedure will help promote the sustainable use of pastureland and development of a stewardship ethic which is important for a stable long-term tenure system (Squires, Chap. 2). These are important procedures and should increase the capability of Mongolia to manage its natural rangeland resources more effectively in the face of the increasing effects of climate change and other natural events (Batima 2006).

The draft law contains many legal elements considered essential for successful rangeland law, but it is considered that additional support systems will be required to enable herder communities and legal entities to achieve a sustainable livelihood and for the state to achieve its national goals for pastureland management

(Hannam 2007; Mongolian Government 2009). Some areas identified include development of operational policy, development of a "national strategy for grassland management", development of land management plans, formation of local stakeholder advisory committees, providing access to finance and credit, developing a comprehensive education and training programme and ensuring stakeholders have access to information and knowledge.

4.1.2 Rangeland Conservation and Climate Change

In an effort to improve its management of the climate change effects on rangeland ecosystems, the Mongolian Government has taken steps to develop a Nationally Appropriate Mitigation Action (NAMA) for grassland and livestock management by following the procedure established under the FCCC (Mongolian Government 2010; ADB 2011). Under these procedures, a NAMA is defined as any kind of activity that reduces greenhouse gas emissions. For Mongolia, the specific grassland and livestock management activities developed under the NAMA procedure would be nationally appropriate and tailored to Mongolia's national circumstances and in line with the FCCC principle of common but differentiated responsibilities. In following the FCCC procedure, the NAMA will be embedded within Mongolia's national sustainable development strategy. The mitigating activities will be measurable, reportable and verifiable and supported. Importantly, by following the procedures set down in the FCCC process and satisfying the standards for national and international registration, this opens up the potential for Mongolia to access climate change funding to implement the NAMA (KPMG 2011). One of the essential requirements for the effective implementation of the grassland/livestock NAMA is a legal and policy framework that in Mongolia's case would include the Pastureland Law. It is considered that the NAMA approach adopted by Mongolia would also be suitable for other developing countries which have extensive rangeland areas to follow.

4.2 Example 2: Management of Dry Land Ecosystems of People's Republic of China (Integrated Law Approach)

The total area of grassland in the PRC is around 400 million ha, accounting for 42% of the country's land area. In PRC, grassland is considered a multifunctional ecosystem that provides ecological and economic benefits. Consideration of the main functions of the different grassland types in each ecological-economic region of the PRC forms a basis for adopting different management regimes (Brown et al. 2008). During 2004–2009, an investigation was undertaken into the legal, policy and institutional framework in PRC's western ecosystems under an international PRC project, covering the three provinces of Qinghai, Shaanxi and Gansu and the three autonomous regions of Inner Mongolia, Xinjiang Uygur and Ningxia Hui (Du and Hannam 2012). Western PRC includes a significant proportion of PRC's rangelands, and land degradation is a serious problem that affects this area. As many vulnerable communities are dependent on rangeland resources for their livelihood, land degradation is closely linked to poverty across these provinces and regions (Ren et al. 2008; Williams et al. 2009). The project employed an integrated ecosystem management (IEM) approach to managing land degradation (GEF 2000) with an emphasis on capacity building and technological support and to strengthen cross-sector coordination and trans-boundary management of the natural resources. The total population of this area is around 120 million people.

The objective of the legal and policy investigation was to improve the policy and regulatory framework for land degradation control as an essential part of strengthening PRC's enabling environment and to build capacity to adopt an integrated approach to sustainable land management. It aimed at improving the policies, laws, regulations and procedures for combating land degradation, and this was achieved using three programmes: (1) improving the law and policy framework to intensify institutional capacity, (2) capacity building to implement the law and policy and (3) a supporting study programme to innovate and reform environmental laws and policies (Du and Hannam 2012).

4.2.1 IEM as a Legal Concept

The concept of IEM has been applied in international environmental law and progressively developed into normative principles and rules (UNEP 1995b). In the PRC investigation, IEM was defined as "a holistic approach to address the linkages between ecosystem functions and services (such as carbon uptake and storage, climatic stabilization and watershed protection, and medicinal products) and human social, economic and production systems" (Jiang Zehui 2007). From a legal perspective, as a comprehensive strategy and method to manage natural resources, IEM is a suitable framework in which to consider the national and provincial legislation (Du and Hannam 2012). By definition, IEM requires taking all components of an ecosystem into account and in consideration of the social, economic and natural environment. The investigation was designed in a manner for each province and region to develop a practical framework to enhance the capability of different groups to implement law and policy of IEM, including legal officers, legal draftsmen, judicial officials, policymakers, government officials and private individuals.

The starting point to assess the capacity of the existing legal and policy framework and policies for land degradation control was the categorisation of legislative materials into nine principal law areas (Du and Hannam 2012):

1. Grassland (includes the *Grassland Law 2002, Regulations on Prevention of Grassland Fires 2005, Administrative Measures for Balance of Grass and Husbandry 2007*); at provincial level, it includes over 20 individual regulations and legal instruments
2. Desertification
3. Water and soil conservation
4. Water resources

5. Forestry
6. Agriculture
7. Land administration
8. Environmental protection
9. Environmental impact assessment

A methodological procedure was developed to accommodate the principles of IEM in the provincial and regional law and policy reform process. The method was applied by officials from the People's Congresses and governmental legislative offices of Qinghai, Gansu, Shaanxi provinces and Inner Mongolia, Xinjiang and Ningxia autonomous regions. Experience from applying IEM in the legal and policy processes of the respective provinces and autonomous regions of western PRC was substantial and includes many lessons which other countries could follow (Du and Hannam 2012):

- IEM provides a scientific approach to fulfil commitments to various multilateral conventions concerning environmental protection and sustainable use of natural resources, and IEM establishes a strategic framework to manage land, water and biological resources for sustainable rangeland development.
- The IEM approach is a cross-cutting mechanism that accommodates multiple scientific means and is a good policy tool to coordinate national implementation requirements of international environmental conventions as they apply to rangelands.
- It is an effective means to achieve sustainable use of rangelands and combat land degradation and is a sound framework to review and solve issues concerning natural resource ownership, use of protected areas, access to resources and benefit sharing.
- The flexible framework of IEM provides multiple options for implementation, including incorporation of IEM principles in national strategies and action plans, regional plans, and applying IEM principles in policymaking, land use and institutional planning. It is a good basis for reform of institutions and organisations to support sustainable use of rangeland ecosystems.
- IEM is a relevant tool for planning, decision-making and evaluation of ecosystem activities associated with all aspects of rangeland management, policy and law.

As an outcome of the investigation, the central government and the governments of the three provinces and three autonomous regions made a commitment to adopt IEM to improve the legal framework. It provided a valuable opportunity for PRC to introduce IEM into its legal procedures by adopting international best experiences and rules of law. As a result, the PRC had revised various laws and rules by December 2007, including the *Measures for Implementation of Grassland Law in Qinghai Province*. A new legislative reform programme was also introduced which included the *Regulations on Administration of Grassland in Tianzhu Tibetan Autonomous County* and *the Detailed Rules on Implementation of Grassland Law of the People's Republic of China in Xinjiang Uygur Autonomous Region*.

5 Improving Legislation for Rangeland

It is considered that there is room to improve the capacity of both international environmental law and national environmental law frameworks concerning rangeland management. In this regard, the following discussion can be used as a basis to improve the environmental law for rangeland management in the Asian region.

5.1 International Level

At the international level, a number of multilateral and regional treaties contain elements and principles that provide for various problems associated with rangeland management, but a formal coordination mechanism would have to be developed to enable this to happen. The fact that rangeland occupies a significant proportion of the earth's terrestrial surface would seem to justify that this is a matter that requires urgent consideration. Two principal framework structures that an international environmental law instrument for rangeland may take are in the form of either a legally binding instrument or a non-legally binding instrument.

As a binding instrument, this could take the form of either a specific treaty with all the essential legal elements for rangeland, a framework treaty which identifies the rangeland elements in existing treaties and links them through a separate binding instrument or a protocol to an existing treaty that creates specific rules for rangeland. The development of a binding instrument under one of the three global treaties discussed earlier in this chapter, with key rules and guidelines for rangeland management, is an option that would benefit all rangelands of the world.

As a non-binding option, this could take the form of an international charter for rangeland or a declaration for rangeland.

5.1.1 Process

It is essential that the promotion of an international legal framework to protect rangeland provides an opportunity for the input of all interested parties, including international environmental organisations, state governments, rangeland science institutions, private sector interests and non-government organisations. Should this proceed, it would be open for the Central Asian states to participate. The experiences involved in the development and introduction of other existing environmental treaties should be examined before embarking on a process for rangeland. In general, for rangeland, such a process could involve:

- Building an adequate understanding of current rangeland degradation processes and issues to establish a clear vision of the benefits of an international legal framework for rangeland

- Assembling existing policy, strategic material and legislation, which have specific or indirect references to rangeland conservation (e.g. biodiversity, environmental planning and natural resources legislation)
- Reviewing appropriate international instruments and strategic material and identifying the relevant instruments that may be accommodated within the political, cultural and physical circumstances for rangeland
- Outlining a capacity-building process, including environmental education for the international community, focusing on the most effective types of technical training for those involved in the development and implementation of state strategies for the legal protection of rangeland

For this process to be effective, it will require cooperation between international rangeland policy and science organisations and an international environmental law institution such as the IUCN Commission on Environmental Law. For Central Asia or even the Asian region, there may be benefits in the respective states pursuing a regional instrument that provides specifically for the social, economic and ecological characteristics of rangeland in the particular geographic region.

5.2 *National Level*

To establish or improve national rangeland legislation in Central Asia or any other region, various international legal principles can be applied to form the philosophical basis on which to select a suitable approach to develop a framework for national rangeland management legislation. It is appreciated that some states may prefer to develop rangeland strategies with a minimum of legal regulation, whereas others may prefer a stronger regulatory law (Hannam and Boer 2002).

5.2.1 Non-regulatory Strategies

Non-regulatory strategies would feature a predominance of elements that concentrate on:

- Education and awareness programmes for sustainable use of rangelands
- Ecosystem research, assessment and monitoring rangeland use
- Financial support for research and extension
- Support and development of participatory community rangeland planning
- Development of ecologically sustainable rangeland standards and practices
- Development of rangeland management and incentive-based programmes

5.2.2 Regulatory Strategies

Regulatory strategies would feature a predominance of elements that concentrate on:

- Development of statutory land use plans that prescribe limits and targets of rangeland use (e.g. maximum number of livestock at particular times of the year, permissible cultivation practices)

- Issue of licences or permits to control rangeland use (these would prescribe use entitlements relating to fencing, stock numbers, access and soil restoration requirements)
- Land use agreements between the state and individuals or groups of land users, which set legally binding land use standards
- The use of restraining notices where sustainable rangeland use limits (as set out in a statutory plan or agreement) are exceeded
- Prosecution for failure to follow prescribed standards of sustainable use

5.3 *Elements of National Rangeland Legislation*

To assist in the development of a national rangeland law framework, with a mix of legal protection and management elements, states may benefit from a set of "generic elements" from which to select appropriate elements for the construction of a national rangeland law (Hannam 2004; Commonwealth of Australia 2010). The general range of legal elements which could be considered by Central Asian states to consider for national rangeland management law, for which the specific procedures would be developed, includes the following types of elements:

- A comprehensive statement of the purposes of the legislation.
- Goals and objectives with a mandate for ecologically sustainable use of rangelands – specific objectives can be formulated from the objectives and principles of global conventions, strategies and policies concerning ecology, the conservation of nature, biodiversity and sustainable land management.
- The preparation of rangeland policy, codes of practice, sustainability indicators and the physical and ecological limits of rangeland.
- The preparation of a rangeland management strategy, outlining national rangeland management policy and policy to manage land use problems.
- That sustainable use of rangeland can be achieved through a mix of regulatory and non-regulatory means, including incentive and support programmes and community rangeland management advisory groups.
- Provision to manage all classes of rangeland that are based on sustainable land use criteria and contain ecologically sustainable standards for implementation at the national, provincial and local levels.
- Procedures to manage natural resources generally, including provisions to protect soil, water and biodiversity.
- An equitable distribution of responsibilities in managing rangeland, including the state, minister, administrators, advisory bodies, officials and herders.
- A facility to enter into legal contracts and agreements with rangeland users and occupiers for the sustainable use of rangeland resources.
- A facility for the community to participate in rangeland assessment, planning and decision-making, including establishing community advisory groups; rangeland management plan preparation; public exhibition of management plans, policies and strategies, and calling for public submissions; and provision for community representatives to sit on rangeland management committees.

- A facility to develop education programmes on sustainable use of rangeland and implementation of technical seminars and conferences.
- A facility to develop and implement a variety of rangeland research and investigation programmes and to relate the research outcomes to state programmes.
- A facility to take formal action where prescribed standards of rangeland management use are not being met and where there has been a contravention of the legislation.

6 Conclusion

In general, legal and policy frameworks can be effective in raising awareness to improve the management of rangelands, and they also present an opportunity to include procedures for adaptation and mitigation of climate change impacts while enhancing livestock productivity and food security. In particular, for the Central Asian region, the development of a sound legal and policy framework by an individual state should contain procedures that improve the documentation and dissemination of information on rangeland management and build capacity and to manage rangelands through incentives, including payments for environmental services and other nonfinancial rewards. Based on experiences from other areas of land use, the use of voluntary and regulatory procedures can help change behaviour towards sustainable and adapted management of rangeland ecosystems. Including incentive mechanisms within legal instruments can lead to adoption of sustainable activities and reverse land degradation in rangeland – activities that will also enhance livelihoods and reduce the vulnerability of pastoral and agro-pastoral people.

National rangeland legislation should also provide the means to develop livestock policies that address the barriers and bottlenecks faced by the inhabitants, including procedures to build local and policy-level awareness and capacity for rangeland husbandry and help secure tenure at community and landscape levels. Legislation should also provide for targeted research in rangeland ecosystems, effective institutions and with a focus on practices that can improve the economic aspects of pastoral management.

At the administrative level, it is essential that integrated multi-sectoral, multistakeholder and multilevel processes be available that address the range of natural resources (land, water, rangelands, forests, livestock, energy and biodiversity) and social dimensions of rangeland environments, with active participation of stakeholders. A holistic approach and partnership can take advantage of the objectives of local, national and global goals.

Importantly, it is also appropriate that the international and national frameworks for rangeland management address climate change impacts and adopt the processes of the United Nations FCCC that are relevant to rangeland management, especially in relation to adaptation and mitigation actions. Consideration should be given by Asian states to support the concept of an international environmental law instrument for rangeland management, in particular, or at the regional level with elements and procedures that provide for the unique aspects of rangeland in this region.

References and Further Reading

ADB (Asian Development Bank) (2011) *Inception report*, ADB R-CDTA 7534, Strengthening carbon financing for regional grassland management in Northeast Asia

African Convention (African Convention on the Conservation of Nature and Natural Resources) (2003) see http://www.intfish.net/treaties/africa2003.htm; extensive analysis of the Convention is available at http://uicn.org/themes/law/pdfdocuments/EPLP56EN.pdf

Batima P (2006) Climate change vulnerability and adaptation in the livestock sector of Mongolia. Final Report Project AS06. Assessments of Impact and Adaptation to Climate Change. International START Secretariat, Washington, DC, 409 p

Brown C, Waldron S, Longworth J (2008) Sustainable development in Western China: managing people, livestock and grasslands in pastoral areas. Edward Elgar Publishing, Cheltenham

Chalifour N, Kameri-Mbote J, Lin Heng Lye P, Nolon JR (eds) (2007) Land use law for sustainable development. IUCN Academy of Environmental Law Research Studies, Cambridge University Press, Cambridge, UK

Commonwealth of Australia (2010) Principles for sustainable resource management in the rangelands. Natural Resources Ministerial Council, Department of the Environment, Water, Heritage and the Arts, Canberra

De Klemm C, in collaboration with Shine C (1993) Biological diversity conservation and the law: legal mechanisms for conserving species and ecosystems. IUCN Environmental Law Centre and IUCN Biodiversity Programme

Du Q, Hannam I (eds) (2012) Law, policy and dryland ecosystems: People's Republic of China. IUCN, Gland, xvi + 140 pp

European Commission (2000) Analysis of national reports on the implementation of the United Nations convention to combat desertification. Turkey, Lebanon, Jordan and Syria, Drylands Program, International Institute for Environment and Development

Fernandez-Gimenez ME (2006) Land use and land tenure in Mongolia: a brief history and current issues. USDA Forest Service Proceedings RMRS-P-39

Friedel MH, Laycock WA, Bastin GN (2000) Assessing rangeland condition and trend. In: Mannetje L't, Jones RM (eds) Field and laboratory methods for grassland and animal production research. CABI, Wallingford, pp 227–262

GEF (Global Environment Facility) (2000) GEF Operational Program 12 Integrated Ecosystem Management, Nairobi

Grossman M, Brussaard W (eds) (1992) Agrarian land law in the western world, essays about agrarian land policy and regulation in twelve countries of the western world. CAB International, Wallingford

Halimova N (2012) Land tenure reform and implications for land stewardship. In: Squires V (ed) Rangeland stewardship in Central Asia. Springer, Dordrecht, pp 305–332 (Chapter 13, this volume)

Hannam ID (2004) A method to identify and evaluate the legal and institutional framework for the management of water and land in Asia: the outcome of a study in Southeast Asia and the People's Republic of China. Research report 73. International Water Management Institute, Colombo

Hannam ID (2007) Report to United Nations Development Programme Mongolia on review of Draft Pastureland Law of Mongolia. United Nations Development Programme Sustainable Grassland Management Project, Ulaanbaatar

Hannam ID, Boer BW (2002) Legal and institutional frameworks for sustainable soils: a preliminary report. IUCN, Gland/Cambridge, UK, xvi + 88 p

IUCN (The World Conservation Union) (2000) Draft international covenant on environment and development, Second Edition: Update Text, Commission on Environmental Law in Cooperation with International Council of Environmental Law, IUCN Environmental Law Program, Bonn, IUCN

Jiang Zehui (2007) To implement integrated ecosystem management to accelerate combating Land degradation. In: Integrated ecosystem management, Proceedings of the international

workshop, Beijing 1–2 November 2004, Global Environment Facility, Asian Development Bank and People's Republic of China Forestry Publishing House, pp 2–6
KPMG (2011) Financing low carbon investment in developing countries: public-private partnership for implementing nationally appropriate mitigation actions. KPMG International Cooperative, Switzerland
Kurucz M (1993) Land protection, property rights and environmental preferences (land use control and land development). Conn J Int Law 8(2):467
Kyoto (1997) The Kyoto protocol to the convention on climate change, the Climate Change Secretariat, UNEP
Miller KR, Lanou SM (1995) National biodiversity planning. Guidelines based on early experiences around the world. The World Resources Institute and IUCN, Baltimore
Mongolian Government (2007) Draft pastureland law 20 July 2007 and brief introduction to the draft law on pastureland
Mongolian Government (2009) Government policy towards herders. 4 June 2009, Resolution 39, Ulaanbaatar
Mongolian Government (2010) Mongolian second national communication under the United Nations framework convention on climate change. Ministry of Nature, Environment and Tourism
Neely C, Bunning S, Wilkes A (eds) (2009) Review for evidence on dryland pastoral systems and climate change, implications and opportunities for mitigation and adaptation. Food and Agriculture Organization of the United Nations, Rome
Nori M, Switzer J, Crawford A (2005) Herding on the brink: towards a global survey of pastoral communities and conflict – an occasional paper from the IUCN Commission on Environmental. Economic and Social Policy, Gland
Ren JZ, Hu ZZ, Zhao J, Zhang DG, Hou FJ, Lin HL, Mu XD (2008) A grassland classification system and its application in China. Rangel J 30(2):199–210
Secretariat CCD (Secretariat of the Convention to Combat Desertification) (2000) Chart of key elements in the CCD National Reports by the Countries in Asia
Secretariat CCD (Secretariat of the Convention to Combat Desertification) (2007) Report, 8th session, Conference of the Parties, Madrid 3–14 September 2007, ICCD/COP (8)/16/Add.1, 23 October 2007
Squires VR (1998) Dryland soils: their potential as a sink for carbon and as an agent in mitigating climate change. Adv GeoEcol 31:209–215
Squires VR (2012) Better land stewardship: an economic and environmental imperative. In: Squires V (ed) Rangeland stewardship in Central Asia. Springer, Dordrecht, pp 31–50 (Chapter 2, this volume)
Squires VR, Glenn E (1997) Carbon sequestration in the drylands: an agenda for the twenty-first century. World atlas of desertification, 2nd ed, Edward Arnold/UNEP
Swift JJ (2007) Case study: institutionalizing pastoral risk management in Mongolia: lessons learned. Prepared under the overall guidance from the Rural Institutions and Participation Service, FAO Project Pastoral Risk Management Strategy, TCP/MON/0066
Tarasofsky RG (ed) (1999) Assessing the international forest regime. IUCN Environmental Policy and Law Paper No. 37
Taylor JL (2006) Negotiating the grassland: the policy of pasture enclosures and contested resource use in inner Mongolia. Hum Organ 65(4):374–386
UN (United Nations Environment Program) (1982) World Charter for Nature
UN (United Nations) (1992a) Rio declaration, New York
UN (United Nations) (1992b) Agenda 21, New York
UN (United Nations) (1995a) Convention to combat desertification, New York
UNEP (United Nations Environment Program) (1995c) United Nations framework convention on climate change
UNEP (United Nations Environment Program) (1995b) Convention on biological diversity, Nairobi

White RP, Tunstall D, Henninger N (2002) An ecosystem approach to drylands: building support for new development policies, World Resources Institute Washington, DC Information Policy Brief No. 1

Wilkes A, Wang S, Tennigkeit T, Feng J (2011) Agricultural monitoring and evaluation systems: what can we learn for the MRV of agricultural NAMAs? ICRAF Working Paper No. 126. World Agroforestry Centre, Beijing, China

Williams A, Wang MP, Zhang M (2009) Land tenure arrangements, property rights and institutional arrangements in the cycle of rangeland degradation and recovery. In: Squires V, Lu X, Lu Q, Wang T, Yang Y (eds) Rangeland degradation and recovery in China's pastoral lands. CABI Publishing, Oxford, pp 219–234

Chapter 18
Conclusions and a Way Forward

John E. Leake

Abstract Based on the preceding chapters and other experience, we seek to review the impacts of the changes of the last 100 years in the Central Asian region and to point a way forward that might be more 'sustainable' and of more value to both the farming and livestock communities who live in the region than the negative impacts of the adjustments to recent political and economic events have been.

This suggested path forward seeks to enhance the symbiotic relationship that has almost always characterised farmer-livestock-grassland relationships, by better adapting to the new institutional and economic circumstances and to make the whole system more resilient to future change as urban populations and demands increase.

The pathway requires a more integrated and decentralised approach than has been the experience through this last century or so and a better community understanding of the interdependencies between the biophysical and social systems that underpin life in the region. The different, but complementary, paths taken by farmers, who are bound to place, and nomadic people and their animals are the subjects of this chapter.

Keywords Resilience-biophysical • Social • Decision-making • Stakeholders • Institutional governance • Integration • Symbiotic • Pastoral • Risk management • Adaptation • East Africa • Ecosystems • LFA • Landscape approach • Panarchy • Mobility • Offtake • Stewardship

J.E. Leake (✉)
The Institute for International Development, Adelaide, Australia
e-mail: jleake@iid.org

V. Squires (ed.), *Rangeland Stewardship in Central Asia: Balancing Improved Livelihoods, Biodiversity Conservation and Land Protection*,
DOI 10.1007/978-94-007-5367-9_18, © Springer Science+Business Media Dordrecht 2012

Key Points

- Farming and livestock keeping have long diverged because of the different imperatives of agriculture and grassland use, particularly in dryer zones, but have always included symbiotic elements that are often overlooked when the systems are looked at in isolation.
- The resulting cultural and institutional practices of both are as important as the prevailing biophysical and economic circumstances in planning to reverse the accepted negative trends of recent times throughout Central Asia.
- Notions of stewardship of land for this and future generations are not well developed in any of the Central Asian nations at official or land holder level; this is partly due to continuing government policies to maximise production of cereals and cotton, lack of communication between ministries responsible for land management and those responsible for land allocation, partly because of insecure or confused land use rights at the land holder level and a widespread lack of appreciation of the market and other linkages between stakeholders in land, water and livestock production and associated processing and marketing within and between states.
- It needs to be more commonly recognized that the people who live in the rapidly degrading areas in populated regions of Central Asia will not change behavior unless they see this will increase profits and improve the predictability of family livelihoods.
- International experience shows that land and water governance systems that enable more immediate stakeholders to participate in decision-making achieve better and more balanced outcomes than remote centralised systems.
- Institutional governance systems have to promote integration between stakeholders vertically and horizontally and act to inform decision-making to reduce risk and to improve biophysical and social resilience in rural areas.

1 Introduction

Interaction between humans and grasslands is as old as humans themselves, beginning with their assumed evolution in the savannah grasslands of East Africa. The co-evolution of agriculture and human communities from this origin of hunter-gathers in forests and grasslands has, and continues to have, transformational impacts on both with, perhaps partly concomitant, changes in climate. These impacts are detectable genetically in the biological make-up of grasslands and forests and at least culturally in humans such that humans are now seen as within these ecosystems and not the low-impact visitors they were until a few thousand years ago.

Humans are increasingly seen as important parts of forest and grassland ecosystems because of their observed impacts, rather than minor players drawing sustenance as they began in the East African savannah, perhaps two million years ago. Over the last 10,000 years or so, these ecosystems can be seen to have changed biologically so they are now seminatural communities when considered from a human perspective

(Olsen 2011). The changes have been less dramatic in dryer areas where cultivation is not possible, but constant adaptation in response to climate change, technological developments and population growth has occurred through a process of stasis between livestock, pasture and people over much of this period. Recent, perhaps accelerated, pressures of reducing pasture area marginalization and political isolation are producing new changes.

The question addressed in this book is how can the stakeholders best influence the process to their benefit or will pastoralism continue the decline of the last 30–70 years. Or, as put to a recent conference, 'is the path forward or back' (Kreutzmann 2012). This chapter presents some options for going forward.

2 Pastoralism as a Way of Life

The evolution of nomadic herding out of this hunter-gatherer beginning may have happened as some people found benefit (or more independence) by following the migratory herds away from the cultivatable and irrigable oasis in dryer regions to exploit the pulsating biological riches that appear as a result of the irregular rainfall (or snow) in grassland regions. This freedom to move had many benefits including the ability to exploit fresh nutritious pastures in different zones seasonally (transhumance) and provided products and opportunities for trade with their more sedentary cousins. The urban dwellers in turn could use these to produce clothing for onward trade, establishing symbiotic links between the two. But it also set up competition for bordering lands and opportunities to raid or to extract tribute instead of raiding (Lattimore 1941). Many writers also see a co-evolution of farming and pastoralism in border regions, for example (Hardesty 2003; Barfield 1992). The system was not ideal, adjustments were often violent and significant losses often occurred through periodic weather extremes, but it was in aggregate sustained over millennia.

It has been suggested that nomads achieved a stasis between themselves, livestock and grasslands because the dominant knowledgeable herders who emerged as leaders managed numbers (and access to game), so they and their families had fat stock to eat. This restricted grazing pressure (to enable animals to remain fat) and maximized the area of grasslands utilised but put pressure on those lower in the hierarchy to adapt to survive, particularly in poor seasons or times of war (Barfield 1992). Importantly, this required good knowledge of the grasslands and a means of communication horizontally and vertically across scales (both spatial and temporal) to facilitate mobility. This anthropological view of the importance of social hierarchy in preserving the resource base is echoed in ecological thinking. For example, Gurney found *'a social structure to be a strongly beneficial influence on population stability so long as the rewards of social dominance are not too extreme'* (Gurney 1979).

A more nuanced version of this idea appears in systems thinking, for example, by C. S. Hollings, who suggested, *'Hierarchies and adaptive cycles comprise the basis of ecosystems and social ecosystems across scales. Together they form a panarchy..a healthy system.. (that) can invent and experiment, benefiting from inventions that create opportunity.. each level is allowed to operate at its own pace, protected from above.. but invigorated from below by faster smaller cycles of innovation*' (Hollings 2001) (see figure 18.6 below).

A system that reproduces the adaptive success of traditional pastoralism in the future will need to forge a new symbiotic relationship with farmers and provide just such a 'panarchy' (see further discussion below).

3 The Love-Hate Relationship Between Pastoralists and Farmers

The farmer and the cowman should be friends, 'Oklahoma' the 1955 film.

This difference in practice and interests between farmers and herders is seen in all pastoral regions and apparently in all historical periods (Galaty and Johnston 1990). As noted, it is also a symbiotic one as the benefits in trade between pastoral products of fiber and leather and manufactured products desired by the nomads are also ubiquitous. The system was well articulated by Lattimore (1940) for Central Asia as a result of travels in north China and Mongolia in the 1930s (Lattimore 1941).

The contrast between the two systems is that farmers could produce and store greater surpluses, which gave them strength in farming areas, while the pastoral nomads retained flexibility and were able to raid and trade, safe in the knowledge they could retreat into the desert if threatened. This was a dialogue of equals of a sort. Lattimore (1940) describes how their relative power fluctuated as either benefited at the expense of the other and exposed weaknesses that were in turn exploited in the other direction. He suggests it was only the rise of railways and machine guns that enabled the power to be permanently projected into semi-desert refuges, leaving the nomads increasingly marginalized and without a means of response.

As Kreutzmann (2012) implies; there is no path back, new forms of interaction are required to retain the symbiotic benefits. Attempts at achieving this are mixed, and with the loss of power and mobility, pastoralists are no longer able to look after their resource base as previously occurred.

4 Stewardship and Ecosystem Service

It is commonly agreed that land degradation is now a very serious problem in Central Asia and that improved stewardship is essential. There are many statistics on the scale of the problem but different opinions on how serious it is and what can be done about it. Many suggest that some land has to be taken out of grazing in most countries, but nobody has proposed a source of income for the livestock owners while this takes place. Others suggest changes in grazing practice and the establishment of improved access to animal feeds by changes in policies towards grain and cotton.

Squires (Chap. 10) has summarised what is known and points out that international practice shows that livestock keepers (and farmers) will not undertake change unless:

1. It is profitable.
2. It conforms to a significant extent with their experience and values.

As Squires points out, 'there really are two clear and fundamental elements of "stewardship" – awareness and action. That means:

- *Recognizing our collective responsibility to retain the quality and abundance of our natural resources;*
- *Putting that awareness into action by making the appropriate decisions for how to best use and manage these resources not only for today but for future generations as well'.*

This has aspects of improved stewardship but needs to be seen against a background necessity for livestock keepers and farmers to provide a livelihood.

5 Recent Pressures for More Offtake

The intrinsic differences between the essentially fixed and more predictable farm production systems and the erratic shifting pastoral production systems of the dryer grasslands remain, but new forms of relationship are not producing the desired results. As urban-based planners, informed by the apparently higher productivity of settled agriculture began to experiment with ways to improve the offtake of livestock products for sale, the inherent resilience of the adaptive traditional nomadic system decayed, particularly as mobility was greatly reduced leading to serious imbalances between people, livestock and grasslands that have yet to be resolved.

In Central and North Asia, this began in the nineteenth century but accelerated with the collectivization of agriculture and livestock production (Rahimon, Chap. 3). This saw the introduction of more intensive production systems based on the availability of feed reserves and imported concentrates to offset the loss of the ability to move (Sedik, Chap. 9) and (McCauley 1994). This made use of an inherent complimentary interest; it gave a market for feed, hay and second-quality grain for farmers and a source of supplementary feed for livestock, but the social hierarchal system of control has not proved durable.

Between the 1950s and the 1980s, this 'scientific' management resulted in an increase in productivity and livestock numbers, but interplay between institutions with different objectives resulted in a reorientation of farming towards grain and cotton production in most of Central Asia and reduced feed inputs available for livestock, at least on a per head basis. This can be seen in a reduction in productivity in each country which mostly began before 1990 and accelerated rapidly after this as the supply of supplementary feed evaporated with the collapse of the state farm sector (Sedik, Chap. 9).

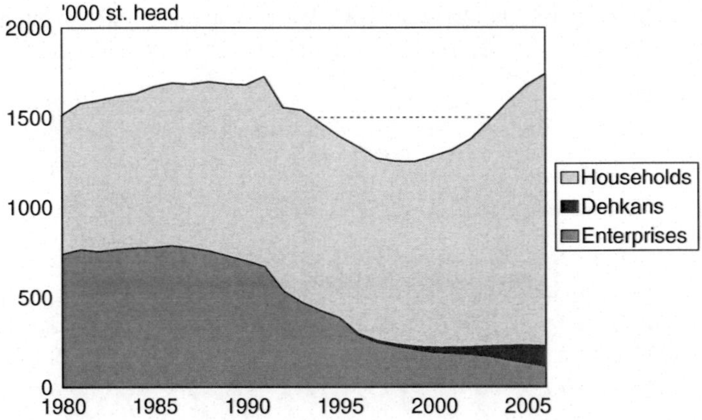

Fig. 18.1 Livestock inventories by farm type, 1980–2007 ('000 standard cow head) (Source: CISSTAT 2008)

It is commonly accepted that rangelands internationally have and continue to have profound influences on the world's carbon balance and on climate because of their ability to sequester carbon into soils (Fisher 2002). Land degradation however reverses this as carbon stores are lost with pasture loss or species change (Squires 1998).

The potential exists to mobilise ecosystem payments to meet the cost of part of this need for improved stewardship but as yet little work to achieve this.

Sedik (Chap. 9) discusses the situation in Tajikistan to show that *'livestock numbers were increasing, particularly the household sector, even back in the Soviet era. By 2007 the share of household plots in livestock had risen to over 90% (measured in standard head). This rapid expansion of livestock inventories, despite the fall in feed availability, has kept feed availability per animal extremely low'.*

Figure 18.1 shows the changing proportion of livestock ownership attributed to different entities. Collectively, households who own a few head own the most of the national livestock inventory in most Central Asian countries as these trends from Tajikistan illustrate.

Sedik goes on to show that the grain area in Tajikistan remains at levels not seen since the 1950s because of state policies that limit areas sown to cultivated feed or rather support areas sown to grain and cotton. This situation of policy distortions in land use allocation affects most countries in Central Asia because of this wish to maximize offtake for export and urban consumption. The break-up of the state farm sector and privatization of most other livestock also resulted in greatly reduced areas being grazed, partly because people brought their animals home for security and partly because of the collapse of watering points in remote areas. This caused a liquidation of herds in most countries and much degradation in the lands around villages from private herds (Shaumarov, Chap. 16). Livestock numbers have recovered since, but the aforementioned policies towards grain and particularly cotton production continue to constrain feed supplies in the autumn and winter in most countries and the ability of farmers and livestock producers to re-establish a complementary trading relationship for feed supply (CACILM preparation report 2006). This imbalance between feed demand in the winter and in periods of stress is noted as the major technical constraint to stabilising the industry.

What are the prospects for change?

6 Feed Resources and Potential

Feed resources vary significantly between states, but there are some similarities. The following is a qualitative assessment of feed resources available or potentially available for livestock over the 5 'stans' (Kazakhstan, Kyrgyzstan, Tajikistan, Turkmenistan and Uzbekistan) based on field investigations in 2006, 2009, 2010 and 2011. (This and other summaries of the situation across Central Asia are based on unpublished consultancies connected to the Central Asian Initiative for Land Management (CACILM), see www.adb.org/projects/cacilm, and other relevant consultancies have taken place in Mongolia, between 1992 and 2002, and in Azerbaijan in 2002, 2004, 2005 and 2010.) This has been divided into categories to suit the discussion on approaches to managing lands allocation to improve productivity:

- Pasture lands available to small holders near water supplies (usually overgrazed)
- Pasture lands in larger farms or farmer groups in former collectives (some of this land is now out of production because of the destruction of water points)
- Feed resources available as a by-product of cereal farming (market distortions and lack of market information are inhibiting the use of much of this)
- Feed resources that would be available if policies permitted present poor-quality cereal land to be redirected towards feed production
- Feed resources that would become available if brackish irrigation drainage water was to be used in feed production (this would require some adaptive research, investment in infrastructure and secure rights to use the water and land rehabilitated for this purpose)

Change in this situation can be addressed at the individual level and by segmenting the industry between these sectors to take advantage of comparative advantages and to mitigate risks that exist in each. There is international experience that can be considered in assessing the potential for this as discussed in the next section.

7 Segmentation of the Industry

The symbiotic links between pastoralists and farmers have evolved since the nineteenth century in various ways in different climates as increased areas of former pastoral land have been taken over by farming and these may be instructive to revisit, for example (Abdeuerfi and ElHassini 2011; Rovere et al. 2005).

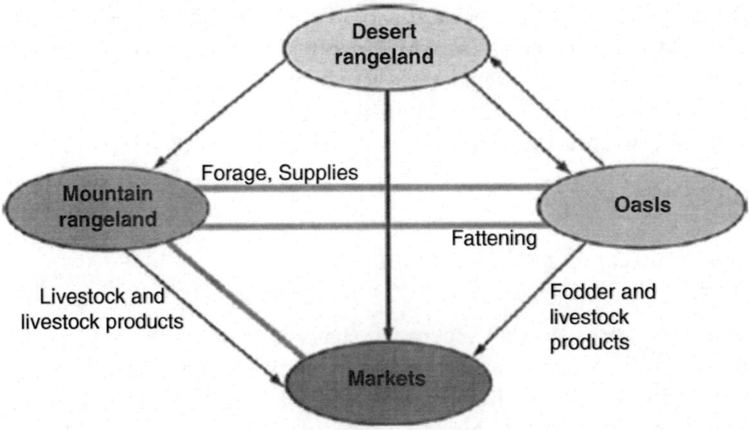

Fig. 18.2 Linkages between uplands (breeding area) and lowlands (markets and fattening areas) (Source: Squires et al. 2010)

The most common segmentation has been between breeding and fattening enterprises with breeding occurring more often in pastoral areas and fattening and processing more often occurring closer to feed supplies and sometimes markets as depicted in Fig. 18.2. This practice is being seriously considered in Kazakhstan (2011) and has considerable promise in other countries and between some countries.

Other developments to improve the quality and quantity of feed resources are possible through investments in such activities as:

- Improving species by adding legumes for ley farming as a cycle between cropping and grazing and adding fertilizer. This occurred during the Soviet period on state farms, notably through the rotation of *Medicago sativa*, grown to use rising brackish groundwater with the concomitant production of hay, followed by cereal production when ground water circumstance permitted this again. There is evidence that this practice is returning in some area, notably Kazakhstan, but this is not widespread.
- Use of fallow land for grazing, this was also a practice in Soviet times but has reduced due to destruction of water points.
- Investments to bring unconventional or waste resources into production, such as systems to use brackish irrigation drainage water and to irrigate salt tolerant species, experiments are occurring in this direction (ICARDA) and this will require further investigation.

However, as discussed in previous chapters, there are formidable barriers to achieving a rationalization of resource use in this way. Although these differ between countries (*see* Robinson, Chap. 11 and Halimova, Chap. 13), some general observations can be made:

- Landholders often have short-term, insecure or confusing rights to use the land or pasture. They cannot use the pastures as they may wish.

- Government policies towards the allocation and use of land and pasture often differ between institutions, and there is often ambiguity about who makes decisions between those allocating land rights and those apparently responsible for monitoring and supervising their use; there is sometimes further ambiguity about whether state, regional or local governments have the responsibility to enforce regulations.
- There are new customs regulations and other barriers to trade and the movement of livestock and feed supplies between the nations.

8 Land Use Rights, Regulation and Law

Halimova (Chap. 13) has discussed the progress with the creation and distribution of land use rights and shown a process in train but with far to go. Each country in the Central Asian region (CAR as defined in Chap. 1) has embarked on land use rights change pursuant to the adoption of a market economy following the break-up of the state farm and collectives sector but progress is mixed. The Kyrgyz Republic may be furthest forward with the passing and partial implementation of *the Law of the Kyrgyz Republic on Pastures 2009*. Laws and regulation governing pasture use are evolving differently in each country even though each has a common history of Soviet law (Robinson, Chap. 11). This is partly because each country has somewhat different land and different requirements. Kyrgyz has no desert and is mostly mountainous, and Tajikistan is about 20 % desert but also mountainous and has more farm land and a mostly Farsi tradition. Kazakhstan is about 32.9% desert and has much less mountain land and large areas of nonirrigated cereal lands, some of which has been taken out of production due to degradation. Uzbekistan is some 42.8% desert and has much more horticultural and irrigated crop lands, while Turkmenistan is 67.2% desert; it is somewhat warmer and has more deserts linked to oasis and irrigated land (Figures for deserts from Babaev 1999).

This process has mostly been planned from above and people's rights are not well understood at the level of the pasture user and confused with rights to cultivate. In some countries, the rights to graze are quite short term, for example, Turkmenistan, and in all countries, there is confusion between the state's duty to monitor and supervise land use under central government laws and the often the regional government duty to allocate land for use.

Contested access to and conflicts over pastures along with ecological degradation has resulted in heightened land tenure insecurity. Tenure insecurity refers to the degree to which land users lack confidence that neither the state nor other people will interfere with their access rights to the land for an extended period of time.

Tenure insecurity has four broad dimensions: first, conflict over rights to pastures amongst groups of village residents and mobile groups; second, differences of opinion about the preservation of pastures between farmers with access to farmland and those without access to farmland, but with a dependence on livestock; third, contradictions between governmental agencies empowered by formal law establishing state ownership of pasture land and local communities which, by custom and necessity, use the pastures; and fourth, land-grabbing by powerful elites who establish control over pastures (Kurbanova, Chap. 7).

Hannam (Chap. 17) discusses international approaches to pasture land regulation and law citing the PRC taking an integrated ecosystem management (IEM) approach to manage land degradation (GEF 2000) with an emphasis on capacity building and technological support to strengthen cross-sector coordination and trans-boundary management of natural resources.

IEM, promoted by GEF since 2000 (quoted in Hannam, Chap. 17), is '*a holistic approach to address the linkages between ecosystem functions and services (such as carbon uptake and storage, climatic stabilization and watershed protection, and medicinal products) and human social, economic and production systems*' (Jiang Zehui 2006).

Different forms of resource use right are required in sedentary and nomadic situations as the requirements for management are necessarily different but are often confused. The former can be fixed boundaries of cultivatable land and water, and production is relatively predictable. With the latter, the valuable resource is where it has rained (or snowed) and this changes as weather is a kind of accident in the semi-desert areas where nomadic practices have evolved; this is a low-cost production system but much less predictable than farming.

The way forward is to create regional resource management institutions made up of government and private stakeholders to 'manage' enforceable resource use rights that are a mixture of fixed and flexible grazing rights. International best practice now that marries local knowledge and interest with some more structured 'supervision' and technical assistance. This should aim to facilitate a better system for benefiting from the strengths of each (herders in breeding and farmers in fattening) by a combination of education, assistance with value chain analysis and monitoring of progress, funded by levies and capital grants from government. This is discussed in more detail under solutions (below).

9 Markets

One of the intended benefits of that change to a market economy was a freeing up of markets, thus improving resource allocation. Unfortunately, this has not occurred very effectively in pastoral and farming zones. Continued policies to cultivate cotton and grain, referred to above, subsidized access to some inputs and lack of access to others distorts farmers and livestock owners decisions although livestock have uniformly increased in price, which is a motive for people to increase numbers (Shaumarov, Chap. 16).

Prior to the period of Russian occupation, in the 19th century there was a good deal of long-distance migration of herds between these lands, particularly between Kazaks and Kyrgyz and free movement of products and trade and this continued to some extent during the Soviet period, until about 1989, subject to Soviet plans (Kerven et al. 2006; Rahimon, Chap. 3). The breakup of the Soviet System has seriously interrupted the exchanges of feed, livestock and products between countries that were possible in Soviet times and remains a barrier to sensible resource allocation.

Perhaps of most importance is that many participants and few in government involved in the livestock/feed markets understand the impediments in the sense of a value chain from breeding animal through fattening, input supply and use, logistics, regulation, processing and marketing to retail and so are unable to collaboratively address the many barriers. There is still a habit of technical planning for high production rather than return to effort and investment.

Sedik (Chap. 9) has discussed the market failures in the feed supply system in Tajikistan, and the remaining large farms, state or private, apparently have plenty of feed in the form of crop by-products, grass, etc., while small holders have very little and degraded pasture lands around settlements. There is still evidently an impression that feed should be free or subsidised, even though prices are so high that fattening livestock can appear uneconomic.

The value chain approach to enterprise planning and upgrading is potentially a valuable tool to assist in making the co-dependencies between the different concerned sectors clear and importantly how each can profit from the linkages. This is because value chains focus on the distribution of power and value across the chain, a framework that is eminently capable of addressing the agency of workers and small producers (Mitchell and Coles 2011).

At the present time, there is not an effective process for achieving this as the different industry sectors are now responsive (or not) to individual institutions of government and to, at the most, only immediate links in the value chain, for example, between livestock producer and trader or between feed producer and trader.

10 Understanding the Changes and Risks to the Pasture Base

The two elements of stewardship referred to by Squires (Chap. 2), -awareness and action-, can also be applied to awareness about the changing state of pasture lands and the knowledge of when to act. There is a general understanding of the need to monitor pasture and land condition but little agreement about how to do this and how much is required to provide useful information. It is little wonder then that there is reluctance to commit to the high cost of doing this that is sometimes planned because of doubts about how the information can be used effectively.

Although much is written about the requirement for monitoring pasture condition as a way to inform action to address decline, there is a reluctance to commit resources to this now. The former Soviet system did invest in monitoring but mostly to estimate production and productive potential. While this baseline of information still exists, memories are fading. New systems are being developed notably under CACILM (Central Asian Initiative for Land Management, see www.adb.org/projects/cacilm), but these are not yet applicable at the national level and work may have slowed since 2010 when ADB reduced its input.

There is weariness in many Central Asian countries over money for research as it is sometimes felt that the previous academies studied for study's sake and would not come to a conclusion about cost-effective action. This phenomenon is not limited to Central Asia and stems from the natural fear of scientists that their work may attract criticism if it is seen as incomplete, and this leads to collecting too much information without useful analysis.

11 A Risk-Based Approach to Monitoring

One approach to this difficulty is to adapt a risk-based approach to monitoring through Ecosystems Functional Analysis (EFA) (this section is adapted from Leake 2012). This involves classifications of natural systems into ecosystem units across which similar actions can be expected to have similar outcomes and similar impacts on other ecosystem units allows another means of analysis that can identify points of risk of breakdown (Tongway and Ludwig 2010). One early writer[1] in this area developed a technique, which he called landscape functional analysis (LFA), to facilitate the analysis of rangeland landscapes affected by wind and water erosion (Tongway 1997). Under this system, landscapes are conceived as a number of components that that form a spatial hierarchy based on size, which also have different functions.[2]

The system also postulates a feedback process that governs the productive functions of the landscape, in both ecosystem goods and services terms, and how these can become dysfunctional when certain thresholds are reached, such as a depletion of reserves through high losses, as illustrated in Fig. 18.2. These thresholds are really a matter of definition along a continuum between a fully functional landscape and a quite dysfunctional 'leaky' landscape (leaky here refers to loss of energy, nutrients or water through run-off and erosion), depending on the nature of the goods and services being looked for.[3] The significance is that it facilitates identification of points of leverage in the landscape for monitoring where damage to its productive capacity is more likely to occur and can more effectively be addressed. These can be seen as risk points in the landscape, and so facilitate a risk-based approach to Natural Resource Management (NRM).

Although LFA was developed with rangelands in mind, it has been used in a great many landscape and marine-scape situations and has been renamed ecosystems functional analysis (EFA)[4] as a consequence, which also reflects its background in systems thinking.[5] Multi-scale ecosystem analysis is increasingly being used for

[1] Notes

Tongway D in "Landscape Ecology; Function and Management" Edits Ludwig J, Tongway, D Feudenberger, Noble J &Hodgkinson K CSIRO publishing 1997.

[2] Ludwig J Tongway D et al 1997 (Ibid.) Page 10–12.

[3] Ludwig J Tongway D et al 1997 (Ibid.) Figure 5.1 and Pages 50–61.

[4] Tongway D in "Landscape Ecology; Function and Management" Edits Ludwig J, Tongway, D Feudenberger, Noble J &Hodgkinson K CSIRO publishing 1997 as elaborated in "Putting Principles into Practice" David J Tongway& John A Ludwig CSIRO Sustainable Ecosystems CSIRO Sustainable Ecosystems SBN: 9781597265812 2010.

[5] This may have been unconscious, as they termed the earlier diagrams 'Frameworks'.

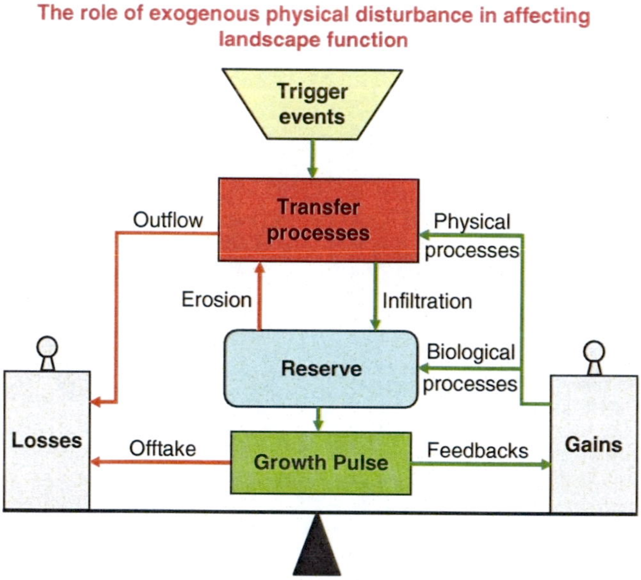

Fig. 18.3 The trigger, transfer and reserve, pulse cycle in ecosystem functional analysis (Ludwig J, Tongway D, et al. 1997 (Ibid.) Fig. 1.1 and Pages 3–9 and as elaborated subsequently in Tongway DJ and Hindley NL 2004. (Ibid.))

natural resource planning as it enables some projections to be made about the likely interactions between components of an NRM system and the possible outcomes (Bailey 1996). For example, the EFA system is increasingly being employed in mine-site rehabilitation and natural reserves management as an objective way to specify objectives and intended outcomes for specifically defined activities directed at defined stresses (Randall 2004) (or strange attracters) in complex natural systems.[6]

In Fig. 18.3, which relates to slow acting 'disasters' or breakpoints, the term 'hazard' as used in disaster and recovery management (DRM) is termed a 'trigger point' for a possible system phase change. In this system, the presence of reserves (of materials and energy) allows the system to restore itself to its previous state after a trigger event, if the reserves are sufficient.

Reserves is an important aspect of system resilience as depleted reserves render a system vulnerable to catastrophic phase change as is described here in biological terms and in elsewhere for social ecological systems (Berkes et al. 2003). The subject is returned to later when concluding similarities between fast and slow disasters and means of minimising risk by focussing monitoring on stakeholders who can mobilise the human and social capital required for this. The shaded triangle in Fig. 18.3 may be interpreted to denote a lack of 'reserves' that effectively changes the 'tipping point'. It should also be noted that 'reserves' can also refer to spiritual reserves as a subset of human and social capital.

[6] http://www.pir.sa.gov.au/__data/assets/pdf_file/0008/11024/mj35_ecosystem.pdf.

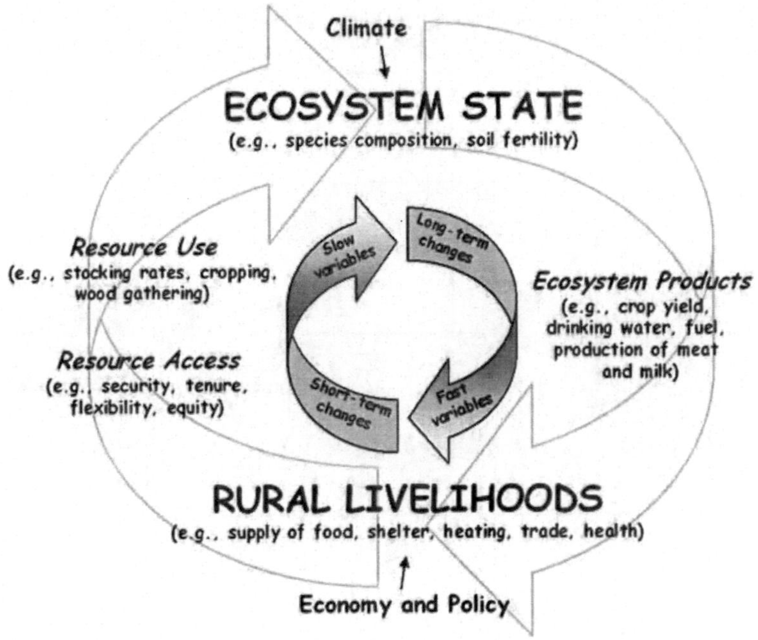

Fig. 18.4 Social and biophysical factors in rangelands are closely linked, difficult to predict and involve a mixture of 'fast' and 'slow' variables. The core of the biophysical system is the 'state of the ecosystem', whereas the core of the socioeconomic system is 'rural livelihood'

12 Solutions

The question posed in this chapter is how to recreate a situation where there is some adaptive process that enables a new panarchy between people, livestock and grasslands in an era in which the livestock keepers have reduced political power when compared with farmers and urban elites.

There are needed changes in grassland governance to improve *awareness* and *capacity for action*:

- Collective awareness on the part of stakeholders at all levels of the need for improved stewardship, information on the evolving state of grasslands and what can usefully be done to address this and reduce the risk of calamity
- Capacity for immediate stakeholders to take action that will improve the capacity for grasslands to provide desired income now and for future generations

These changes have different temporal scales; collective awareness of the evolving ecosystem state is a slow long-acting process, whereas actions in pursuit of rural livelihoods is a faster process that acts on the slower process, much as the adaptive actions of nomadic herders prior to industrialisation acted on the longer process of the evolving state of grasslands, resulting in the stasis referred to

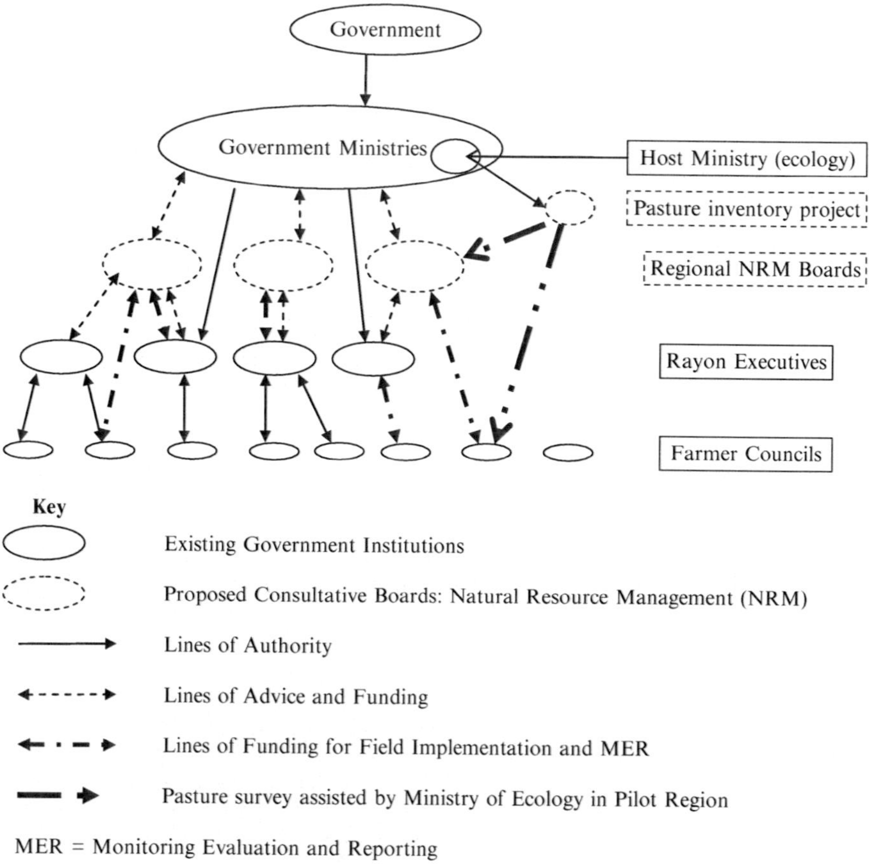

Fig. 18.5 A schematic of the linkages between various actors in a proposed restructuring of agencies land management

in the section on 'pastoralism as a way of life' (see above). Figure 18.5 shows the interrelationships between elements within an ecosystem and the implications for rural livelihoods.

As stated above, improving collective awareness is essentially data gathering, information generation and promulgation to appropriate stakeholders. These data and information requirements are relevant to biophysical information about grasslands (and associated water resources) and importantly information about financial and economic potential and performance consistent with sustainability as it is only by improving profits that changes in action on ground will occur.

Enabling changes of this nature will require an institutional framework that will enable communication horizontally and vertically between relevant stakeholders, so decision-making processes can consider all of the relevant circumstances across disciplines and sectors.

Fig. 18.6 Schematic showing the relationships between fast and slow variables in natural resource management

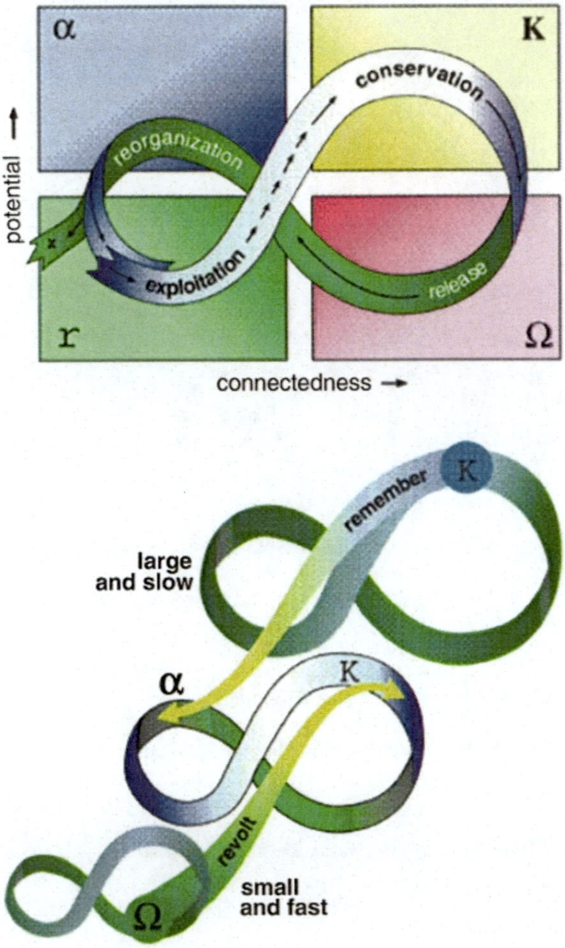

A source of funding for activities that meet a test of public good for such activities as:

- Applied research into ecosystem services and ecosystem goods
- Monitoring
- Promotion
- Education
- Price information and value chain analysis

Such a system (Fig. 18.5) would mostly operate at national and regional levels in each country, as in this possibility (drafted for Azerbaijan). The system to be adopted would need to be negotiated in each country to reflect local circumstances. Ideally, there would also be a multi-country structure to facilitate international discussion, learning and liaison, as intended under CACILM.

Panarchy (see above) places great emphasis on the interconnectedness of levels, between the smallest and the largest and the fastest and slowest (http://www.sustainablescale.org/ConceptualFramework/UnderstandingScale/MeasuringScale/Panarchy.aspx). The large, slow cycles (e.g. soil formation, nutrient change) set the conditions for the smaller, faster cycles (like forage production) to operate. But the small, fast cycles can also have an impact on the larger, slower cycles. There are many possible points of interconnectedness between adjacent levels (Fig. 18.6).

The multiple time scales and slow variables (Slow variables like soil development define long-term change by constraining the fast variables like forage production. http://sedac.ciesin.columbia.edu/tg/guide_glue.jsp?rd=lu&ds=4.4) of ecology have profound implications for forward-looking human behavior towards the environment. We must recognise that current issues in ecosystem management in the various countries in Central Asia differ in many ways. Nevertheless, a lack of understanding of slow ecological variables and the concomitant poor integration of such variables into policy choice must be recognized as a barrier to better land management (Carpenter and Turner 2001). It is easy to see how failures of foresight, explainable, in part, by insufficient knowledge of slow variables, can lead to environmental collapse and why it is difficult to reverse land degradation problems and instil a philosophy of better land stewardship.

References and Further Reading

Abdel gueryfi A, Hassani AE (2011) Interactions between cereal cropping systems and pastoral areas as the basis for sustainable agriculture. In: Lemarie J, Hodgson J, Chabbi A (eds) Mediterranean countries grasslands productivity and ecosystems. CAB International, Wallingford

Babaev AG (1999) Desert problems and desertification in Central Asia The Desert Research Institute. Springer, Berlin/Heidelberg/New York

Bailey RG (1996) Multi-scale ecosystem analysis. Environ Monitor Assess 39:21–24

Barfield TJ (1992) The perilous frontier. University of California Press, Berkeley

Berkes F, Colding J, Folke C (2003) Navigating social–ecological systems: building resilience for complexity and change. Cambridge University Press, Cambridge

Carpenter SR, Turner MG (2001) Hares and tortoises: interactions of fast and slow variables in ecosystems. Ecosystems 3:495–497

Carbon storage by introduced deep-rooted grasses in South American savannas Fisher MJ et al. Letters to Nature No 371, 236–238 15 Sept 2002 doi: 10.1038/371236a0

Fisher MJ (2002) "Carbon storage by introduced deep-rooted grasses in South American savannas Fisher M J Et al. Letters to Nature No 371, 236-238 15 Sept 2002 doi: 10.1038/371236a0

Galaty J, Johnston D (1990) The world of pastoralism, herding systems on comparative perspective. The Gilford Press, New York

Gurney WSC (1979) Ecological stability and social hierarchy Department of Applied Physics. University of Strathclyde, Glasgow, UK

Hardesty DL (2003) Archaeology. In: Hardesty DL (ed) Encyclopedia of Life Support Systems (EOLSS), Developed under the Auspices of the UNESCO. Eolss Publishers, Oxford

Hollings C (2001) Understanding the complexity of economic, ecological, and social systems. Ecosystems 4:390–405

Kerven C, Alimaev I, Behnke R, Davidson G, Smailov A, Temirbekov S, Wright I (2006) Fragmenting pastoral mobility: changing grazing patterns in post-Soviet Kazakhstan. In: Bedunah D, McArthur E, Fernandez-Gimenez M (eds) Rangelands of Central Asia: transformations, issues and future challenges. Rocky Mountain Research Station Serial, RMRS-P-39 2006. US Department of Agriculture, Fort Collins

Kreutzmann H (ed) (2012) Pastoralism: a way forward or back? In: H Kreutzmann (ed.) Pastoral practices in High Asia, Advances in Asian Human-Environmental Research. Springer

La Rovere R, Hiernaux P, Van Keulen H, Schiere JB, Szonyi JA (2005) Co evolutionary scenarios of intensification and privatisation of resource use in rural communities of south-western Niger. Agr Syst 83(2005):251–276

Lattimore O (1940) Inner Asian frontiers of China. American Geographical Society, New York

Lattimore O (1941) Mongol Journeys. New York: Doubleday Doran, New York

Leake JE (2012) (in press) Disaster and recovery. Thaksin University, Songkhla, Thailand

McCauley M (1994) Agriculture in Central Asia and Kazakhstan in the 1980s. In: Akiner S (ed) Political and economic trends in Central Asia. British Academic Press, London

Mitchell J, Coles C (eds) (2011) Markets and rural poverty; upgrading in value chains. Earthscan Park Square, Milton Park/Abingdon/Oxon

Olsen G (2011) Forests and grasslands as cradles for agriculture. In: Squires V (ed) Food, agriculture, forestry and fisheries in human nutrition. Encyclopedia of life support systems Developed under the Auspices of the UNESCO. Eolss Publishers, Oxford

Randall J (2004) Ecosystem function analysis – a tool for monitoring mine-site rehabilitation success. MESA J 35:24–27

Squires VR (1998) Halophytes: their potential as new crops in coastal deserts and saline inland regions using brackish water irrigation. Desertification Control Bull No. 33, pp 50–56

Squires V, Hua L, Zhang D, Li G (2010) Towards sustainable use of rangelands in North West China. Springer, Dordrecht, p 384.

Tongway D (1997) In: Ludwig J, Tongway D, Feudenberger D, Noble J, Hodgkinson K (eds) Landscape ecology; function and management. CSIRO publishing, Canberra, Australia

Tongway D, Ludwig J (2010) Restoring disturbed landscapes: putting principles into practice. Island Press, Washington, DC. Canberra, Australia

Zehui J (2006) Integrated ecosystem management-proceedings of the international workshop. China Forestry Publishing House, Beijing

Glossary of Terms

Ailoq	Term meaning summer highland pasture, *also called Yaylag* (Russian), *yaylak* (Turkish), *jaylaw* (Kazakh), *jayloo* (Kyrgyz), *yeilâq* (Persian) is a Turkic
Akims	Local leaders
Amlyak	According to Sharia law, land belongs to the state in the name of the Supreme governor-khan, all the others were users and not owners
argali	The mountain sheep (species *Ovis ammon*) that roams the highlands of Central Asia
Ayilok motu	In Kyrgyzstan, the lowest level of local government
Badachi	In Kyrgyzstan, is used for those herding collective herds of cattle; however the term *badachi* is often used generically to cover all shepherds and others responsible for herding transhumant livestock
bai	Rich men in the village, chiefdom in the nomadic unit
bag	Local political units
Bogara land	Is land found in a horizontal belt which covers the foothills of most mountain ranges. It is arable only in spring and early summer because of the increased presence of water and produces only one harvest per year
"byednyak"	A Russian word used to describe poor Russian peasants and/or poor locals
Chekene	A little like the Kyrgyz *bada* system in which a professional shepherd is paid to herd the animals of others
chon manap	In Kyrgyzstan, a chieftain but holding no more honours than other authorities of the clan. Manap is like a Russian Duke
Chorvodori va Baitori	In Tajik meaning livestock husbandry and veterinary
dayhan birlishik	In Turkmenistan, associations of leaseholders, or peasant associations
Dehkan farm	Private or cooperative farm
Dehkan farmer	(Literally, "peasant")

V. Squires (ed.), *Rangeland Stewardship in Central Asia: Balancing Improved Livelihoods, Biodiversity Conservation and Land Protection*,
DOI 10.1007/978-94-007-5367-9, © Springer Science+Business Media Dordrecht 2012

Dzud	Extremely severe winter leading to large widespread death of livestock
el-jurt	In Kyrgyzstan, a league had a common territory composed of various types of seasonal pastures and recognized the authority of a single ruler
Farsi	Persian language, spoken in Iran and parts of Central Asia
Fazendas	In Kazakhstan, livestock operations where most herds grazing outside village pastures are owned by single households holding individual title to pasture land
fermerskie khozyaistva	Peasant farmers also known *fermer*
Gastarbeiter	A somewhat derogatory German word that refers to immigrant labourers
Giprozem	The State Project Institute for Land Management
Glavgidromet	Uzbek Meteorological Organization
Gosregister	In Kyrgyzstan, a land titles office
Goskomzem	State land committee
Gur	A portable home (see yurt) common among nomads in Central Asia
halal requirements	Method of slaughter and food handling dictated by the Koran
Hakim	In Uzbekistan, a local leader
Hukamat	The administrative body of a *raion/rayon*
Irrigable land	An area where water can be delivered to allow irrigation
Intensivnie	Pastures, leased from the raion administration
jailoos	In Kyrgyzstan, summer pasturelands
Jamoat	Throughout the former Soviet Union countries, the local representative organ administrative unit below *raion/rayon*, comprising a group of villages
Jayit	In Kyrgystan, consists of representatives of pasture users, deputies of local self-government bodies, and heads of executive local self-government bodies.
jilgachi	In Kyrgyzstan, a word used for those who herd horses
Jut	In Uzbekistan, unusually thick layer of ice on top of snowfall
kristianskikhozyaistva	In Kazakhstan, independent land units designated as (peasant farms) on former State land
kezu system,	In Kyrgyzstan, group members take it in turns to herd animals according to a daily rota
Khashar	In Tajikistan, which is the peoples' assembly that identifies problems and their solutions
Kishlak	A village
koichi	In Kyrgyzstan, a term used for sheep herders
Kolkhozes	In the former USSR, cooperative organizations of peasants who have come together voluntarily for the joint management of large-scale, livestock farms

Kuchi	In Afghanistan, transhumant pastoralist nomads. A Farsi word meaning 'move'
Kyzyk	In Kazakhstan, running common herds on a rota basis similar to the *kezu* system in Kyrgyzstan
Latif	A national NGO in Tajikistan
Leskhoz	State Forestry department in Soviet era
Mahalla	Village Council
Madzhlisi Oli	In Tajikistan, parliament
markhor	Wild goat *Capra falconeri*
mardikhors	In Uzbekistan, informal labour markets
metayage system	Islamic law provides that the decision made in establishing the lease of a piece of land must be derived from the decision to what use the land shall be put
Mission East	Danish relief and development *NGO*
Negdels	Herding and grazing collectives
Oblast	Throughout the former Soviet Union countries, the local government level equivalent to a county
Otgon/otgonnye	Remote pastures, to be leased from the oblast administration
Panarchy	Hierarchies and adaptive cycles comprise the basis of ecosystems and social ecosystems across scales. Together they form a panarchy
prisel'nie	In Kyrgyzstan, village pastures
qarakul (karakul)	In Afghanistan, a breed of sheep for producing karakul lamb, a type of folded hat
Qrut	In Afghanistan, a dried yogurt product
Raion/Rayon	A district
rain-fed cultivation	Opportunistic cropping entirely dependent on rainfall
Scaling-up	Taking results from experiments and field trials and expanding the area and scope of the approach
sens lat.	In the broadest sense
Sharia	Islamic religious law
Shirkat	In Uzbekistan, mainly joint-stock shareholding companies. Large-scale pastoral land areas in central and NW parts of Uzbekistan are allocated to Karakul farms
Shoats	Sheep and goats run together and managed as a single flock
Sheriyat/Shariat	A body of sharia law relating to land
shura	In Afghanistan, a meeting of clan and district leaders
soum	Local political unit
Sotki	In Azerbaijan, is a measure of land area equal to 0.10 ha
Sovkhoz	State-operated agricultural estate in the USSR organized according to industrial principles for specialized large-scale production
Tunduk –	In Kyrgyzstan, refers to the central part of a pastoralist's
Turkmen Mallary	In Turkmenistan, the state agency responsible for livestock
Tugai vegetation	Riparian vegetation on flood plains of Central Asia

Uzkhlebprodukt	In Tajikistan, state-controlled system to establish feed storage facilities and sale outlets in rural areas
Vakuf	Literally, the dedication of an object to the way of Allah
wakuf lands	Donated land that passes in time to eternal possession of clergy
Uigur/Uyghur	Are a Turkic ethnic group living in Eastern and Central Asia. Most Uyghurs live in the Xinjiang Uyghur Autonomous Region in China
Urf	In Afghanistan, community property rights based on customary rights
urial	*An upland wild sheep (Ovis vignei) of southern and central Asia*
Yurt	*A portable home (see Gur) common among nomads in Central Asia yurt and is* meant to root its society in pastoral traditions

Index

V. Squires (ed.), *Rangeland Stewardship in Central Asia: Balancing Improved Livelihoods, Biodiversity Conservation and Land Protection*,
DOI 10.1007/978-94-007-5367-9,